T0074299

The Application of Emerging Technology and Blockchain in the Insurance Industry

RIVER PUBLISHERS SERIES IN MANAGEMENT SCIENCES AND ENGINEERING

The "River Publishers Series in Management Sciences and Engineering" aims at publishing high quality books on management sciences and engineering. The Series provides discussion and facilitates exchange of information on principles, strategies, models, techniques, methodologies and applications of management sciences and engineering in the field of industry, commerce, government and services. The Series aims to communicate the latest developments and thinking on the management and engineering subject world-wide. Thus, the main aim of this book series is to serve as a useful reference and to provide a channel of communication to disseminate knowledge between academics, researchers, managers and engineers and other professionals.

The Series seeks to link management sciences and engineering disciplines to helping organizations in private and public sectors addressing technological, business and societal challenges in a context of complexity, uncertainty and change. The Series stimulates the development of new approaches to management and engineering in a human, organizational and societal context, from a socio-technical systems perspective, and in aspects as design, innovation, planning, coordination, communication, engagement and decision-making. Examples of societal challenges requiring new approaches to management and engineering include the sustainable, secure and responsible development of energy, transport, water, infrastructural and informational resources, infrastructures and services. The Series is highlighting cultural and geographical diversity in studies oriented to transformation in organizations and work environments. It emphasizes the role of human resource management, the role of organizational decision-making and collaboration, empowerment of stakeholders, and the management of innovation and engineering activities in a context of organizational change. This way it reflects the diversity of human, organizational and societal conditions.

Books published in the series include research monographs, edited volumes, handbooks and text books. The books provide professionals, researchers, educators, and advanced students in the field with an invaluable insight into the latest research and developments.

Topics covered in the series include, but are by no means restricted to the following:

- Human Resources Management
- Socio-technical Systems Change
- Large-scale Systems in a Human and Societal Context
- Culture and OrganizationalBehaviour
- Higher Education forSustainability
- Management in SMEs, large organizations, networks
- Strategic Management
- Entrepreneurship and Business Strategy
- Interdisciplinary Management
- Management and Engineering Education
- Knowledge Management
- Management of Innovation and Engineering
- Operations Strategy, Planning and Decision Making
- Sustainable Management and Engineering
- Production and Industrial Engineering
- Materials and Manufacturing Processes
- Manufacturing Engineering
- Interdisciplinary Engineering
- Management of System Transitions (energy, water, transport)

For a list of other books in this series, visit www.riverpublishers.com

The Application of Emerging Technology and Blockchain in the Insurance Industry

Editors

Kiran Sood

Postdoc Researcher, University of Usak, Faculty of Applied Sciences,
Dept of Finance and Banking
Chitkara Business School, Chitkara University,
Punjab, India

Simon Grima

Department of Insurance,
Faculty of Economics Management and Accountancy, University of Malta,
Msida, Malta

Ganga Sharma

School of Computing Science and Engineering, Galgotias University,
Greater Noida, India

Balamurugan Balusamy

Associate Dean-Student Engagement, Shiv Nadar University,
Delhi-National Capital Region (NCR),
India

River Publishers

NEW YORK AND LONDON

Published 2024 by River Publishers
River Publishers
Alsbjergvej 10, 9260 Gistrup, Denmark
www.riverpublishers.com

Distributed exclusively by Routledge
605 Third Avenue, New York, NY 10017, USA
4 Park Square, Milton Park, Abingdon, Oxon OX14 4RN

The Application of Emerging Technology and Blockchain in the Insurance Industry / Kiran Sood, Simon Grima, Ganga Sharma and Balamurugan Balusamy.

Routledge is an imprint of the Taylor & Francis Group, an informa business

ISBN 978-87-7004-006-8 (hardback)
ISBN 978-87-7004-060-0 (paperback)
ISBN 978-10-0381-147-3 (online)
ISBN 978-1-032-63094-6 (ebook master)

Contents

Preface

The technological revolution has transformed how companies interact with customers, creating an environment where marketing, information and technology must work together. Retailers and industry giants have made significant strides in adopting digital platforms to deliver a satisfying customer experience. Insurance companies must adjust their business models and strategies to remain competitive and take advantage of technology.

Insurance companies are increasingly investing in IT and related technologies to improve customer experience and reduce operational costs. This book will help the insurers strengthen customer interactions, target potential customers, provide usage-based insurance and optimise the overall insurance business.

Innovation through new technologies is a key driver of change in the financial sector accompanied by uncertainty and doubt; this book will play a pivotal role in risk management through fraud detection, regulatory compliances, and claim settlement leading to the overall satisfaction of customers. Besides, it answers readers' questions in real, everyday business terms, tailored specifically to the insurance industry's unique needs, challenges, and targets.

This book is a unique guide to the disruptions, innovations and opportunities that technology provides the insurance sector and acts as an academic-industry-specific guide for creating operational effectiveness, managing risk, improving financials, and retaining customers. This book focuses less on the architecture and technical details, rather more on practical guidance on translating technology into target delivery. The discussion examines implementation, interpretation, and application to show what technology can do for business, with insights and examples targeting the insurance industry and with a special outlook into blockchain. From fraud analytics in claims management to customer analytics, from risk analysis to solvency management, comprehensive coverage presented from a layman's perspective makes this guide an invaluable resource for any insurance professional. It also contains the current philosophy and actionable strategies from a wide range of contributors who are experts on the topic. The speciality of this book rests on the fact that it logically explains why the traditional ways of doing business will soon become irrelevant and, therefore, provides an alternative to thriving in the present competitive world by embracing technology by the insurers. Further, it intends to discover new ways of grasping and defining growth opportunities and explore the business models, value chain, customer engagement strategies and organisational structures that can transform the insurance industry.

This book offers a valuable resource for undergraduate and postgraduate students, Academic Researchers, and MBA/Executive Education. Besides, it intends to provide insurance practitioners and academics with a high-level overview of the main research topics and to encourage future academic work in this field.

This book caters for research work in many fields of application of technology, predictive analysis, artificial intelligence, robotic process automation, blockchain etc., and it provides an in-depth knowledge of the fields covered. Besides, it will also cater to the needs of professionals/practitioners working at insurance companies, insurance agents, brokers, bankers and consultants, Policymakers working at insurance companies, Insurance underwriters Underwriter managers Insurance associations Bankers, Consultants Attorneys, Risk managers, Financial planners, Business owners, Policymakers.

The book's content will be most valuable to the readers in terms of understanding how technology has brought innovation and introduces modern methods to measure, control, and price risk in the insurance business.

Foreword

Rebecca Dalli Gonzi

Head of Department, Construction and Property Management, Faculty for the Built Environment, University of Malta, Malta

Affiliate member: Faculty of Economics, Management and Accountancy; Institute of Sustainable Energy, University of Malta

The appropriate application of data to support the industry is constantly expanding. When we consider the benefits that this offers, we can start to reap the reward of good judgement being made in the upper echelons of society. In the areas of insurance, regulation, compliance, and other activities, we can observe well-informed decisions that satisfy the demands of the public when the gap to achieve data bottom-up is secured. This book provides readers with answers to questions in practical, everyday business terminology specifically adapted to the needs, goals, and challenges of the insurance sector. However, with a view towards implementation and transitioning technology into desired delivery, it focuses on the pragmatic implications that serve such demands. A never-ending probe into how data is used, intentionally or inadvertently, will always be required to safeguard the public domain. Offering solutions to social segments that were previously unreachable and helping the vulnerable towards leading solutions could be realised if data acts as a foundation to bridge the gap between the two.

This book makes us aware that we currently live in a time where we must reassure the public that the data being harnessed can be used well. The academic body of knowledge has an opportunity to publish more about the debate concerning technology and its effects on citizen choice and rights across the globe. In this sense, we must encourage greater transparency in big data debates to dispel common myths and disclose underlying currents as people increasingly rely on technology to make decisions. Instead of acting as a controlling instrument, technology may empower individuals in everyday life. Still, to ensure that underlying assumptions are rational and that unanticipated consequences are minimised, regulators must openly discuss how algorithms are handled.

The fact that AI was developed by humans and is subordinate to human intuition or self-expression promotes conversation about how procedures are handled and give developers a chance to construct frameworks that honour the intelligence that was poured into such a creation. The motivation provided by this book encourages the reader to reflect upon emerging markets that may still need to experience the success of its use, thus laying the groundwork for stirring fresh inquiry.

List of Contributors

Ajmani, Prerna, *Vivekananda Institute of Professional Studies-TC, GGSIP University, India*

Baldacchino, Peter J., *University of Malta, Accountancy Department, Faculty of Economics Management and Accountancy, Malta*

Bansal, Rajni, *Chitkara Business School, Chitkara University, Punjab, India*

Bassi, Payal, *Chitkara Business School, Chitkara University, Punjab, India*

Batra, Reenu, *Computer Science and Engineering, SGT University, India.*

Dahiya, Manju, *Galgotias University, India*

Dangwal, Aarti, *Chitkara Business School Chitkara University, Punjab, India*

Dhatterwal, Jagjit Singh, *Department of Artificial Intelligence & Data Science (AI&DS), Koneru Lakshmaiah Education Foundation, Andhra Pradesh, India*

Diego, Norena-Chavez, *Escuela de Posgrado, Universidad de Lima, Lima 70001, Peru*

Gagandeep, *Chitkara Business School, Chitkara University, Punjab, India*

Garg, Girish, *School of Finance and Commerce, Galgotias University, India*

Goel, Cheenu, *Chitkara Business School, Chitkara University, Punjab, India*

Grima, Luke, *Department of Insurance and Risk Management, Faculty of Economics, Management & Accountancy, University of Malta, Malta*

Grima, Simon, *Department of Insurance and Risk Management, Faculty of Economics, Management and Accountancy, University of Malta, Malta*

Gupta, Monica, *Chitkara Business School, Chitkara University, Punjab, India*

Gupta, Sumeet, *UPES, India*

Iwendi, Celestine, *School of Creative Technologies, University of Bolton, UK*

Jindal, Priya, *Chitkara Business School, Chitkara University, Punjab, India*

Kaswan, Kuldeep Singh, *School of Computing Science & Engineering, Galgotias University, Greater Noida, Uttar Pradesh, India*

Kathuria Pankaj, *University School of Business, Chandigarh University, India*

Kaur, Jasmine, *Chitkara Business School, Chitkara University, Punjab, India*

Kaur, Simranjeet, *USB, Chandigarh University, Mohali, India*

Khanna, Rupa, *Department of Commerce, Graphic Era Deemed to be University, Dehradun, India*

Kumar, Naresh, *Dean Applied Computational Science and engineering, G L Bajaj institute of technology and management , Greater Noida, India*

Kumar, Sanjay, *Computer Science Engineering, Galgotias College of Engineering and Technology, Greater Noida, India*

Lal, Sandeep, *Faculty of Law PDM University, India*

Maple, Carsten, *Secure Cyber Systems Research Group (SCSRG), WMG, University of Warwick, UK*

Maurya, Ashok, *Galgotias University, India*

Mishra, Naman, *Galgotias University, India*

Misra, Neeti, *Uttaranchal University, India*

Mohit kukreti, *University of Technology and Applied Sciences, Ibri, Oman*

Morina, Fisnik, *Faculty of Business, University "Haxhi Zeka" in Peja, Kosovo*

Noja, Graţiela Georgiana, *Faculty of Economics and Business Administration, West University of Timisoara, Romania*

Ozen, Ercan, *University of Usak, Faculty of Applied Sciences, 64200 Usak- Türkiye*

Rani, Seema, *Amity Institute of Information Technology, Amity University, India*

Rao, T. Joji, *O.P. Jindal Global University, India*

Rawat, Nishtha, *IIT Roorkee, India*

Samridhi, *Department of Management, Indira Gandhi University, India*

Sharma, Aradhana, *Punjabi University, Patiala, India*

Sharma, Bhavna, *School of Finance and Commerce, Galgotias University, India*

Sharma, Vandana, *CHRIST (Deemed to be University), Delhi-NCR, India*

Sharma, Vimal, *Chitkara College of Sales and Marketing, Chitkara University, Punjab, India*

Sidhu, Gurpreet Singh, *Department of Management, Punjab Institute of Technology, Rajpura, District Patiala, Punjab, India*

Singh, Amandeep, *Chitkara Business School, Chitkara University, Punjab. India*

Sood, Deepak, *Chitkara Business School, Chitkara University, Punjab, India*

Sood, Kiran, *Postdoc Researcher, University of Usak, Faculty of Applied Sciences, Dept of Finance and Banking; Chitkara Business School, Chitkara University, Punjab, India*

Taneja, Sanjay, *Department of Management Studies, Graphic Era Deemed to be University, Dehradun, India*

Thalassinos, Eleftherios, *University of Piraeus, Greece*

Todorović, Igor, *Faculty of Economics, University of Banja Luka, Bosnia and Herzegovina*

Trivedi, Sonal, *School of Management, Birla Global University, Orissa, India*

Verma, Jyoti, *Chitkara Business School, Chitkara University, Punjab India*

Zammit, Mark Laurence, *Department of Insurance and Risk Management, Faculty of Economics, Management and Accountancy, University of Malta, Malta*

List of Figures

List of Tables

List of Abbreviations

AGCS Allianz Global Corporate and Specialty
AI Artificial intelligence
AIG American International Group
ALM Asset-Liability Management
AMCA American Medical Collection Agency
AML Anti-money laundering
ATAWAD Any Time, Any Where, Any Device
B.I. Business interference
B3i Blockchain Insurance Industry Initiative
BC Blockchain
BCT Blockchain technology
BIDAAS Blockchain based Deep Learning as a Service
CAGR (compound annual growth rate)
CIO Chief Information Officer
CIS Center for Internet Security
DABS Distributed Autonomous Banks
DAI Decentralised autonomous insurer
dApps Decentralized applications
DEA Data envelopment analysis
DLT Distributed ledger technology
DoS Denial-of-Service
DPO Data protection officer
DSCSA Drug Supply Chain Security Act
DWWP Digital World Working Party
EHR Electronic health record
EMR Electronic medical record
ERM Enterprise risk management
EU European Union
FD Fraud detection
FDI Foreign Direct Investment
FFIEC Financial Institutions Examination Council
FI Fire insurance

GDP	Gross domestic product
GDPR	General Data Protection Regulation
GLM	Generalized Linear Model
GRC	Governance, risk, and compliance
GWP	Gross written premium
HRM	Human Resource Management
HVAC	chap-18
IAIS	International Association of Insurance Supervisors
ICO	Initial coin offering
ICP	Insurance core principle
ICT	Information and communication technology
IdM	Identity management
IdMS	Identity management system
IDV	Insured declared value
IFoA	Institute and Faculty of Actuaries
ILS	Insurance linked securities
IoT	Internet of Things
IPR	Intellectual property right
ISO	International Standards Organization
IT	Information technology
ITU	International Telecommunication Union
KM	Knowledge management
KYB	Know Your Business
KYC	Know your customer
LTC	Long Term Care
M2M	Machine-to-machine
MGI	Massive group insurance
MHYD	Manage how you drive
ML	Machine learning
MoU	Memorandum of Understanding
NAIC	National Association of Insurance Commissioners
NARDL	Nonlinear autoregressive distributed lag
NASA	National Aeronautics and Space Administration
ORSA	Own Risk and Solvency Assessment
P&C	Property and casualty
P2P	Peer-to-peer
PAYD	Pay as You Drive
PBFT	Practical byzantine fault tolerance
PCA	Principal component analysis
PHYD	Pay How You Drive

PII	Personally identifiable information
PMFBY	PM Fasal Bima Yojana
PMJAY	Pradhan Mantri Jan Arogya Yojana
PoA	Proof of Authority
POC	Proof of Concept
POS	Proof of Stake
POW	Proof of Work
PRISMA	Preferred Reporting Items for Systematic Reviews and Meta-Analysis
PWS	Perceived wellness survey
ROI	Return on investment
RPA	Robotic process automation
S&T	Science and technology
SC	Smart contracts
SDP	Single disease payment
SMA	Survey of monetary analysts
SME	Small- and medium-sized enterprises
SSO	Single sign on
UBI	Usage-based insurance
UBI	Usage-based insurance
UBI	Usage-based insurance
UHN	University Health Network
ULIP	Unit linked insurance plan
VOS	VOSviewer
WBAN	Wireless body area network
WEF	World Economic Forum

SECTION I

1

Technological Developments in the Insurance Industry: An Overview

Monica Gupta[1], Rajni Bansal[1], Aradhana Sharma[2], and Norena-Chavez Diego[3]

[1]Chitkara Business School, Chitkara University, Punjab, India
[2]Punjabi University, Patiala, India
[3]Escuela de Posgrado, Universidad de Lima, Lima 70001, Peru

Email: Monica.gupta@chitkara.edu.in; Rajni.bansal@chitkara.edu.in; aradhanasharma.as@gmail.com; Dnorena@ulima.edu.pe

Abstract

The application of cutting-edge technologies to enhance business operations in the insurance sector is known as insurtech or insurance technology. This is an essential step for preventing and predicting the model. Thus, the data is shared among the parties and insurers.

This chapter analyses the newest insurance technology trends, the top five technologies impacting in insurance industry in the present days, an analysis of the use of technology by insurance companies, the top tech trends upending the insurance industry, how the insurance industry will be improved by implementing new technologies, and the top 13 insurrection innovations.

Technology can minimise the total operational costs. The insurance industry must adequately utilise all the technological tools to develop the insurance company. Unmanned drones are a technology tool for insurance that carriers will use more in the future. Insurance companies can use IoT to examine client data and determine their needs and dangers thoroughly. Wearables allow for a more precise assessment of customer health hazards. Service to policyholders will switch from being a customer- or insurer-initiated activity. There are five advantages of technology in the insurance sector: lower insurance rates, fraud prevention, lower underwriting costs, billing efficiency, specialised insurance, and data-driven pricing.

1.1 Introduction

The insurance industry is known for being conservative and traditional. This sector of the economy changes slowly. Innovations in technology are the industry's new frontiers. According to the FinTech Developers, several drivers permit the interchange and collection of data. As a result, insurers can predict the risk. The activities of insurers can be made more efficient overall and in terms of pricing, risk selection, and pricing. Transparency is increased, and underwriting risk is decreased through data technology. Insurance companies can assess risk and provide customers with tailored plans by obtaining personalised information.

In recent years, the advancement of technology in all industry segments is also evident in the insurance sector. For example, blockchain technology diffusion is used widely for insurance technology, which essentially helps analyse data and provides future scope for diffusion possibility as opposed to actual diffusion [1]. The issues of robust data management related to top insurance processes are also discussed in the chapter, outlining the variabilities of usage and utilisation based on specific requirements of insurance companies. Improving operational efficiency can be aimed at implementing insurance technology and providing customers with better support and transparency [28, 29].

1.2 Insurance Technology Overview

The entire insurance industry will benefit from technology improvements such as AI technology, and big data will also help them effectively. The implementation of AI technology and the significant data promises cost savings and the opportunity to provide customers with services that are too effective and well-organised [2]. However, new legal and regulatory obstacles are paralleled by this new potential. This article covers issues with Australian data privacy regulation with a specific emphasis on insurers' (possible) gathering of consumer data from non-traditional sources. We look at instances where customers might need to be made aware that the information gathered may be used to calculate insurance rates. We give two practical instances of non-traditional data sources in our analysis: consumer loyalty programmers.

"Using Porter's (1985) value chain and Berliner's (1982) insurability criteria," we analyse the importance of AI technology on the insurance industry using data from '91 papers and 22 industry studies. We also provide directions for future study from both academic and professional perspectives. The findings show that when the insurance industry transitions, it can identify the total loss and possible loss for the business; it also identifies the cost savings and all the revenue-related information that can be generated through this technology [3]. Furthermore, we pinpoint two potential changes in the way risks are insured.

There are numerous examples of how artificial intelligence (AI) changes the insurance sector in the business press and at industry conferences. However, only some thorough conceptual analyses and evaluations of AI technology put AI in a strategic perspective and consider how various AI applications fit together and make a coherent picture [4]. A thorough case study of the BGL company, a renowned insurance provider in Europe, is provided in this essay. The analysis is organised using a broad model of insurance company business processes. The marketing influence of AI technology is demonstrated, as well as the nature of the business value generation process, through the description of five AI applications utilising an insurance firm–customer data flow diagram.

The survivorship principle postulates that a firm's level of business diversity is correlated with its likelihood of enduring over time. This study looks at this problem in the context of long-term regulatory efforts to harmonise fragmented European regulatory environments and the factors that affect the technical effectiveness of acquiring insurance companies before and after the financial crisis [5]. Data envelopment analysis modelling is used to evaluate technical efficiencies throughout significant consolidation and harmonisation of monetary and insurance regulating laws (DEA). Both technical efficiency and the survivorship principle support our prediction that the likelihood of being an acquirer will be high.

Industry adoption of Internet of Things (IoT) technologies has increased. However, studies on IoT in knowledge management (KM) still need to be made available. By implementing the IoT technology, the industry can understand the critical issue based on the IoT technologies. This paper reveals the supporting role and impacts of IoT-based technologies. This technology also helps in the decision-making process and improves claim accuracy and efficiency [6]. The critical responsibility or role of the IoT system is to enhance the drivers' attitudes. It also provides high satisfaction, good behaviours, and resources for data analysis using the socialisation–externalisation–interaction model.

The importance of insurance markets in economic growth has been well-documented by economists for many years. For five highly polluting economies, this study examines an effective relationship between the insurance company's growth and CO_2 emissions using all the data from 1990–2019 [7]. In this study, a panel and time series NARDL framework is used. We discover that the growth of the insurance industry unevenly affects CO_2 emissions. The results suggest that a possible shock to insurance sector expansion in high-polluting economies enhances the CO_2. In contrast, a nasty shock and the development process of the insurance sector also significantly decrease.

The primary goals of insurance businesses is to see the innovative activities, the nature of the innovation environment, and how it affects the efficient insurance operations. The peculiarities of innovation introduction in the insurance sector have been determined [8]. Innovation for determining business activities and predicting the scope for efficient insurance operations is related to efficiency improvement. On a larger scale of utilisation, innovative insurance processes hold significance for being compatible with dynamic environmental conditions.

They examine the dynamic effects of the "CO_2 emissions in emerging economies from 1991 to 2018." The study adds to the body of empirical research on the subject. The results demonstrate that when we are using the "non- linear autoregressive distributed lag (NARDL) approach", The development process in the life insurance sector has significantly increased due to the CO_2 emissions in China, South Africa, and Russia [9]. However, the overall development process has been impacted due to the enormous carbon emissions in Russia. Moreover, it is quite a big shock for the Russian insurance companies, mainly impacting their long-term business. Additionally, findings demonstrate that when the non-life insurance market develops, CO_2 emissions rise in Russia and China; thus, South Africa is falling in that time.

An analysis of the macro innovation input–output link between the "science and technology insurance (S&T insurance) and regional innovation" is presented in this chapter. Using a dynamic panel regression model, the influence of S&T insurance on regional innovation is investigated using the "provincial panel data from 2010 to 2019" [10]. According to the findings, S&T insurance significantly promotes inputs of the innovation, but it hurts the outputs of the innovation. Additionally, the function of S&T insurance is subject to changes. S&T insurance significantly influences innovation in the eastern and central regions; thus, it is applicable or impacted in the western regions.

1.3 Newest Insurance Technology Trends

There is a change in the P&C insurance industry because of the omnichannel approach to customer service and digital insurance offerings. Insurance companies are looking for technology that can change as per the changing

Figure 1.1 Artificial Intelligence, Chatbots and Internet of Things represent the newest technology trends in the insurance industry [11].

demands and help them stay positive and ahead of their competitors [11]. Insurance companies have already employed these technologies. Some of the emerging insurance tech trends are shown in Figure 1.1.

1.3.1 Predictive analytics

The customer's behaviour can nowadays be predicted by predictive analytics. It can improve the accuracy and perfectness of the data too efficiently. Thus, proper utilisation is required in order to improve data accuracy. Most insurance companies use predictive analysis to anticipate trends, identify fraud risks, and risk selection. It is also used to increase the revenue and direct written premiums of many P&C insurers.

1.3.2 Artificial intelligence (AI)

Implementing AI or artificial intelligence technology is significantly increasing in the present-day world. The requirements of the customers are also changing day by day, and AI technology can provide personalised experiences to customers. It enhances the overall infrastructure of the insurance sector [12]. It changes as per the consumers' requirements, habits, and behaviours. AI technology also enhances efficiency, and this technology is focusing on automation, and it will have a significant role in the upcoming day's world.

1.3.3 Machine learning (ML)

Machine learning, or ML, is generally a bunch of artificial intelligence used to improve data validity and quality. It is improving the algorithm of the insurance industry through its digital technological support. An SMA survey found that more than “66% of P&C insurance executives believe that machine learning has a high impact potential for commercial lines of business, while 53% of executives believe it has a high impact potential for personal lines.”

1.3.4 Internet of Things (IoT)

Customers are always providing their essential information for insurance purposes, and the role of the IoT is to save their information and the insurance plan effectively. Insurance users can use IoT devices’ data and prevent losses [13]. The data can be used for various IoT products, such as wearable technology and smart homes. It also helps identify the possible risk and outcomes for the users. This technology is generally improving the techniques of data saving and enhancing the techniques of insurance saving.

1.3.5 Insurtech

Insurtech is generally an insurance technology that helps reduce the overall costs for the consumers and the insurance company and enhances the entire customer experience. The investment in Insurtech reached $4.9 billion in 2018, and it helps identify all the possible risks of the insurance effectively and helps mitigate all kinds of risks.

1.3.6 Social media data

The role of social media is evolving beyond marketing strategies in the insurance industry. It is significantly improving the risk assessment for P&C insurers [13]. It also helps the insurance company by enhancing the consumer’s experiences by investigating the insurance company’s fraud.

1.3.7 Telematics

Telematics capabilities will have an impact on auto policies. Telematics will benefit both the insureds and the insurers. It will effectively encourage better driving habits, and it also helps to change the relationship between the consumers.

1.3.8 Chatbots

As per the future estimates, “95% of the customer interactions will be done through chatbots by 2025.” The chatbots are based on artificial intelligence (AI) and machine learning (ML), which helps to save consumers' insurance time with their advanced technological support. The capabilities of chatbots are expected to increase in the upcoming days.

1.3.9 Low code

The insurers are responsible for managing the software platforms, which helps the new products effectively. This process uses skilled developers or an IT team and has made this process more accessible than before. It allows business stakeholders to manage apps and software using a user-friendly drag-and-drop functionality [14]. By implementing this, the insurers will be able to apply completely different, and it enhances the efficiency of the software by using a user-friendly user interface. This also helps to enhance the market speed, and it also helps to develop the organisation by implementing more existing features for the insurance company.

1.3.10 Drones

The drone is also a new and existing technological instrument that most sectors use and basically collects all the data. It collects all kinds of data by calculating all the risks and identifying them before starting the policy [15]. It saves the consumers from a significant loss and informs them about the possible risks or losses. It flies in the sky, can collect all kinds of data, and is used in the many stages of the life insurance company. Generally, it is used to improve the efficiency of the insurance and helps the carriers stay ahead of their competitors.

1.3.11 Robotic process automation (RPA)

Robotic process automation is also an advanced technology that depends on the automation process and is used in document verification. Consumers will have more options for self-service, and, over time, the requirement for back-office jobs will be less. People with a background in machine learning and data analytics will be required more. It ensures a practical collaboration between humans and machines.

1.3.12 Blockchain technology

Blockchain is generally a record including all the digital events that are conducted among the parties, and it is conducted for the participant. Information

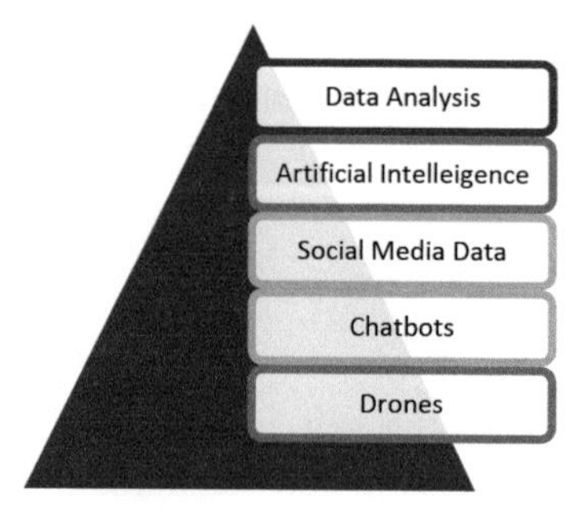

Figure 1.2 The technologies impacting the insurance industry include data analysis, Artificial Intelligence, Social Media Data, Chatbots and Drones [13].

on it cannot be erased, and people expect more in the case of blockchain in terms of faster services. It helps in risk assessment and fraud prevention.

Technology has changed the entire picture:
Technology is significantly increasing in the present-day world, and it is improving the overall lives of humans with its advanced technology [15]. The insurance industry is also dependent on technology and is expanding its technological innovation daily. Driving innovation is the most improved technology; this industry already completes a huge milestone through technology. Technology generally improves the overall experiences of the consumers, and it provides more customisation to the consumers.

The insurance market of the world is also implementing a lot of technology in the companies to improve their overall work practice. Technological innovation includes digitalisation and is possible due to improved technological support. The new technology generally helps identify all kinds of fraud activities and provides a good risk analysis to the insurance company. Companies are implementing it to improve their business for the upcoming days worldwide [16]. Thus, much improvement is needed for the insurance industry to help it increase its overall efficiency quickly in the upcoming days.

1.4 The Top Five Technologies Impacting the Insurance Industry

Implementing advanced technology in the insurance sector has seen growth over the years. Advancement in technology is now benefiting insurance agencies and customers. The insurers are using these technologies to do something better than the competitors, to improve the workers' compensation coverage and the coverage for other business-related insurances (Figure 1.2).

- Predictive analysis of data

- Artificial intelligence
- Data from social media
- Chatbots
- Drones

1.5 The Innovations that are based on the Technology in the Insurance Industry

1.5.1 Increase distribution and penetration process in the existing market

The fundamental goal of the insurance company is to expand their business towards the world, and the enhancement rate should be high. The insurance industries are taking some emerging strategies with high growth rates and increase in size, and now it can be done quickly with the help of technology.

1.5.2 Customer expectation

Customer satisfaction is also a significant factor for the insurance industry, and they always need to provide good consumer service. Mainly, they need to focus on the communication process and update it with digitisation [5]. As per the Global Consumer Insurance Survey, "80% of customers are willing to use digital and remote channel options for fulfilling different tasks."

1.5.3 Technology at its peak

Technologies like digital platforms, cloud computing, and telematics are delivering a lot of new and effective ways to control insurance risks. It also helps the insurance company identify all kinds of fraud and can improve the overall transaction in the insurance industry.

1.6 Systems Optimised by Implementing Technology

By implementing technology, some of the systems can be optimised, which are given below.

1.6.1 Internal workflow automation

Insurance companies can transfer their work with full automation by implementing new and advanced technology. Robots can be involved in the business process to eliminate all risks identified through automation [8]. Automation is required to increase the data quality and identify all the risks in the insurance industry.

1.6.2 Hassle-free insurance

The new technology also enhances the insurance's reliability; the ageing process needed to be more time-consuming. Hence, the new technology can minimise the period of its exceptional technology.

1.6.3 Telematics insurance

It is one of the crucial innovations that are based on digitisation, which is having an impact on the insurance industry. It can be used to increase customer interaction and increase profit due to specialisation. Risk assessment and calculating the premium cost can also be done through it.

1.7 Top Insurtech Innovations

The rate of customers around 61% prefers digital tools to monitor their applications' status. The insurers spent around 225 billion dollars on IT in 2019. According to the report of McKinsey, among the insurance companies, 90% of them have identified that infrastructure is a barrier in the way of digitalisation [16]. Therefore, Insurtech innovation depends on technology adoption. Some corporations make changes by going for start-ups and establishing innovation labs. They are trying to address problems like the complicated internal process of insurance and solving the problem of service providers. As per the PWS survey, 41% of consumers are willing to switch to the insurance companies they are dealing with to achieve more digitalisation.

1.7.1 Internal automation of workflow with machine learning and RPA

The cost of routine work can be reduced with the help of automation, which will help employees to focus more on creative tasks. As per the sources, the company shall be able to save 1.25 million dollars within the first year if AI is used. Using "robotic process automation" (RPA) addresses routine tasks; however, it does not involve machine learning regularly.

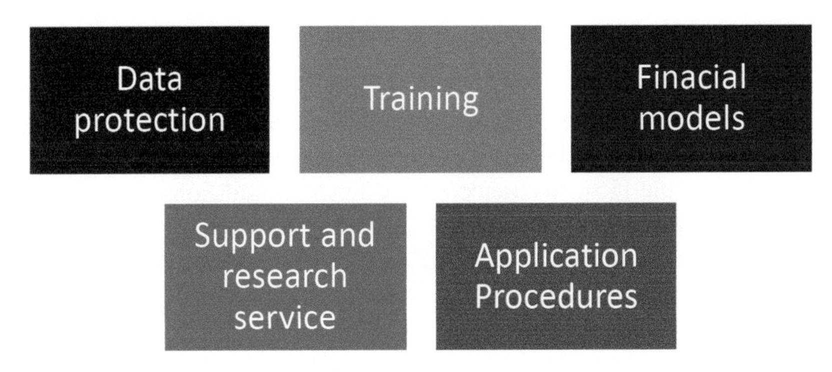

Figure 1.3 The data sharing framework includes data protection, training, financial models, support and research services and application procedures.

1.7.2 Digitisation of paper records with the help of optical character recognition

The document can be reviewed and rejected automatically if there are any sorts of inconsistencies among the digitised files. The management of online documents is emerging as a more common factor increasingly. However, the agents are detected to deal with various documents [5]. These documents can be in email, paper, scans, and other formats. It is time-consuming for humans to translate the data into structured records. That is why there is a need to digitise paper records.

1.7.3 Redefinition of the traditional ways of policy management and claims during the age of digital insurance

The broader spectrum of software solutions impacts claims management besides AI-driven automation. Claim management can be denoted as a necessary business process. This generally begins with the registration of claims but ends with insured party payments. The usage of claim management software reduces manual workflow. The number of man-to-man interactions also gets affected by this factor.

1.7.4 Usage of IoT and social media in personalised insurance pricing

Social media and endpoint devices can provide large amounts of confidential data. This approach can help both customers and insurers (Figure 1.3).

1.7.5 Disruptive business model – P2P insurance

Due to the available businesses related to technology, it is gaining popularity as one of the most disruptive business models. Creating a single pool consisting of premium shares, the risks would be covered and agreed upon by the network of people [15]. Insurance companies and traditional intermediaries are not required for the P2P model. The main milestones of insurance distribution, along with the insurance carrier and self-governance, have already been passed by P2P insurance.

1.7.6 Innovation through insurance APIs

Industry innovation can be slowed down by the lack of technology and over-regulation; it can be denoted as a harmful procedure for the customers' experience. Innovative and flexible experiences can be denoted as the primary demand of the customers. According to a survey by IBM, the carrier of 41% of insurants gets abandoned when trying a new one as the provider needs to adjust to customers' needs. Insurance APIs that insurrectos can utilise are as follows:

- AXA Insurance
- The Lemonade Public API
- NAIC Registry

1.7.7 Industries are brought to a new level by insurance fraud detection software

Fraud detection software is rising. Cloud and mobile technologies support insurance agents in dealing with duplicated claims along with fake diagnoses and internal employee scams in real time [2]. Claim data is compared with the system's database to identify fraud. The shift technology solution is an example based on probability analysis that goes beyond traditional claim scoring.

1.7.8 Product distribution is brought to online space by the insurance marketplace

The web is helping people to be more aware of the offerings within the market, and they easily compare the products and find plans that match particular requirements. Insurance shopping platforms are actively redefining distribution models. Marketplaces help insurers in terms of cutting distribution costs along with bringing in the most targeted leads.

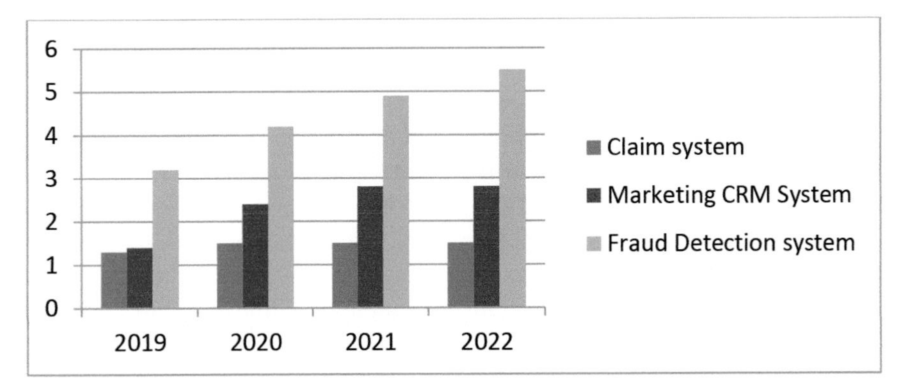

Figure 1.4 The importance of IT in the insurance industry is represented by claim system, Marketing CRM system and fraud detection system [2].

1.7.9 Insurance data analytics: data pipelines and ingestion

The importance of data analytics is acknowledged by 50% of insurance executives. According to the survey, the second most critical digital innovation is analytics after cloud-based technologies.

1.8 What Should the Insurance Industry Expect?

The insurance industry expects much improvement regarding the insurance policy and all the related work practices. The new technology can rectify all the errors of the insurance and make a more competitive advantage in the upcoming days in the world. It allows cutting of premium costs by 10% and 8% in claim expenses; it helps in more accurate risk assessment, better customer experience, move to proactive decision making from reactive decision making and expanding portfolio [17].

However, the insurance industry should properly implement technology because consumer demands are changing daily [2] (Figure 1.4). Moreover, it is challenging to meet all the requirements of the consumer; hence, if the insurance industry implements the new technology correctly, the insurance industry will be able to improve its business in the upcoming days.

1.9 Literature Survey

Insurance technology's utilisation helps improve different operational sectors within the insurance industry. Technology for communication, for example, can be identified as one of the primary functions for using insurance technology to demonstrate transparency and effective communication between the client and the insurance company. Chatbot technology is used for insurance-related activities, demonstrating ambivalent perceptions, beliefs, and attitudes in alignment with the users or potential users [18]. On the other hand, considering the formative structure of IoT and its potential usage in the insurance sector instils confidence among organisations. Technological investment in IoT can establish an interconnection between different data storages and systems, enabling efficient marketing and communication between consumers and enhancing engagement [19]. The impact of advanced technology is highly significant in insurance management. Therefore, reviewing the prospects and advantages of insurance technology is illustrated in the paper, highlighting the scope for future implementation.

1.10 Advantages of Implementing Insurance Technology

Entrepreneurial business welcomes technological changes that impact the entrepreneurial ecosystem. In the present age, the development of fintech companies indicates technology's potential advantages in financial activity management. Advancement of financial technology contributed to a substantial increase in the USA regarding the entry of new entrepreneurial businesses in the market [20]. Amplifying economic or financial activities is a prominent advantage of technology, supporting the management of multifarious activities. During the COVID-19 pandemic, the position of technology usage acquired greater significance. The advantages of remote operations helped the insurance industry aid organisations and customers during the severe economic crisis and professional stagnation. Insurtech was used during the COVID-19 pandemic to establish communication between the policyholders and accommodate the changes in the insurance ecosystem by facilitating data processing operations, remote data analysis and identification, acceleration of participation, and so on [21]. Enhancing the capacity for accommodating the market's growing demands is a critical advantage identified with insurance technology implementation.

The prospect of strategic alliance and management can be correlated with insurance technology and its utilisation in different areas of activities. The use of technology optimally can guarantee the management of asset–liability matching and strategy implementation [22]. The involvement of financial technology for insurance businesses can increase the overall business prospects with specific protection for policyholders. In essence, the distribution of resources can be optimally analysed and measured with financial technology used for insurance activities, supporting the customers effectively. Using big data analytics technology for insurance activities helps conduct usage-based insurance activities [23]. The scope for management and monitoring data using this technology can provide sustainable and long-term benefits, especially for risk management. Big data analytics conducts a prediction of market trends along with the considerations of behavioural and emotional factors among customers while using insurance services. The scope for beneficial technology usage for insurance activities helps convert traditional methods into optimised modern operations methods that benefit both the customers and the insurance companies. Therefore, the projected advantages of insurance technology for practical usage demonstrate the ability to create a smooth flow of operations with optimal risk management.

1.11 Conclusion

Insurance companies can adopt advanced technologies for innovative and flexible practices, ensuring that customers are provided with sufficient support and optimum experience. Insurance activities are related to extensive scale data management and the development of proper frameworks for protecting the interest of policyholders. On the other hand, the emergence of insurance technology helps to manage data remotely, and the documentation process is simplified with the adoption of automated information management systems. Implementing automation for business processes helps eliminate risk factors, and it can store and manage confidential data without high-risk factors. The volatility of the current market space is also considered in the paper to demonstrate that the insurance industry requires support and safety for transactional activities. Customer satisfaction is addressed with the incorporation of financial technology that helps to develop transparent and effective communication channels between customers and insurance companies. The communication process also helps the customers to conduct insurance-related activities with a user-friendly interface. Therefore, considering the current market trends and the advancement of technology regarding insurance activities marks the advantages of insurance technology in terms of data analysis, AI implementation, and social media data management for marketing, chatbots, and drones.

References

[1] Kar, Arpan Kumar, and L. Navin. "Diffusion of blockchain in insurance industry: An analysis through the review of academic and trade literature." Telematics and Informatics 58. 2021: 101532.

[2] Ashwini, D. N. (2022). Progress of banking and insurance sector in India–an economic overview.

[3] Eling, M., Nuessle, D., &Staubli, J. (2022). The impact of artificial Intelligence along the insurance value chain and on the insurability of risks. The Geneva Papers on Risk and Insurance-Issues and Practice, 47(2), 205-241.

[4] Li, M., & Gao, X. (2022). Implementing enterprises' green technology innovation under market-based environmental regulation: An evolutionary game approach. Journal of Environmental Management, 308, 114570.

[5] Klumpes, P. J. M. (2022). Consolidation and efficiency in the major European insurance markets.

[6] Liu, L., Li, W., He, W., & Zhang, J. Z. (2022). Improve enterprise knowledge management with the internet of things: a case study from the auto insurance industry. Knowledge Management Research & Practice, 20(1), 58–72.

[7] Li, X., Ozturk, I., Ullah, S., Andlib, Z., &Hafeez, M. (2022). Can top-pollutant economies shift some burden through insurance sector development for sustainable development? Economic Analysis and Policy, 74, 326–336.

[8] Matiyazova, S. R. (2022). The Process of Formation of an Innovative Environment in Competitive Conditions in the Insurance Market. Oriental Journal of Economics, Finance and Management, 2(1), 7–11.

[9] Rizwanullah, M., Nasrullah, M., & Liang, L. (2022). The asymmetric effects of insurance sector development on environmental quality: challenges and policy options for BRICS economies. Environmental Science and Pollution Research, 29(7), 10802–10811.

[10] Wang, Z., Yin, H., Fan, F., Fang, Y., & Zhang, H. (2022). Science and technology insurance and regional innovation: evidence from provincial panel data in China. Technology Analysis & Strategic Management, 1–19.

[11] Xiong, Q., & Sun, D. (2022). Influence analysis of green finance development impact on carbon emissions: an exploratory study based on fsQCA. Environmental Science and Pollution Research, 1–12.

[12] Feng, S., Zhang, R., & Li, G. (2022). Environmental decentralization, digital finance and green technology innovation. Structural Change and Economic Dynamics, 61, 70–83.

[13] Seo, D., & Kwon, Y. (2022). An Analysis of Productivity and Efficiency in Indian Non-Life Insurance Companies: DEA-Based Approach. Journal of the Korea Convergence Society, 13(3), 217–225.

[14] Cummins, J. D., & Rubio-Misas, M. (2022). Integration and convergence in efficiency and Technology gap of European life insurance markets. Annals of Operations Research, 1–27.

[15] Apella, I., & Zunino, G. (2022). Technological change and labour market trends in Latin America and the Caribbean: a task content approach. CEPAL Review.

[16] Bednarz, Z., &Manwaring, K. (2022). Hidden depths: The effects of extrinsic data collection on consumer insurance contracts. Computer Law & Security Review, 45, 105667.

[17] Born, P., & Bujakowski, D. (2022). Economic transition and insurance market development: evidence from post-communist European countries. The Geneva Risk and Insurance Review, 47(1), 201–237.

[18] Rodríguez Cardona, Davinia, et al. "A mixed methods analysis of the adoption and diffusion of Chatbot Technology in the German insurance sector." (2019).

[19] Liu, Lixin, et al. "Improve enterprise knowledge management with internet of things: a case study from auto insurance industry." Knowledge Management Research & Practice 20.1. 2022: 58–72.

[20] Berman, A., Cano-Kollmann, M., &Mudambi, R. (2022). Innovation and entrepreneurial ecosystems: fintech in the financial services industry. Review of Managerial Science, 16(1), 45–64.

[21] Volosovych, Svitlana, et al. "Transformation of insurance technologies in the context of a pandemic." Insurance Markets and Companies 12.1. 2021: 1–13.

[22] Li, X., Xie, Y., & Lin, J. H. (2022). Life insurance policy loans, technology choices, and strategic asset-liability matching management. Emerging Markets Finance and Trade, 58(7), 1838–1847.

[23] Arumugam, Subramanian, and R. Bhargavi. "A survey on driving behavior analysis in usage based insurance using big data." Journal of Big Data 6.1. 2019: 1–21.

[24] Shalender, K. (2021). Building Effective Social Media Strategy: Case-Based Learning and Recommendations. In *Digital Entertainment* (pp. 233-244). Palgrave Macmillan, Singapore.

[25] Gupta, M., Bansal, R., Hothi, B. S., & Shashidharan, M. (2021). Overview And Growth Of Micro, Small And Medium Enterprises In Punjab. *Academy of Entrepreneurship Journal*, *27*, 1–7.

[26] Rana, A., Bansal, R., & Gupta, M. (2022). Emerging Technologies of Big Data in the Insurance Market. In *Big Data: A Game Changer for Insurance Industry* (pp. 15–34). Emerald Publishing Limited.

[27] Rana, A., Bansal, R., & Gupta, M. (2022). Big Data: A Disruptive Innovation in the Insurance Sector. In *Big Data Analytics in the Insurance Market* (pp. 165–183). Emerald Publishing Limited.

[28] Sood, K., Seth, N., & Grima, S. (2022). Portfolio Performance of Public Sector General Insurance Companies in India: A Comparative Analysis. In Simon Grima, Ercan Özen, Inna Romānova (Ed.), In *Managing Risk and Decision Making in Times of Economic Distress, Part B*. (pp.215–229). Emerald Publishing Limited,UK.

[29] Grima, S., & Marano, P. (2021). Designing a Model for Testing the Effectiveness of a Regulation: The Case of DORA for Insurance Undertakings. *Risks*, *9*(11), 206.

2

Analysing the Impact of Emerging Technologies on the Existing Insurance Sector

Gagandeep, Jyoti Verma, and Amandeep Singh

Chitkara Business School, Chitkara University, Punjab, India

Email: profgaganabdc@gmail.com; jyotiverma17@gmail.com; amandeep.garai@chitkara.edu.in

Abstract

Insurance can be considered as a societal scheme to minimise the threat of damage of life and non-life items. Many emerging technologies like blockchains, big data, Internet of Things (IoT), and Insurtech have created industrial revolution in the insurance sector. The existing insurance companies face a dilemma due to mismatch between the rates of digital transformation and rates of scientific advancements in the insurance trade. This study analyses the impact of emerging technologies on the existing insurance sector. Many databases like Ebscohost, Emeraldinsight, and Jstor were explored for the relevant literature using keywords. This study identified that the emerging technologies directly impact the insurance sector based upon the factors like competency of their processes, focus on customers, and advantages over competitors, whereas there are many factors that are creating many challenges for the insurance companies. These challenges can be handled with the help of human capital, better leadership, and corporate innovations. The insurance companies can understand the impact of emerging technologies on the existing insurance sector and can tailor strategies to enhance the overall insurance penetration.

2.1 Introduction

The term insurance can be considered as a societal scheme to lessen or reduce the threat of damage to life and non-life items. In India, insurance is generally considered a tool to save on tax rather than the rest financial aid. Due to the entrance of many companies supported by FDI, the financial marketplace has to turn out to be the backbone of the economy. Every modern economy, whether developed or not, is in the need of a strong and emerging insurance sector. The insurance sector provides major avenues for public welfare like motivating the public to save money, acting as a safeguard against many enterprises' losses and acting as a long-term source of funds for the overall development of the nation. The contribution of the insurance sector towards a healthy economy can never be ignored. Insurance provides protection as well as coverage against many natural calamities and tragedies. Insurance also contributes to minimising the risk associated with the individual's health, life, and property. To support all the above benefits, the growth of the insurance business has become mandatory.

2.2 Background: Origin and Growth of Insurance

In 1818, life insurance in India originated when the oriental life insurance company was started in Kolkata. It was considered as earnings to offer for English widows. During that time, the premium for Indian lives was higher than non-Indian lives. The origin of non-life insurance in India can be traced to the Triton Company Ltd., the first non-life insurance company established in 1850 in Calcutta by the British. In 1870, the Bombay mutual life insurance society started its business [1]. Regulation of insurance was legally initiated in India with the approval of the Life Insurance Companies Act of 1912 [2]. The insurance industry expanded at a rapid pace after freedom. Despite its growth, insurance remained an urban phenomenon [21, 22].

In the past few decades, the rate at which technology is changing has been multiplied and targets to expand by considerably widening the possibility of one or more of its dimensions like customer group, customer function, and technology [3–5]. The factors like convergence, innovation, effectiveness, speed, and buyer preferences indicated the emergence of technology [6]. The emerging technology and digitalisation have already had a far-reaching impact on the insurance service industry. Many new market entry opportunities are being created by the growing penetration of the internet along with emerging technologies, thereby increasing the pressure to withstand the market competition. These emerging technologies in capsule form may be formulated as new technological goods and facilities designed for customers through the application of scientific knowledge.

The insurance business constitutes the biggest and most significant business segment in the customer services industry [7]. The performance of the insurance sector can be evaluated based on factors like their level of penetration as well as density. The division of the premium underwritten in one year and the respective GDP of the same year resulted in penetration value. The Indian insurance industry is based upon 57 insurance firms out of which 24 companies are dealing with life insurance businesses including products such as life, health, debility, and death while the remaining 34 are dealing with general insurance consisting of motor, commercial, property, and casualty. This trade is estimated to rise at a CAGR of 5.3% from 2019 to 2023. In the financial year 2021, the rate of penetration of Indian insurance was estimated at 4.2% out of which 3.2% and 1.0% are contributed to long-term and short-term insurance, respectively. According to S&P Global Market Intelligence data, in Asia-Pacific's insurance technology market, India acts like a challenger against the market leader. Due to the advancement of technologies, many companies entered into strategic alliances for providing cyber insurance. The persistent changes in the regulatory framework can lead to a promising future for insurance companies [8].

2.3 Research Problem

No doubt, insurance is a potential as well as a profitable business, but most of the rural public is still deprived of insurance services due to the slow rate of penetration, the demand for its services, and lack of awareness in such areas. Typically, the emerging technology in India is mainly focused on enclosure efforts directing the under-deprived marketplaces that are now disrupting the traditional way of insurance with cutting-edge technology through internet and mobile technologies to increase the reach, frequency, and impact of affordable insurance to the target market. The existing insurance companies face a dilemma due to a mismatch between the rates of digital transformation and the rates of technical advancements in the insurance trade [9]. The current insurance companies are left with just three options either to adapt or collaborate or eliminated the emerging technologies [3, 10, 11]. The present study attempts to study the challenges faced by insurance companies, the potential level of emerging technologies, the gaps available, and the abilities to close these gaps. The purpose of the study is to help insurance companies in understanding the areas mostly affected by emerging technologies.

2.4 Literature Review

This literature review is based upon the critical appraisal of information about the introduction of the digital age to insurance services and explores the latest emerging technologies and their uses in the existing insurance industry. In order to improve performance and efficiency, many tools like high-tech technology as well as digital data are required to ease the digital transformation of insurance companies [12]. To become a leader in the market, they need to adopt digital technology to revise the market presentations. Along with the ways to generate more profits and opportunities, market players must be aware of the issues that arise due to digitalisation. Only with the help of a hybrid combination of unique research, latent opportunities, and the issues related to digitalisation, the insurance business can transform itself to suit the era of online business.

Many new market opportunities have been generated for technology providers due to the growing penetration of the internet and emerging technologies. This not only increased the level of competition but also increased the market pressure to perform [4].

Some researchers explored how digitalisation and digital knowledge affect marketplaces, companies, and commercial organisations based on an international perspective [13]. They explored the technical aspects of digitalisation like in what way digitisation generates and maintains significance in a business, how corporate based on digitalisation generates novelty, how multinational firms can grow the efficiency inside the digital economy, and in what way commercial organisations and institutions changed because of new scientific knowledge. Some highlighted the prospects and threats of digitalisation for the worldwide economy at both the micro and macro levels [14]. They highlighted new advances in technology in areas such as blockchain, cryptocurrencies, and the Internet of Things, and examined the issues such technologies can create for rendering public services. The researcher illustrated how the effect of digitisation can influence public welfare areas like health, education, and infrastructure.

The investments made by leaders in the insurance market are going to give threats to the market challengers, and market followers as this sector are on the edge of transforming itself due to the digital era. The digital vision of insurance leaders is going to change the business definition, alternative technologies, and primary cum support activities, events, trends, issues, and expectations of customers. Due to big data and the Internet of Things digital revolution, companies need to pass through vicissitudes of vectors like technology, competition, or management fashion [15].

In a few decades, due to the digital revolution, many latest scientific knowledge-based tools like blockchain, big data, and the Internet of Things have become popular due to their wider acceptance across financial industries including insurance. Due to the digital design of these technologies, more value can be added to the primary as well as support activities of insurance by reducing the level of managerial efforts required for them [16].

2.5 Research Methodology

The exploratory research was applied to research as it helps to recognise the problem, explain the problem, develop a methodology for the problem, identify the key variables, and interpret primary data more insightfully [17]. For this study, various electronic databases like bibliographic databanks are based on citations to articles. Numeric databanks comprise statistical and numerical facts. Full-text databanks comprise the thorough manuscript of the source papers consisting of the databank and directory databanks deliver facts on persons, establishments, and services that are being referred [17, 18].

Exploratory research is referred to as building a healthier understanding of a business issue at the times when a researcher is unable to frame an elementary statement of the problem under study [18]. Exploratory research gives valuable insight and generates ideas and hypotheses rather than measuring or testing them [19].

2.6 Results

The present study finds that there are challenges facing existing insurance companies that are related to change caused by the emerging technologies in the insurance business sector. The insurance companies need to re-frame their existing revenue model due to changing consumer expectations of the consumers [4]. The existing legal constraints as well as obsoleted infrastructure within the existing insurance companies leading to inflexibility and inactiveness in terms of their performance are a thoughtful weakness for the insurance industry. The inability of the insurance companies to adapt fast enough to leverage the technological advantage is jeopardising their business operations.

The emerging technologies have a positive impact on the insurance companies on the basis of factors like efficiency of their operations, focus on customers, and advantages over competitors [20]. For this, the insurance companies need to frame customised services to satisfy each customer preferences and improve insurance services such that it generates more value for their customers. This would build a strong relationship between the external and internal communication, motivating the customers from the buying to policy renewal with better reach, frequency and impact rates, new distribution channels, and high claims focus. The emerging technologies can create many rewards for withstanding the market pressure for existing insurance companies, as with the help of such technique's companies can dynamically choose superior options, create additional correct customer profiles, and work creatively. Based upon latest technologies, insurance companies can save their costs through speed and simplicity mainly in claims processes.

In order to fill the prevalent gaps to in-cash the opportunities of latest technologies, the companies need to address the core elements of their existing business which are related to target market, value proposition, and revenue [3]. With this, the existing insurance player can move closer to the revenue model of a player based on emerging technology. Thus, addressing the said core elements is necessary for the insurance companies to design suitable response to emerging technology.

In order to minimise the knowledge gaps between traditional insurance trade and business based on cutting edge technology, the companies need to focus on their inherent capabilities like organisation structure, leadership styles, and innovation. In the technology space, the innovative human resources would be required by the insurance companies based on innovative technology. The synergic effects of human resources and leadership style can lead to innovation in order to support the insurance companies to minimise the knowledge gaps in the present business definition [3].

2.7 Conclusion

The study identified that the present insurance companies are knowing emerging technologies related to the insurance market. So, insurance companies covering most of the market share in the insurance sector need to invest more into emerging technologies to generate innovative methods of generating and maintaining worth for clients. The issues faced by insurance companies due to emerging technologies are drivers propelling the growth of technology. So managers need to study their traditional ways of insurance business to understand the origin of such challenges. The managers need to build the core as well as distinctive competency to reap the benefits of emerging technologies through human capital and corporate innovative efforts. As a customer is a king, leaders are obliged to focus on developing an emphatic association with the client to recognise their core need, want, and demand. Managers should create an environment that fosters risk-taking and experimentation with new technologies. To leverage the benefits of emerging technologies, managers need to recruit, select, and train people with familiarity and expertise in the field of data science, data analytics, etc.

References

[1] Srivastava, A., & Tripathi, S. Indian life insurance industry-The changing trends. *Researchers world*, *3*,2, 93, 2012.

[2] Venkatesh, M. A study of trend analysis in insurance sector in India. *The International Journal of Engineering and Science (IJES)*, *2*, 6, 01–05, 2013.

[3] Catlin, T., & Lorenz, J. T. Digital disruption in insurance: Cutting through the noise. McKinsey & Company. Retrieved from https://www.mckinsey.com/industries/financial- services/our-insights/time-for-insurance-companies-to-face-digital-reality, 2017.

[4] Lee, I., & Shin, Y. J. Fintech: Ecosystem, business models, investment decisions, and challenges. *Business horizons*, *61*,1, 35–46, 2018.

[5] Verma, J. Application of Machine Learning for Fraud Detection–A Decision Support System in the Insurance Sector. In *Big Data Analytics in the Insurance Market*. Emerald Publishing Limited, 251–262, 2022.

[6] Shim, Y., & Shin, D. H. Analyzing China's fintech industry from the perspective of actor–network theory. *Telecommunications Policy*, *40*, 2-3, 168–181, 2016.

[7] Pollari, I. The rise of Fintech opportunities and challenges. *Jassa*, 3, 15–21, 2016.

[8] Varma, P., Nijjer, S., Sood, K., Grima, S., & Rupeika-Apoga, R. (2022). Thematic Analysis of Financial Technology (Fintech) Influence on the Banking Industry. *Risks*, 10(10), 186.

[9] Lai, R., Chuen, D. L. K., & Deng, R. Handbook of blockchain, digital finance, and inclusion. *Volume*, *2*, 145–177, 2018.

[10] Puschmann, T. Fintech. *Business & Information Systems Engineering*, 59,1, 69–76, 2017.

[11] Li, Y., Yan, C., Liu, W., & Li, M. A principle component analysis-based random forest with the potential nearest neighbor method for automobile insurance fraud identification. *Applied Soft Computing*, *70*, 1000–1009, 2018.

[12] Mezghani, K., & Aloulou, W. (Eds.). *Business Transformations in the Era of Digitalization*. IGI Global. https://doi.org/10.4018/978-1-5225-7262-6, 2019.

[13] Ratajczak-Mrozek M. & Marszałek P. *Digitalization and firm performance : examining the strategic impact*. Palgrave Macmillan. https://doi.org/10.1007/978-3-030-83360-2, 2022.

[14] Herberger, T. A., & Dötsch, J. J. *Digitalization, Digital Transformation and Sustainability in the Global Economy*. Springer International Publishing. https://doi.org/10.1007/978-3-030-77340-3, 2021.

[15] Vander Linden, S. L., Millie, S. M., Anderson, N., & Chishti, S. *The insurtech book: The insurance technology handbook for investors, entrepreneurs and fintech visionaries*. John Wiley & Sons, 2018.

[16] Avdeev, E. Application of the Blockchain Technology in the Insurance Industry (Doctoral dissertation, Universität St. Gallen) ,2018.

[17] Nunan, D., Malhotra, N. K., & Birks, D. F. *Marketing research: Applied insight*. Pearson UK, 2020.

[18] Hair Jr, J. F., Babin, B., Money, A. H., & Samouel, P. Essentials of business research methods: Johns Wiley & Sons. *Inc., United States of America,* 2003.

[19] Naveenan, R. V., Madeswaran, A., & Arun, K. R. Green Banking Practices in India-The Customer's Perspective. *Academy of Entrepreneurship Journal*, 27, 1-19, 2021.

[20] Baden-Fuller, C., & Haefliger, S. Business models and technological innovation. *Long range planning*, 46, 6), 419-426, 2013.

[21] Sood, K., Seth, N., & Grima, S. (2022). Portfolio Performance of Public Sector General Insurance Companies in India: A Comparative Analysis. In Simon Grima, Ercan Özen, Inna Romānova (Ed.), In *Managing Risk and Decision Making in Times of Economic Distress, Part B.* (pp.215–229). Emerald Publishing Limited,UK.

[22] Grima, S., & Marano, P. (2021). Designing a Model for Testing the Effectiveness of a Regulation: The Case of DORA for Insurance Undertakings. *Risks*, *9*(11), 206.

3

The Impact of Technology in the Health Insurance Sector

Manju Dahiya[1], Ashok Maurya[1], Naman Mishra[1], and Ercan Ozen[2]

[1]Galgotias University, India
[2]Department of Finance and Banking, Uşak University, Türkiye

Email: manju.dahiya@galgotiasuniversity.edu.in;
ashok.maurya@galgotiasuniversity.edu.in; mnaman225@gmail.com;
ercan.ozen@usak.edu.tr

Abstract

The main aim of this study is to focus on the impact of technology on the health insurance field. The change in the growth rate of the health insurance sector concerning its work efficiency and time management, which is affected by the inclusion of technology, is anatomised and is therefore used to show the boost technology provides to this sector economically. Including new technologies like artificial intelligence, big data, and cloud computing would help lessen people's workload and increase the time efficiency of the industry. After successfully conducting the research, the inclusion of technology is proving to be the boost that the health insurance industry requires, as it reduces unwanted economic burdens and makes the process less time-consuming. We can show the positive impact of technology in terms of workload efficiency and time management over the past years. The study is significant because it highlights the importance of technology in health insurance. Since there are many doubts regarding the inclusion of technology, this study will clear them as it shows the positive impact in terms of work efficiency, which has been reduced as well as time consumption, which will significantly decrease with the inclusion of new technologies.

3.1 Introduction

We live in a society where every step we take is bolstered with the help of technology, and whatever task we do; we use the service of emerging technologies to ease our efforts and make our lives easier. The new age technologies usher an era with many opportunities for ordinary people in every sector, including but not limited to the agriculture, health industry, textile industry, and insurance sector. This study mainly focuses on the fact that efficiency, as well as the quality factors of the industry, are impacted by the inclusion of technology.

Before delving deep into the topic, what is essential for us is to understand the topic on which the study is being conducted, which is the basic meaning of the term insurance. Insurance is a contract in which a corporation or the government agrees to give a reimbursement guarantee for defined loss, damage, disease, or death in exchange for a certain premium [6]. Including insurance with intricate technologies such as artificial intelligence, predictive analysis, big data, etc., has been slow but is very important and impactful in this new era [23, 24].

There has been a tremendous change in the inclusion of technology in this sector, resulting in the fruition of a new type of sector, named the Insurtech sector, which is the amalgamation of emerging technologies and the insurance sector. These types of companies aim to make their business grow with the help of new and emerging technologies as well as develop new products and services and reduce the financial burden, which is upon the shoulder of both the insurer and the client [3]. It is the advent of the fourth industrial revolution, which focuses not only on the efficiency factor of the industry but also on the inclusion of the economy, which is helmed by knowledge; this acts as a catalyst for the growing use of technology in the insurance sector [5]. The pillars of interconnection characterise this revolution (allowing people to be able to connect easily over the internet), technological support (helping people to make decisions with the help of technology), transparency (allowing people to keep faith in the system), and unbiases (allowing technology to decide without any bias leading to a more just outcome). This revolution is heralded with the help of linking everyone with the internet, the use of mobile telecommunication devices, the Internet of Things (IoT) to connect devices to the internet, and a vast sea of computing resources. These extravagant opportunities have found their use even in the insurance sector, which includes an intricate balance of its banes and boons.

3.2 Literature Review

Many opinions are floating in the research on how technology impacts the insurance sector and which technologies are currently being used. Some of the pieces of literature are given below.

Roberto Rocha focused on how the insurance sector has been performing in different developed or developing countries. They have also examined the reasons that drive the development of this industry [9]. Bill Chen studied how mandatory insurance affects the daily life of a person and how it will lead to long-term effects that pertain to the growth of the country's economy. He also weighs the pros and cons of having such a mandatory thing [10]. Vikas Sharma contributed to this field by studying the effects of privatisation on the insurance industry and what effect it will have on the pockets of the general public. He also highlighted what it would mean to the country when there is more push for privatisation [12]. Balamurugan Muthuraman and Karthik Mohandoss studied the trends in the insurance industry with a primary focus on the Indian insurance industry. They compared the different aspects used to determine the industry's growth, like GDP, growth ratio, etc., to gauge the trends in the Indian insurance industry. Sukanta Sarkar et al. studied the insurance industry's journey from its time in British India to the current era, wherein the sector has become more liberal and flexible. They focus on the journey the insurance sector has seen and the changes it has undergone [8]. Matthias de Ferrieres studied how the insurance industry is lacking in implementing technology and what factors majorly digress in its implementation [1]. He also studied the impact digitalisation would leave on the already existing major players in the field, whether they would embrace the change gracefully or would it be too harsh for them to shift their fundamentals. Alexander Bohnert, Albrecht Fritzsche, and Shirley Gregor analysed the relationship between the economic growth using technology and the business performance of numerous European insurance companies [2]. They were able to establish that technology's role is crucial when it is implemented with utmost care in both the internal as well as external functioning of the companies. Klapkiv Lyubov reviewed the various technological innovations in the investment industry. She has also studied the driving factors in the field of the insurance industry as well as the effect of technology on the insurance value chain; she also pointed out the balance between the positive and negative effects the insurance industry will experience with technological innovations. Nandhini Muniappan studied the impact of emerging technologies on the insurance sector, primarily focusing on the LIC. She also highlighted the different factors impacted by the use of technology and the fields in which technology can be implemented. Dr. Pooja Choudhary, in her study, highlighted the impact that the insurance sector has on the financing aspect of the country [11]. She examined how this industry drives the country's economic processes and how any change in the same creates a ripple effect in the country's economy. Vivek Nagarajan and Samuel Selvan studied the different technologies and how their incorporation into the insurance sector would help the sector's growth and provide the necessary economic boost [13]. Christian Eckert and Katrin Osterrieder studied the different technologies used in the insurance sector and the impact they have been creating. They could find the links between different technologies and how they depend on each other for smooth functioning; they also studied the requirements the company must fulfil to implement these technologies. Antonella Cappiello directed her study to understand the management of the industry and how the dynamics will change when we include technology into the mix. She has also suggested that the companies will have to change their business model to be able to compete in this ever-changing, dynamic race for survival. Deniz Guney Akkor and Suna Ozyuksel analysed the effect of new technologies on the already existing business industries with a primary focus on the insurance sector. They also studied the effect of digital transformation on the workforce that is already present in the insurance sector; the impact of industry 4.0 is also clearly highlighted in this study. Lili Zheng and Lijun Guo directed her study to one of the most known emerging technologies, Big data, and its impact on the insurance industry [4]. They highlighted the importance of data in this tech-savvy era; data is considered the pulse of the new revolution, and therefore being able to analyse this provides an upper edge to any industry; therefore, the insurance industry is not far from this aspect [7]. Wojciech Szymla stressed their study on how the insurance sector made way for innovation after being affected by the 2019 global pandemic and how the industry rebased itself to function effectively and thereby be productive [14]. The above research has already helped us to clearly understand the importance of technology in the field of insurance, but the fact that the trust of the masses is not paid any heed is the main point of my study. To assess the minds of the masses, we have used this study to instil confidence in them about its effect by portraying the different aspects of technology and showing them the numbers that prove technology's power in the insurance field.

3.3 Emerging Technologies and the Insurance Sector

Over time, we have been bombarded with the implementation of many new-age technologies; this section is centred on those technologies and their performance in the insurance field. The different technologies are as follows:

3.3.1 Big data

Whenever the word Big Data is thrown at us, the primary meaning that the human mind can comprehend is that it relates to data that is humongous in its size. This definition is near the ballpark but does not hit the bull's eye; what we aim for in big data is the analysis part of the information, wherein we can extract useful information from the data that is present before us.

Data is the food on which the insurance industry survives, making its presence felt in the customer base and attracting them to their purpose. Data is not just limited to just a sector of the insurance industry but has found its uses in almost every field, be it travel, health, or life; every area leans on data to analyse its customers as well as segregate different groups of people as well as policies that are used to develop their business model, which helps them to grow and compete in this field. Data also allows companies to judge which policies are redundant and which to discard; it helps them mitigate their losses and control their expenses.

The inclusion of technology has paved the way for many new opportunities, and it has ushered us to a new era where this information is even used to make relations between basic human behaviour and their decision-making power, which is crucial to understanding the customer base and thereby forming policies accordingly. For becoming influential policymakers, these techniques help a lot in determining the mood of the customers, which in turn helps the sector to grow.

3.3.2 The cloud computing

Cloud computing can be defined as the sharing of resources over the internet by the use of well-established connections. It is generally a form of sharing infrastructure and computing power where there is no need for physical access to the infrastructure but through the internet. Cloud computing reduces the need for people to invest in infrastructure and stresses the maintenance of such extensive IT infrastructure. It is based on the pay-per-use model, which makes it easier for anyone to afford it without putting a strain on their budget.

The need to catch up with the ever-expanding time is very significant, and in this scenario, cloud technology helps significantly in allowing businesses to scale up without any hiccups. In conjunction with big data, cloud computing can be used as a tool; it provides the necessary resources to critically analyse the terabytes of data and make meaningful relations between them.

Another aspect of cloud computing that fascinates the industry is that it introduces automation, which can be a time saver for any organisation. Insurance companies can use cloud computing to automate data analysis and make meaningful suggestions when the need arises. The main advantage of cloud computing is that it allows companies to focus on policymaking and does all the hard work for them, making it an irreplaceable part of any organisation.

3.3.3 The Internet of Things

This is a network wherein all the electronic gadgets we use can interact with each other using the web, allowing them to transfer data without needing any human being to aid them in the process. The objects are generally embedded with sensors and communicate with other devices using the internet.

Consider the case of automobile insurance; to establish the premium, insurers examined information such as age, gender, mileage, and automobile model. However, owing to the Internet of Things, it is now feasible to evaluate facts such as driving speed, duration, number of full braking per kilometre, drivers' classic routes, and mobile phone involvement.

With the positives come the negatives; this technology can even help companies and take some of their customers away. This scenario includes, for example, usually ensuring our residences in case of a fire. Telematics, however, can detect a gas leak and cut off the gas flow before a fire starts due to intelligent houses. Similarly, a smart factory can monitor the operation of working machinery and take preventative measures before a crisis happens, obviating the need for company insurance.

3.3.4 Artificial intelligence

Artificial intelligence (AI) is a digital computer or computer-controlled robot's capacity to accomplish activities generally associated with intelligent creatures. AI aims to develop systems that can work intelligently and autonomously.

When we look at insurance, it may be extensively employed to enhance the review process without requiring human interaction by reporting the complaint, capturing the loss, auditing the model, and communicating with the consumer.

Furthermore, machine learning enables insurers to use artificial technology more by significantly reducing time consumption and uncovering illicit behaviours. Many insurance businesses have adopted artificial intelligence in settlements and screening, automating operations and improving uniformity and effectiveness. However, it is continually moving from clever mechanisation in the rear to chatbots in the consumer forefront. With touch interface and voice-activated technologies, AI is growing sharper and quicker. Communication is becoming effortless and more intuitive. Many companies use intelligent virtual support to provide consumers with quick customised assistance. With its growing benefits to consumers, artificial intelligence can be a potential risk for insurance companies as it might suggest that people stray from taking unnecessary insurance, thereby threatening the company's customer base to a great extent. This kind of disadvantage can be a big hurdle for people to invest in a technology that might even harm their business, but the fact that the benefits out-weigh the negative is too great to be ignored.

3.3.5 Blockchain

Blockchain is an open and decentralised technology that uses encryption to connect metadata such as purchases and stores it in a shared register. It is a system that consists of an ever-growing list of records known as "blocks" connected by encryption to secure the security of data transmitted, payments, etc. Blockchains hold data over a network of devices, rendering them decentralised and dispersed. This implies that no person, organisation, or anyone who knows the system may be able to change it, but everybody may use it to contribute to its operation.

Blockchain opens up new avenues for reducing expenditures by automating the authentication of disputes or any other type of financial information via foreign entities, such as the following.

- Smart contracts: Financial institutions also establish contracts when marketing their assets and insurance contracts to consumers. Smart contracts are self-enforcing, making them verifiable, public, and irreversible. Smart contracts can computerise the claim resolution process, protecting insurance firms and consumers from fraudulent schemes.
- Identity tracker: Blockchain will enable everyone to build a worldwide identification encrypted and distributed with approved persons and organisations as needed. This identification will be one-of-a-kind, preventing fraud and illegitimate claims.

- Digital record keeping: Another use of blockchain is by using it as a digital record keeper; in this, the details of an insurer and the relevant documents are stored in blocks. Since these blocks are not affected by anything, be it time or storage issues, these last long and can only be accessed on the required time, preventing the spread of fraudulent people and claims.
- Customised insurance products: One size fits all is not a saying that we can justify to the insurance sector as, in this case, every person has different needs and therefore requires different types of insurance. To ease this issue, we can use blockchain technology to store the real-time data variables of the consumer, allowing the companies to look into the future and create customised policies for their customers, keeping up with the customer satisfaction policies and expanding their customer base.

Blockchain integration also allows for faster payment processing, bolstered with complete verification. It will enable the trust to build between the company as well as the consumers, and allows people to refrain from fraudulent practices. Blockchain can be stated as an integral part of the technological machinery, which helps the insurance industry to grow and prosper.

3.4 Theoretical Concerns

The only thing constant in this fast-paced era is change, which is termed inevitable; it seems complicated to embrace change, but it is required. It is the same case in this scenario wherein the public is hesitant to move to technology when it comes to technology; they still believe in the old orthodox systems that are trusted over the years and give them a sense of comfort; for they rely on a machine to make decisions regarding the situations of life and death makes them uneasy. This hesitation and reluctance must be removed to bring out the effectiveness of the technology. In this study, we focus on the impact that technology leaves behind when it is incorporated into the insurance sector; we have focused on whether the inclusion of technology has improved the effectiveness for both the consumers and the companies. From the company's perspective, they are gauging whether their work will be eased with technology or will it be an added burden to hire specialists to keep the machinery working.

On the other hand, customers want to know whether they can lean on technology to carry out their work. The added burden of money is also a factor that has been given importance in our study as implementation costs can be high, dissuading many from implementing such things, which burn a hole in their pockets. The intricate balance of money versus effectiveness has been given importance in this study as it does put a financial strain on changing the structure of a company to include new technologies. In turn, the burden is passed down to the customers.

3.5 Research Questions

- Can technology predict the risk that would compel you to take insurance?
- Do you trust technology enough to give away life-changing in its hands?
- From the point of view of the insurer, does technology act as a boon or a bane?
- What, according to you, would be the future of technology in the insurance field?

3.6 Research Objectives

- To analyse the effectiveness of technology in determining the risk, which may be the potential factors for taking up insurance?
- We aim to gauge the level of trust the masses have in technology as a prime factor determining the course of their lives.
- To understand the insurer's point of view when it comes to implementing technology.
- To analyse the future of technology in the context of insurance.

These research objectives were beneficial in highlighting the scope of my study. These revealed the impact of technology in the insurance field and the trust they can place in technology for making impactful decisions. The inclusion of technology is also highlighted with these objectives, as it is crucial to determine whether or not it is practical to implement such technology. Analysing the masses' thoughts helped us to know what course technology will take in the future regarding insurance.

3.7 Working Hypothesis

The working hypothesis can be stated as follows: "Technology is too great a factor to be ignored, and the inclusion of it in the field of insurance will help in the growth of the sector." Since ancient times, we have been the workers of the time, and, therefore, change with time is essential; therefore, the inclusion of technology is the need of the hour for the insurance sector. The inclusion of technology will remove the unnecessary blockages, which hinder the smooth functioning of the insurance sector.

3.8 Methodology

While performing the study, we used our primary and secondary data, which we have collected from different small insurance companies and the beneficiaries of the health insurance sector, to perform tests to gauge work efficiency. We have used the Chi-squared test to prove the hypothesis of our study, as the data is not normally distributed. We have studied the companies' effectual output over the years, allowing us to see the benefits of technology over the years. The data we used amassed around 500 people, but it was curtailed to approximately 150 people to be able to conduct hypothesis testing feasibly.

3.8.1 Sample

While scouring through the databases, we found numerous research documents, which amounted to 1158. After a thorough review of each article and a clear understanding of their relevance to our study, we selected nearly 16 papers to be included in our study. The PRISMA (Preferred Reporting Items for Systematic Reviews and Meta-Analysis) was our driving factor in choosing the different research articles from numerous sources, including but not limited to Scopus, Researchgate, Google Scholar, Web of Science, and Academia. These databases gave us an extensive number of articles, which surmounted 1158. After manually screening each piece, we found that close to 664 were duplicates and redundant for our use; therefore, we rejected them. The remaining number of articles, 494, were thoroughly examined to determine their relevance to our study. Therefore, the irrelevant ones, which did not make any significant change in our research, around 239 in number, were discarded. The entire process was repeated to increase the number to a reasonable 255, basing our search on new keywords and ideologies. Furthermore, we analysed the methodology and toned down the number of articles; at the very least, we could select 16 pieces that complemented our study. We have used the Preferred Reporting Items for Systematic Reviews and Meta-Analysis flow diagram (Figure 3.1) to show how these articles were

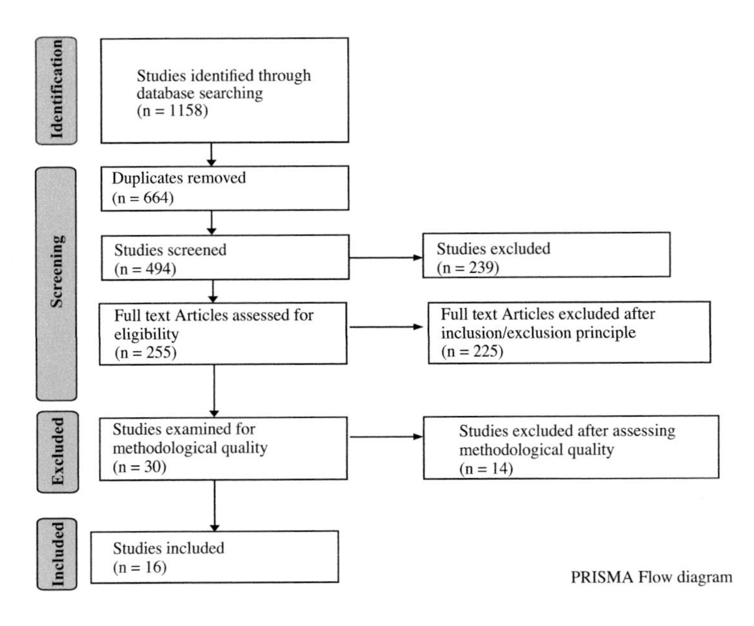

Figure 3.1 PRISMA flow diagram. Source: Author's compilation.

selected and how the final number of 16 was reached, which met all the criteria we proposed for our study.

3.9 Significance of Study

I am performing this study to highlight that technology is bolstered, which should be incorporated in every field, be it insurance. There have been studies in the past, but they have yet to be able to vehemently state the importance of technology and how its implementation would boost economic development. We have already seen the role that technology plays in shaping the future, but the fact that people are still reluctant to give into the hands of technology entirely remains at large; this study is aimed to eliminate those doubts. Seeing the change in economic trends would help people better understand the importance of technology.

3.10 Research Analysis

To gauge the thought process of the people as well as their views regarding the convergence of technology and the insurance sector, we conducted a

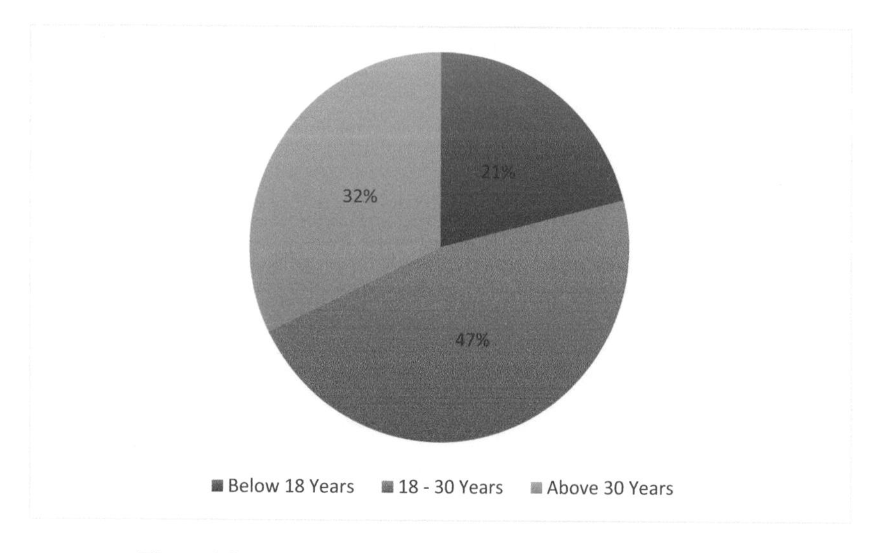

Figure 3.2 Graph denoting age. Source: Author's compilation.

short survey, the data for which is available on GitHub (Mishra, 2022), which resulted in the following.

3.10.1 Your age?

This question was posed to know about the age group of the people and determine what they think regarding technology (Figure 3.2). The three subcategories were then divided based on the fact that the people below 20 years are the most techno-savvy when it comes to the modern world; the age group of 20–30 years is the people who have just ventured into real life and are equally aware of the impacts of technology; on the other hand, the group of people in the bracket of above 30 years old are the people who are less technically oriented but are more focused on the securing income and leading a good life. This question clears the air around the fact that most of the population is aware of technology and, therefore, even pays heed to its effects on the different sectors of the economy.

3.10.2 Do you have insurance?

Posing this question was essential to know what the people think about the importance of insurance in their life and how much they value their own life and belongings compared to other worldly factors. Many people (Figure 3.3) have responded with a resounding "yes," showing that they are very

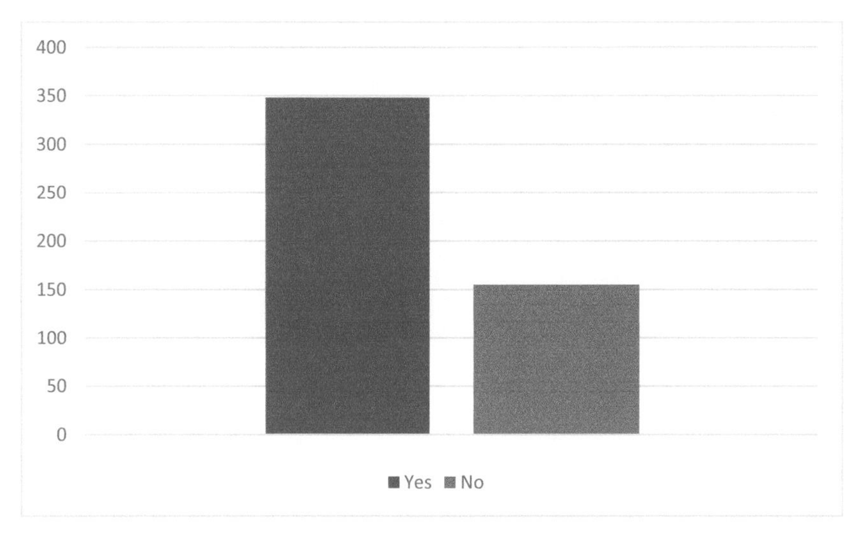

Figure 3.3 Insurance choices. Source: Author's compilation.

concerned about themselves and what they hold dear. This question cleans the air regarding the fact that people consider insurance important regardless of their age group.

3.10.3 Do you believe in technology to let them make your decisions like self-driven cars?

The ushering of artificial intelligence taking over everything essential and helping us with almost everything in life is not hidden. This question (Figure 3.4) shows that people do not shy away from taking up technology as a life-driving factor and are ready to give in to technology to drive them through the ever-crowded cities and roads where accidents happen almost every second. This trust can only be formed when the people are sure of the technology supporting them and are ready to give in to it. This question successfully highlighted that the masses have already come to terms with the inclusion of technology in emerging fields and therefore do not shy away from accepting its help when required.

3.10.4 Seeing the advancement in technology, will a machine advise you on some decisions? Will you pay heed to it?

This question was there to gauge what people think when it comes to thinking of technology as the guiding light in their day-to-day life. Most people are

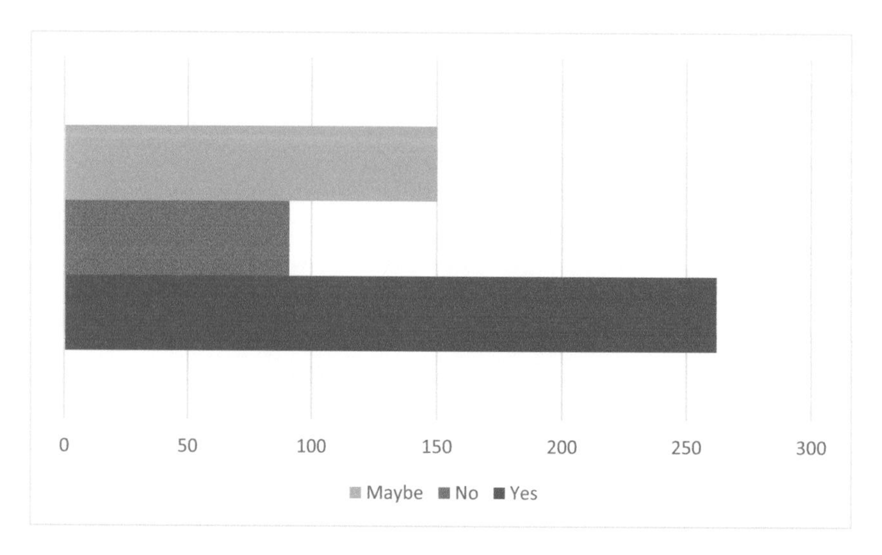

Figure 3.4 Graph denoting technological preference. Source: Author's compilation.

aware that technology is critical for them to thrive and move forward; people showing interest in the fact that decisions can be made with the help of technology are very hopeful to see, and they even pay heed to them. Recently, we have started listening to the decisions of machines; in the case of electronic maps, people are ready to believe the route shown by a machine rather than their own; this indicates that people are prepared to accept that technology can lead the right path and even help them when they are unable to make any decision. This question (Figure 3.5) was important in highlighting the importance of technology in this modern era and how it impacts the daily decision that we make.

3.10.5 Does the convergence of technology and insurance make you feel comfortable?

This question cleared the base of whether people like the inclusion of technology, and many responded positively, showing their level of trust in technology. It (Figure 3.6) indicates that people are inclined to include more technological advancements in the field of insurance as they are aware that technology is the oil that will help the machinery flow quickly without any blockages. Throughout time, the trends in technology have been a significant factor in steering the ship of the economy, and it has allowed the boat to sail without any discomfort since the advent of the technological era. This question highlights the fact that even though people are comfortable with technology, they still need time to understand and process it.

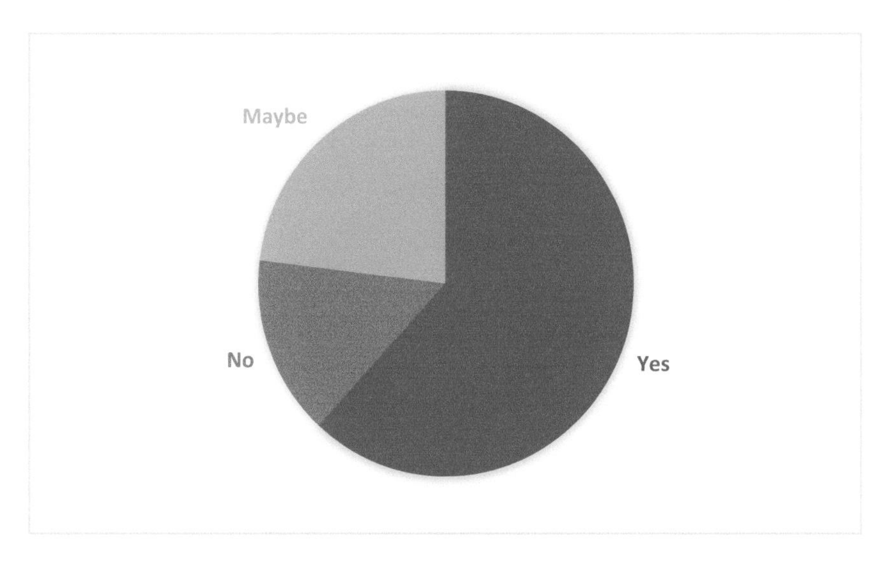

Figure 3.5 Mind of people corresponding to technology. Source: Author's compilation.

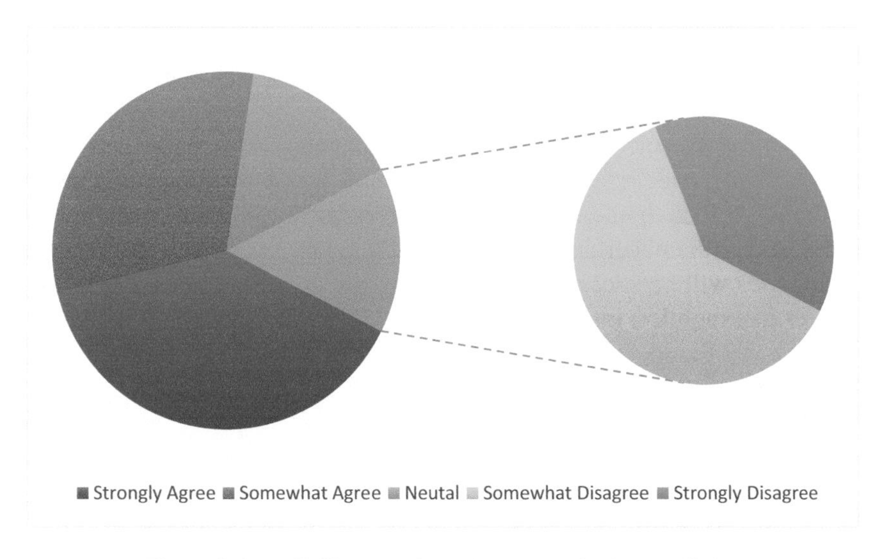

Figure 3.6 Belief in technology. Source: Author's compilation.

3.10.6 With the advent of online technology in the insurance sector, would emerging technologies like AI, ML, big data, etc., be successful?

This question (Figure 3.7) focused on people's mindset regarding the new age technologies and what they think about them. As people have already

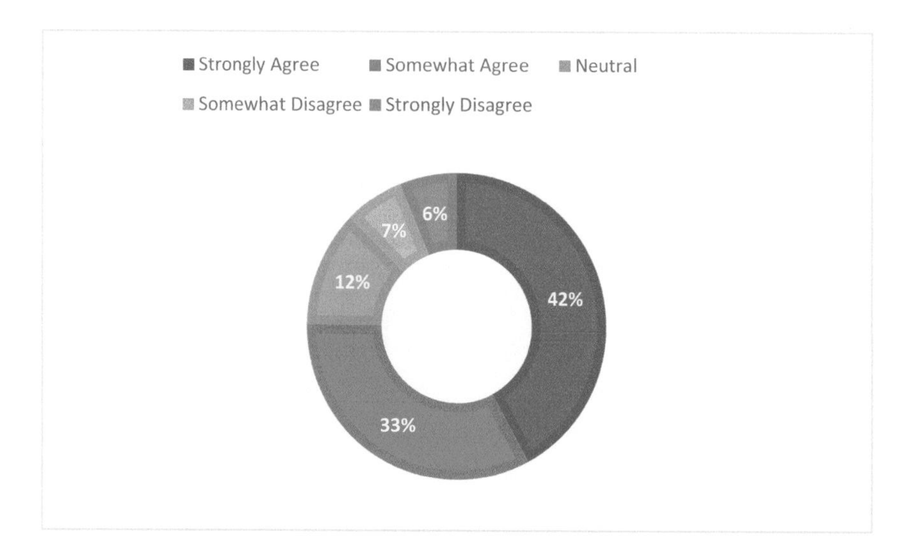

Figure 3.7 Graph showing the mind of the people regarding the future of technology. Source: Author's compilation.

been using technology in combination with the insurance field, their opinion on emerging technologies has been a liberal one. People have already found themselves surrounded by a web of technologies, including the internet, which has made way for an easy workflow in the insurance sector; this has already helped them believe in technology and even allows them to clear the air around the uncertain future. As predicted by the people, the new age technologies will be successful as they have already tasted the current technologies like mobiles, internet, etc.

3.10.7 Assuming that you own a company that is well established, will you take the risk of rebasing your company to include the new age industry 4.0?

This question (Figure 3.8) is the one that restricts the fine line between reality and hypothetical. In this, people believe in technology, but the fact they might need to shake their company to the core to include these new emerging technologies gives them chills. They are not ready to place a bet on their fortune company and give it all away to have things that may or may not help them develop in the future. The reason for people being reluctant can be cleared with the following reasons that people are still in the trance of the old ways, and to make a sudden shift can make anyone feel uncomfortable and uneasy; second, people are more trustworthy of their colleagues and in such a case to let go off such confidants can be very tough which is another reason for them

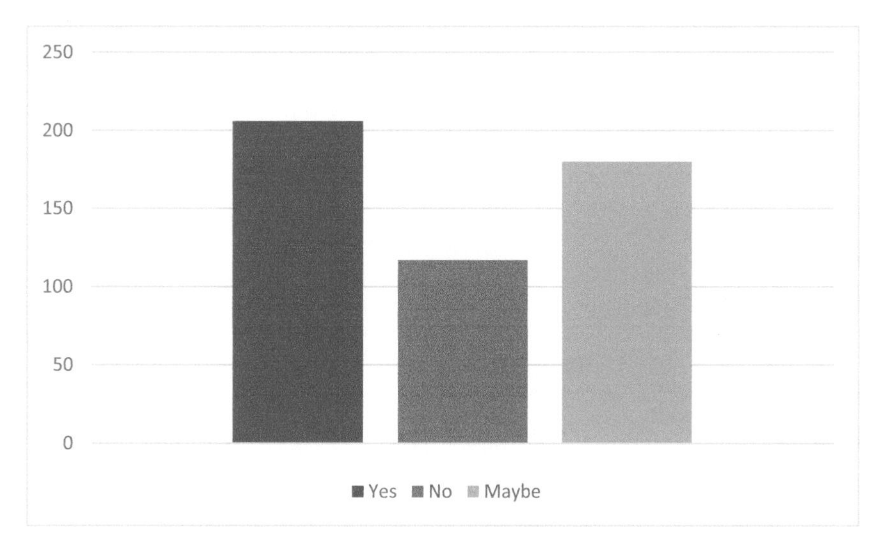

Figure 3.8 Graph showing the mindset of people regarding industry 4.0. Source: Author's compilation.

not to embrace the change. Some people are doubtful, waiting for others to start the procedure, and want to avoid being the front runners. They are here to remain and gauge the effects of technology and will only begin when they feel satisfied. The line between reality and expectations is brought out in this question, and people can voice their opinions on the uncertainty in implementing new technologies like AI, ML, big data, etc.

3.10.8 Has the inclusion of technology reduced the time required for paying premiums and purchasing policies (Table 3.1)?

3.10.8.1 Calculations

H0: There is no relation between work efficiency and technology in the insurance sector.

H1: With the inclusion of technology, time efficiency has improved dramatically.

Degree of freedom = 3 – 1 = 2.

Level of significance (α) = 5%.

The *p*-value is 0.0008652.

Since the *p*-value is very small, the null hypothesis is rejected; therefore, we can say that people already favour the inclusion of technology in the insurance field as they have experienced its positive effects. This makes the inclusion of technology an essential factor in the growth of the insurance sector.

Table 3.1 Hypothesis testing.

Question	Observed frequency (O^i)	Expected frequency (E^i)	$O^i - E^i$	$(O^i - E^i)^2$	$\frac{(O^i - E^i)^2}{E^i}$
Yes	52	38	14	196	5.1578
No	20	38	−18	324	8.5263
Maybe	34	38	−4	16	0.4210
Sum					14.10515

Source: Author's compilation.

Table 3.2 Hypothesis testing.

Question	Observed frequency (O^i)	Expected frequency (E^i)	$O^i - E^i$	$(O^i - E^i)^2$	$\frac{(O^i - E^i)^2}{E^i}$
Yes	88	68.3	19.7	388.09	5.6821
No	63	68.3	−5.3	28.09	0.4112
Maybe	54	68.3	−14.3	204.49	2.9939
Sum					9.0872

Source: Author's compilation.

3.10.9 Have the inclusion of technology helped you to lessen your company's workload and save time (Table 3.2)?

3.10.9.1 Calculations

H0: In an insurance company, there is no need for technology.

H1: Technology is equally vital in the field of insurance.

Degree of freedom = 3 − 1 = 2.

Level of significance (α) = 5%.

The p-value is 0.01064.

Since the p-value is very small, the null hypothesis is rejected; therefore, we can say that people already favour the inclusion of technology in the insurance field as they have experienced its positive effects. This makes the inclusion of technology an essential factor in the growth of the insurance sector.

3.11 Inference

After successfully conducting the research and comparing our results, we were able to draw out a few crucial conclusions and points, which are the highlights of our study. We see that the effective output (Figure 3.9), determined

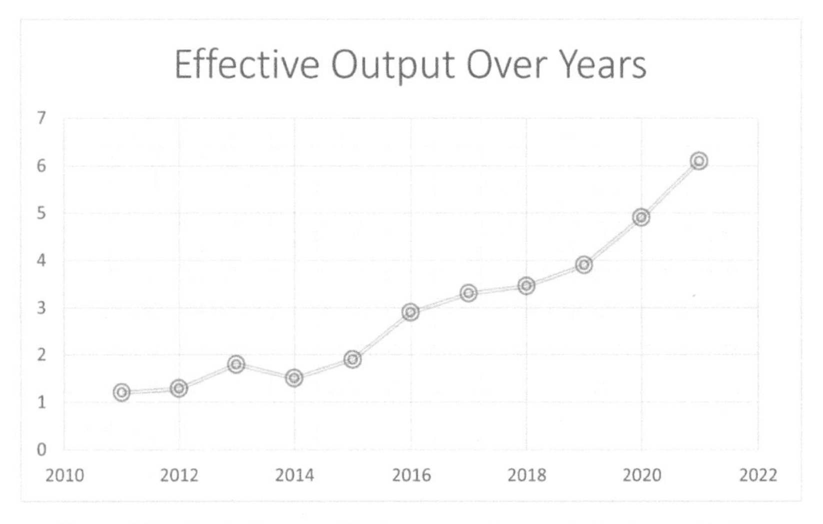

Figure 3.9 Graph showing effective output. Source: Author's compilation.

by the company based on the number of insurance claims handled successfully by the company employee daily and the number of policies they are to sell out, has improved dramatically with the inclusion of technology. To be efficiently able to show our results, we take the help of the following graph.

In this graph, we see that during the beginning of industry nearly a decade back in the 2010s, we see that the efficiency of the companies was relatively very low on a scale that surmounted to 1/7 times what we have today. In the old times, we saw a lot of manual labour involved; therefore, people's abilities needed to be improved, making the insurance process time-consuming. On the other hand, with the intervention of technology near the 2016s, we see that the productive output of the companies increased significantly to about three times the original; this can be attributed to shifting to paperless means and involvement of the internet. Over time, we see that this growth continued at a constant speed, but with the 2020 global pandemic, the growth rate boosted because the world changed to online mode. This era led to a boom in this department, thereby people relying more on technology rather than usual means; this era saw a growth of around seven times from which we started off proving the significant advantage that technology provides to this sector.

3.12 Conclusion

After carefully studying the opinions of the masses and comparing the data that we have found with the growth rate of the previous years of insurance companies, we can conclude that technology is the need of the hour. It is one of the revolution's spearheads and, therefore, a force to which every industry should bow and embrace. It has been proved that the inclusion of technology helps not only the general public but also the insurance companies; we were able to demonstrate how the public will benefit from the efficiency that is provided by the inclusion of technology, leading to the process becoming less time-consuming and hectic. The same can be said for insurance companies, as the overhead costs for maintaining many workforces can be mitigated, and the room for human error is reduced. All these things point in one direction only, that being that technology is one of the prime factors that shall never be ignored, be it the case of any sector of the human revolution and industry; its impact on the economic condition of any sector is too significant to be ignored, therefore making it the pulse of the machinery of the world.

Acknowledgement

To begin with, I would like to thank all the people around me, including but not limited to my mentor, teachers, and colleagues, for helping me and constantly bolstering me. I want to dedicate my gratitude toward some significant advisors and contributors.

First, I would like to thank my guide and mentor, Dr. Manju Dhaiya, for her support and kind words of encouragement. She peered through my research and gave her valuable advice on many aspects of this topic.

Second, I want to extend my warm gratitude towards Dr. Ashok Maurya for his constant bolster and guidance throughout the academic year, which help me steer this research of mine.

In conclusion, I sincerely thank my colleagues and parents who made me able enough, and without their support, this research would not have been possible.

References

[1] Zheng, Lili & Guo, Lijun. (2020). Application of Big Data Technology in Insurance Innovation. 10.2991/assehr.k.200401.061.

[2] Iliev, Anton & Rahneva, Olga & Pavlov, Nikolay. (2014). Insurance and Insurance Information Systems.

[3] Kędra, Arleta & Lyubov, Klapkiv & Lyskawa, Krzysztof & Klapkiv, Yuriy. (2019). Digitalization in insurance companies. 10.3846/cibmee.2019.086.

[4] Wu, Jihong & Wang, Junmei & Liu, Yanjun. (2019). Design and Research of Insurance Survey Claims System Based on Big Data Analysis. 211–214. 10.1109/ICVRIS.2019.00059.

[5] Mittal, Varun. (2020). Insurance as a Service.

[6] Bodla, Bhag & Seema, & Suman,. (2015). Growth of Indian Insurance Sector in Post Irda Era.

[7] Hebbar, C. Kusumakara & Shenoy, Sandeep & Rao, Guru & Rao, Abhishek. (2013). The Role of Information Technology & Insurance Penetration: A Study.

[8] Sarkar, Sukanta & Suman, Kalyan & Chaudhury, Suman. (2015). Insurance Industry in India: A Journey from British India to Independent India. iv. 819–825.

[9] Feyen, Erik & Lester, Rodney & Rocha, Roberto. (2011). What Drives the Development of the Insurance Sector? An Empirical Analysis Based on a Panel of Developed and Developing Countries. Journal of Financial Perspectives.

[10] Chen, Bill & Chen, Dongmei. (2012). The Review and Analysis of Compulsory Insurance. 10.13140/2.1.3751.2964.

[11] Choudhary, Dr-Pooja. (2019). Role of Insurance Sector in Infrastructure Financing in India: A Brief Landscape.

[12] Sharma, Vikas. (2012). "Recent Trends in Life Insurance Sector– A Study after Privatization.

[13] Nagarajan, Vivek & Selvan S.C.B, Samuel. (2019). "Technology in Insurance."

[14] Volosovych, Svitlana & Zelenitsa, Iryna & Kondratenko, Diana & Szymla, Wojciech & Mamchur, Ruslana. (2021). Transformation of insurance technologies in the context of a pandemic. Insurance Markets and Companies. 12. 1–13. 10.21511/ins.12(1).2021.01.

[15] Dionne, Georges & Harrington, Scott. (2017). Insurance and Insurance Markets. SSRN Electronic Journal. 10.2139/ssrn.2943685.

[16] Uddin, Mohammad. (2020). Insurance Industry.

[17] Outreville, J. & Preker, Alex. (2002). Introduction to Insurance and Reinsurance.

[18] Desai, Bhavesh & Desai, Ravi & Algotar, Gaurang & Desai, Kanan & Bansal, Rajkumar. (2013). Health Insurance: Effects and Awareness. Journal of the College of Community Physicians of Sri Lanka. 18. 30–32. 10.4038/jccpsl.v18i1.6078.

[19] Lippert, Ingmar. (2010). Insurance. 10.4135/9781412972093.n179.

[20] Grima, Simon. (2019). The Impact of Technology Innovations on the Governance of Insurance Firms: A Literature Review 44 The Impact

of Technology Innovations on the Governance of Insurance Firms: A Literature Review.

[21] de Ferrieres, Matthias. (2021). An Introduction to Insurance Effective Digital Disruption and Impact of Insurtechs in the Insurance Economy.

[22] Lin, Lin & Chen, Christopher. (2019). The Promise and Perils of InsurTech. Singapore Journal of Legal Studies. 10.2139/ssrn.3463533.

[23] Sood, K., Seth, N., & Grima, S. (2022). Portfolio Performance of Public Sector General Insurance Companies in India: A Comparative Analysis. In Simon Grima, Ercan Özen, Inna Romānova (Ed.), In *Managing Risk and Decision Making in Times of Economic Distress, Part B*. (pp.215–229). Emerald Publishing Limited,UK.

[24] Grima, S., & Marano, P. (2021). Designing a Model for Testing the Effectiveness of a Regulation: The Case of DORA for Insurance Undertakings. *Risks*, *9*(11), 206.

4

Adoption of Artificial Intelligence to Revolutionise the Insurance Industry

Samridhi

Department of Management, Indira Gandhi University, India

Email: samridhi.tanwar@gmail.com

Abstract

The use of artificial intelligence (AI) has exploded across all industries. The availability of more more processing power and shifting customer expectations have all accelerated AI development. Across many industries, including insurance, it is disrupting and improving organisations. Forconsumers, partners, and staff, insurance companies are utilising AI platforms and solutions. The article focuses on how artificial intelligence (AI) is altering the insurance sector. With this in mind, the author will examine how artificial intelligence (AI) is being used to tailor insurance business processes and decisions to better meet the needs and expectations of customers, ultimately increasing the efficiency of the overall insurance production process.

Using descriptive approach, the study revealed that artificial intelligence plays preventive as well as curative role in insurance sector. The paper highlights the way artificial intelligence has eased the process of underwriting, claims processing, fraud detection, and quality customer service. The paper offered practical advice on how insurance companies, regulators, and supervisors can maximise the benefits of this technology.

4.1 Introduction

After the trade market, the insurance industry is another sector where it is difficult to forecast the next great paradigm shift. Insurance firms are often on shaky ground and face tremendous hurdles when it comes to adopting smooth and intuitive new technologies. Considering the shaky scenario and natural disasters, the insurance industry is slowly but steadily adopting digital solutions to solve some of its key pain issues [1]. Numerous examples of digital developments can be observed in the industry, for instance, the use of NLP technology to improve customer experience in health insurance to combat insurance fraud [2]. Likewise, the adoption of AI in life and non-life insurance sectors leads to fast claims processing and better customer services. Further, the image processing component of AI in insurance is used to do property damage analysis. The same machine can be utilised to make a well-informed financial decision using sophisticated algorithms [3].

Indeed, AI can speed up and simplify procedures like underwriting, fraud detection [4], discovering policyholder needs, sales management [5, 6], and customer support. As several benefits are linked with the incorporation of artificial intelligence in the insurance field, a number of insurers utilising artificial intelligence to develop risk models for decision-making and eliminating manual input are rising [20, 21].

Artificial intelligence, like convolutional neural networks, has great potential to meet its promise to replicate the observation, reasoning, learning, and problem-solving of the human mind with a new wave of deep learning approaches. As a result of this advancement, insurance can move from detection and correction to prediction and prevention, shifting all parts of the transaction within methods [7].

4.2 Key Benefits of Artificial Intelligence in the Insurance Industry

Organisations are planning to incorporate more and more AI into day-to-day insurance workflows. Insurers can benefit from machine learning and natural language processing in a variety of ways, including time-saving, cost-cutting, and increased profitability. AI can also assist in eliminating human errors in addition to saving time and money. Stakeholders in this ecosystem seem to understand the benefits of adopting AI in insurance. Some investors believe that AI would transform the insurance industry while others believe that AI would enhance staff productivity [2, 8, 9].

Market leaders are hopeful and confident about reaping the benefits AI brings in insurance. It was predicted that AI will enhance labour productivity by 10%–40% in 11 Western developed countries and Japan by 2035. If this optimistic forecast holds, economic growth will almost certainly quadruple by 2035. Not only insurers, but consumers are also rapidly seeing the benefits of AI. Consumers in the United States are comfortable utilising chatbots to apply for insurance. Insurers utilising AI to give more personalised plans are also acceptable to consumers [2].

a. Insurer investments in AI are increasing:
   Greater investment in AI has accompanied increased understanding of its benefits; that is why insurers are investing heavily in AI technology each year. It is a market where AI spending is expected to skyrocket [4]. In addition, corporate interest in insurtech businesses has also resulted in some significant funding.

b. Startups have a lot of potential:
   There are now around 1500 active Insurtech startups around the world but worth mentioning are the initiatives of Inspekt Labs, Cognitiv+, and Cytora. Inspekt Labs, one of Early Metrics' rated businesses, creates software that uses computer vision and machine learning to automate the asset inspection process for automobiles, property, and consumer products using smartphone-generated photographs and videos. Another ranked startup, Cognitiv+, creates an AI platform that collects data from contracts, policies, and legislation to provide businesses with legal data insights. Cytora is working on an AI-based underwriting solution. The system assesses risks to determine fair pricing and boost profits by lowering loss ratios [2].

c. Use of unstructured data:
   Presently, insurance companies have a lot of money. They have amassed mountains of information about people, houses, and businesses over the years. However, it is frequently locked away in silos and unavailable to those on the front lines. If the data is integrated and available for study, AI can help change that. It can pull together this plethora of unstructured data and use it to improve customer engagement, service personalisation, and the high impact of marketing messages [4, 10].

4.3 Artificial Intelligence-related Trends Shaping Insurance

Companies that have quickly and responsibly adopted artificial technology are enjoying an advantage [5]. To stand out and rise over competitors, it becomes critical to analyse the key technology trends strongly linked to AI and revolutionising the insurance industry. The following section discusses the four main technology trends.

a. Data from connected devices is exploding:
   Devices with sensors have been around for a long time in the industry, but the number of connected consumer products will explode in the

next few years. Existing products like automobiles, home assistants, fitness trackers, smartwatches, and smartphones will continue to gain traction, with new, fast-growing categories including home appliances, eyeglasses, clothes, shoes, and medical devices joining the party. Experts predict that 1 trillion devices will be connected by 2025. Huge data generated by gadgets will allow carriers to better understand their customers, develop new products, customised pricing, and increase in real-time service offerings.

b. Physical robotics is becoming more popular:
Recent advances in robotics are fascinating, and these advancements will continue to revolutionise the way human beings interact with the environment. Additive manufacturing, also known as 3-D printing is another process that is reshaping the future of manufacturing and commercial insurance products. By 2025, 3D-printed buildings will become commonplace, and to remain in competition, carriers have to consider that how the development will affect their risk assessments. Moreover, improved surgical robots, autonomous agricultural equipment, and programmable drones will become commercially available within the next decade. By the end of 2030, the proportion of standard-duty vehicles with autonomous functions, such as self-driving capabilities, will increase significantly.

c. Open-source and data ecosystems:
Open-source protocols will evolve as data becomes more widespread, ensuring that data may be used across industries. Various private and public institutions will amalgamate to form ecosystems where data can be shared for several purposes while adhering to a standard cybersecurity and regulatory framework. For instance, wearable data can be immediately imported to insurers, while connected-home and auto data may be easily available via Google, Apple, Amazon, and several other producers.

d. Cognitive technology advancements:
Deep learning technologies, convolutional neural networks, and others are primarily utilised to process speech, images, and unstructured text but will evolve to be used in a wide variety of applications. Based on the human's ability to learn and understand through decomposition and reasoning, these cognitive techniques will enable "active" insurance products to process the vast and massive data streams generated in association with individual actions and activities. As these technologies become more and more commercialised, insurance companies will

have access to models that continuously learn and adapt to the world around them, responding in real time to potential risks and changing behaviours [11, 12].

4.4 Application of Artificial Intelligence in Insurance Industry

AI has many applications in the field of insurance. Processing of claims, underwriting, customer service, and fraud detection are just a few of the areas where AI assists insurance agencies with efficiency and quality [4]. AI adoption can make it easier to:

- analyse damage caused by a natural disaster that occurred unexpectedly;
- calculate and assess risk tolerance for trading desks;
- insurance agencies can use transaction analysis to help them make better decisions;
- better investment choices based on preferences, risks, and spending habits;
- customer investments and insurance coverage are consistently optimised;
- analyse claims, asset management, risk assessment, and prevention (OECD, 2020).

The necessity for incorporating AI in the insurance sector might generate sky touching growth in this industry as the financial sector is flooded with more financial, insurance, and investment data than ever before. AI-powered data management technology helps people assess and navigate data pyramids, as well as organisations in creating intuitive and dynamic consumer experiences [13, 14].

The following section covers, in a nutshell, the areas/sectors where artificial intelligence makes a huge impact and transforms the insurance industry as a whole.

a. Customer intimacy and cost:
   Adopting these technological, process-oriented changes entails much more than simply installing new technologies. It is also critical to think creatively about improvising customer service in order to match quickly changing market expectations while staying within budgetary limits.

b. Market pressure and competition:
Due to the rising market pressure and competition from new entrants in lean and agile insurtech, established organisations should be prepared to fight back and enhance their approaches to get closer to their customers while reducing their total operational cost.

c. Future opportunities, challenges, and innovations:
It is difficult to drive higher growth and revenue in a digital business world by relying solely on innovation and shared margins to reduce costs. Insurers, meanwhile, are rising to the challenge as industry leaders strive to embrace new AI-driven opportunities and believe that creativity and innovation are critical to the future of their businesses.

d. Humans and machines are in perfect coordination:
When it comes to acquiring competitive advantages from the synchronised functioning of humans and computers applying their combined efforts in a common workforce, CEOs are standing shoulder to shoulder. These new-age advancements and AI capabilities have the potential to revitalise the traditional elements of the insurance industry as we know it today, fostering a well-organised, intelligent, and confident "growth-accelerating platform" that extends far beyond the frontiers of cost, competition, and customer satisfaction [6].

CEOs are standing shoulder to shoulder when it is about acquiring competitive benefits from the synchronised capabilities of humans and computers, leveraging collaboration in a shared workforce. These innovative advancements and artificial intelligence capabilities have the potential to revitalise the traditional insurance sector as we know it today, fostering a well-planned, organised, and "growth-escalating platform" that extends far beyond the aspects of competition, cost, customer satisfaction, and loyalty [6].

e. Enhancing the customer experience:
The main area of AI success in insurance is improving customer experience. Companies are increasingly utilising AI to design products for customers (both retail consumer and business customers), to establish an ongoing connection with customers for upselling, and loyalty, and to analyse data from multiple sources in order to make more accurate estimates.

f. Chatbots:
Many insurers have already invested in virtual assistants such as chatbots, according to the National Association of Insurance Commissioners

(NAIC). These chatbots provide digital services and can naturally converse with humans. The goal is to answer questions, route calls, limit human traffic to only higher-level requests, and provide guidance, billing information, and transactions 24 × 7. Insurance companies like Geico, Allstate, and Lincoln Financial were among the first to use chatbots. Today, almost every big corporation makes use of them. Chatbots are now being used to resolve cybersecurity password difficulties as well as deliver copies of policies and other basic documentation. This saves a lot of time and effort [15].

g. Claims management:
Lemonade, a startup, has used machine learning and chatbot technology in various stages of the claim handling process. Data, photographs, and sensors are used to measure the extent of damage, anticipate costs of repairing, and resolve basic claims using machine learning algorithms. Lemonade claims that its chatbots, Jim and Maya, can get customers a policy in as little as 90 seconds and settle a claim in as little as three minutes.

h. Agent interaction:
Liberty Mutual created a means for its artificial intelligence apps to work with Alexa to handle a variety of tasks. This includes answering customer queries, giving insurance quotations rapidly, and connecting users with the nearest agent who can suit their needs. This AI technology is also used by the corporation to provide risk management recommendations.

i. On-demand insurance:
Consumers nowadays expect rapid gratification. Consumers do not want to dial a 1-800 number, be referred to a local agent, schedule an appointment, drive to the agent's office, wait for paperwork to be prepared, and then sign all of the paperwork. Instead, they prefer to do it on their smartphones or online. For example, the insurance business Trov offers an app that insurers can licence, which simplifies several parts of insurance administration. Consumers can activate or deactivate coverage with a single swipe of their phones, and chatbots are used to automate claim processing. Trov has worked with companies like Slice Labs to provide homeowners, renters, and small business owners with on-demand insurance coverage. Bind Benefits, meanwhile, allows customers to customise their health insurance coverage based on their current requirements or life events. As part of its service model, Bind collaborates with United Healthcare [4].

4.5 How Can Insurers Prepare Themselves for Rapid Change?

Automation, machine learning, deep learning, and external data ecosystems will be widely adopted and integrated, accelerating the industry's rapid transition. While no one can predict how insurance will appear in 2030, providers can begin planning now in the following ways:

a. Become knowledgeable about AI-related technology and advancements: Although the insurance industry's tectonic transformations will be driven by technology, tackling them is not the IT department's responsibility. Board members and customer-experience teams, on the other hand, should invest time and resources in learning about AI-related technology. This effort will include investigating hypothesis-driven scenarios to identify and emphasise where and when disruption can occur, as well as what it means for certain business lines. Small-scale IoT pilot initiatives in specific regions of the organisation, for example, are unlikely to teach insurers much. Instead, they must proceed with purpose and a clear understanding of how their organisation can participate in the IoT ecosystem on a global scale. Pilots and proof-of-concept activities should be developed not only to test the capabilities of the technology but also to test how well network operators can perform in a given position within IoT-based ecosystem.

b. Create a cohesive strategic plan and its proper implementation: Carriers must decide regarding the deployment of technology to achieve their objectives. The senior leadership team's long-term strategy plan will need a multi-year transition in operations, talent, and technology. Some insurers are already experimenting with novel strategies, such as creating strategic ties with prominent academic institutions and purchasing prospective insurtech startups. Insurers should assess what areas they want to invest in to match or beat the market, as well as what strategy is best for their organisation, such as creating in-house strategic capabilities or incorporating a new business. The approach should cover every facet of any large-scale, analytics-based endeavour, from data to people to culture. A road map for AI-based pilots and proofs of concept should be included, as well as information on which parts of the organisation will require change management or focused skill development. Most essential, a clear calendar of milestones and checkpoints is required to allow the organisation to determine how the strategy should be updated on a regular basis to deal with any shifts in AI technology evolution as well as significant changes or disruptions in the industry.

Carriers must develop strategic responses to anticipated macrolevel changes in addition to understanding and applying AI technologies. Carriers will need to reconsider their consumer engagement and branding, product design, and core revenues as many lines adopt a "predict and avoid" strategy. Vehicles with self-driving capabilities will reduce automobile accidents, IoT devices will aid in the reconstruction of structures following a natural catastrophe, and improved healthcare will save and prolong lives while preventing in-home floods. Similarly, when a loved one dies, people will demand effective medical care and comfort, natural disasters will continue to wreak havoc on coastal areas, and vehicles will break down. Profit pools will shift when these changes take effect, new product kinds and lines will arise, and how consumers interact with their insurers will change substantially. All of these efforts can lead to a well-coordinated analytics and technology strategy that spans the entire business while focusing on value generation and differentiation.

c. Develop and implement a thorough data strategy:
Data is increasingly becoming one of every company's most valuable assets. The insurance industry is no exception: the volume and quality of data carriers collect during the policy's life cycle influences how they identify, assess, place, and manage risk. Most AI solutions perform effectively when given a vast amount of data from many sources. As a result, carriers must have a well-structured and actionable plan for both internal and external data. Internal data must be organised in such a way that new analytical insights and capabilities may be developed quickly. When it comes to external data, carriers must prioritise access to data that complements and improves their internal datasets.

The main challenge will be gaining access at a reasonable rate. The external data ecosystem will certainly get more fragmented as it increases, making it difficult to find high-quality data at a reasonable price. A data strategy should include numerous techniques for getting and safeguarding external data, as well as strategies for merging this data with data from internal sources. Carriers should be ready to adopt a comprehensive procurement strategy that involves the use of data APIs, data source licencing, data broker agreements, and direct acquisition of data assets and providers.

d. Build the proper talent and technology infrastructure:
Carriers must make small but consistent investments in people to ensure that advanced analytics is regarded as a must-have ability across the

organisation. In the future insurance organisation, talent with the necessary mindsets and abilities will be required. The next generation of successful frontline insurance workers will be in high demand, and they must be a unique blend of technologically adept, creative, and willing to work at something that is not a static process, but rather a mix of semi-automated and machine-supported tasks that are constantly evolving. To produce value from future AI use cases, carriers will need to mix talents, technology, and insights from across the company to create unique, comprehensive client experiences. To achieve this, most carriers will need to execute an intentional cultural shift, which will require CEO buy-in and leadership. To keep up, a concerted effort will be needed to attract, cultivate, and maintain a diverse workforce with critical skill sets. This team will include data scientists, data engineers, cloud computing professionals, technicians, and experience designers. Many firms will create and implement reskilling programmes to conserve knowledge while also ensuring that the company has the necessary new skills and competencies to compete.

As the final component of developing the new workforce, organisations will seek out external resources and partners to supplement in-house talents and assist carriers in securing the necessary assistance for business growth and execution. Similarly, future IT architecture will be radically different from today's. Carriers should start making strategic investments to enable the shift to a more forward-thinking technology stack capable of supporting a two-speed IT architecture.

During the next decade, rapid technological improvements will generate major disruption in the insurance business. Carriers that use new technology to create unique solutions use cognitive learning insights from new data sources, improve processes and decrease costs, and satisfy customer expectations for personalisation and dynamic adaptation will win in AI-based insurance. Most significantly, carriers that focus on generating opportunities from disruptive technologies rather than viewing them as a threat to their current business will be successful in the insurance industry in 2030 [11].

4.6 Risk and Compliance Implications of Artificial Intelligence

Artificial intelligence (AI) is now so common and incorporated into our daily lives that it is difficult to imagine a new product or service that does not include artificial intelligence. Insurance is no exception. AI is increasingly being used to locate potential clients, assess risk, determine prices, and improve claim experience. However, if artificial intelligence is used poorly or with bias, it can have detrimental implications. As a result, insurers must be aware of the plethora of legal and ethical difficulties that may arise. There are three types of dangers/risks that can occur as a result of noncompliance:

a. Risks of training: If the models are not properly trained, their findings may be skewed, affecting just a limited number of people based on protected traits. As a result, the model may deny claims or charge higher premiums by error.

b. Security risks: Improper testing can affect computer vision algorithms that evaluate claims, text classifiers that detect document fraud, and speech recognition systems in call centres [16].

c. Data privacy risks: Using sensitive components of customer data or personally identifiable information (PII) without customer's permission is illegal. Meanwhile, facial recognition systems are viewed as a danger to personal privacy and a violation of fundamental rights.

These risks can be a legal nightmare, and they raise numerous concerns about insurers' use of artificial intelligence [17].

4.7 AI Implementation Challenges in Insurance

Throughout the insurance ecosystem, a subset of the industry stands out for adopting new technology ahead of the pack. The insurance industry has traditionally shown a larger proclivity for digitisation, albeit slowly but steadily. Insurers require big data-driven insights today more than ever to assess risk, decrease claims, and provide value to their customers. In the insurance industry, artificial intelligence has brought about revolutionary benefits. AI-enhanced systems can break down barriers to collaboration, eliminate manual dependency, and flag problems. However, enterprises today face numerous obstacles in realising AI's full potential.

a. Scarcity of good training data:
   Training datasets can help AI become more effective and make better decisions. Even with the data they had at hand, data scientists discovered that AI training was more challenging than they had anticipated. They even ran into difficulties while identifying and analysing the training data.

b. Clean vision, process, and executive leadership support:
   AI is a continuous process. The insurers may be able to make the most of AI with the support of a clear strategy, concentrated time, patience, and directed leadership from industry leaders and AI thought leaders.

c. Data in-silos
   Silos within an organisation are unwise and restrict the effectiveness and productivity of operations. The majority of businesses find it challenging to collaborate, implement, and measure more general objectives when data is stored in silos.

d. Selecting technology and vendors:
   AI has matured to the point where it can now penetrate organisations. As the number of AI success stories grows, so does the amount of money invested in AI. The major concern is whether AI deployment suits the insurance business process or not, no matter how large the hype is. The insurtech industry grew at a steady pace in 2019, raising $6 billion in funding. Insurance companies have made progress with the support of these insurtech service firms, solving age-old insurance ailments with AI-powered solutions.

e. People, expertise, and technical competency:
   "Skills and talent" in the field of AI are the primary impediments to AI transformation in their company. People are lacking in core skills required for the efficient implementation and utilisation of AI in the insurance sector [18, 19].

4.8 Regulations Governing Artificial Intelligence in Insurance Industry

4.8.1 Regulation and guidelines on AI usage

For a long time, the insurance industry has been investing in artificial intelligence (AI). In reality, at least on a fundamental level, AI in insurance may be called mature. The sophistication of AI-based insurance is improving with each passing year. With the increasing implications of AI, it becomes imperative to regulate it carefully. While a few AI applications have been controlled, the vast majority are still unregulated and guided only by guidelines. Regulations differ from one location to the next. Following are the different regulations and guidelines available to control the AI application:

a. Data related:
   The number of regulations relating to AI-related data privacy is the highest. The GDPR (2018) of the European Union, the California Consumer Privacy Act (2020), China's Personal Information Protection Law (2020), Singapore's Personal Data Protection Act (2020), and Canada's Digital Charter Implementation Act (2020) are just a few of the regulations that protect citizens' and consumers' privacy and personal data.

b. Algorithm related:
   Even though AI prejudice has caused a lot of controversies, most nations only have rudimentary regulation ideas. The Algorithmic Justice and Online Platform Transparency Act, presented in May 2021, is currently pending in the United States, but the Algorithmic Accountability Act, introduced in 2019, was never approved by the Senate. The upcoming AI Bill of Rights will protect consumers by giving them the right to openness and justification. The bill also aims to make AI-enabled technologies more accountable. Contrary to the federal government, many US states have adopted AI legislation. Just two come to mind: the Illinois Artificial Intelligence Video Interview Act and the Automated Decision Systems Accountability Act in California.

   In August 2020 and April 2021, industry bodies such as the National Association of Insurance Commissioners (NAIC) and the Federal Trade Commission produced guidelines aimed at ensuring openness, justice, equity, accountability, and security in the use of AI by businesses.

c. Application related:
   A comprehensive proposal called the Artificial Intelligence Act of the European Union (EU) (April 2021) forbids the use of artificial intelligence (AI) for manipulation, dark patterns, social scoring, and facial recognition. The law mandates that high-risk AI adhere to EU health and safety standards. Additionally, transparency and a code of conduct are required for low-risk groups. In August 2021, the Chinese Cybersecurity Administration announced stricter guidelines for the usage of recommender systems. It exercises control over the internal operations of the model, requests permission before use, limits usage to promote positive energy, and stops the transmission of hazardous content. Further, all 193 members of UNESCO adopted a historic agreement on the ethics and use of artificial intelligence to promote human rights and address important global concerns in November 2021.

4.8.2 Impact of AI regulation on the insurance industry

As a consequence of increased regulation, the insurance industry anticipates some shifts in the insurance sector. Carriers will require the agreement to create risk profiles of customers based on protected attributes and to identify customer interactions and company activities using AI in order to maintain transparency. AI models used by insurers may also be frequently checked for compliance with algorithmic accountability and security. Carriers would expect the same from manufacturers of AI technology.

The cost-effectiveness of implementing these restrictions, as well as their influence on carrier combined ratios, must be considered. If AI proves to be more expensive and comes with legal ramifications, insurers may be hesitant to adopt it. Compliance with multiple AI legislation across geographies is another difficulty for insurers. As a result, the adoption of AI in the insurance industry would be aided by a comprehensive, global regulatory framework [17, 20].

4.9 Things to Consider While Establishing Artificial Intelligence in the Insurance Sector

Concerns about data privacy, security, and prejudice related to AI adoption in the insurance sector are rising. Insurance companies all need to adopt a responsible AI strategy when implementing AI. The underneath section discusses the three things insurers should think about when implementing AI responsibly.

a. Transparency:
   While implementing AI, it is critical to understand how models are created and how they are employed. When insurers plan about it, especially in a business like insurance, which is built on fundamental ideals like equality, mutualisation, and mutual care, its vitality increases. While planning and building an AI-first solution, insurance companies need to be confident in their ability to explain how and why things are done. They also need to know what data was used to train the model and how to check the dataset's legitimacy.

b. Regulation:
   To regulate the insurance industry and hold players accountable, various stakeholders will be required, including business owners, data scientists, agencies, the government, and others. As regulators and compliance teams continue to investigate the usage of AI in the insurance

sector, the number of rules that have been enacted has increased. The Insurance Distribution Directive in Europe, the Explainable Artificial Intelligence initiative in the United States, and the California Consumer Privacy Act are all examples of this. The National Association of Insurance Commissioners has also produced a general AI Code of Conduct.

These top-down approaches can preserve consumer data while also increasing public confidence in AI. While government regulation is essential for monitoring the ethical use of AI, it is not the only option accessible to businesses wishing to help manage their AI. It is critical to check within your company to see if it is cultivating a culture of transparency and encouraging people to come forward and report biases.

c. People:
While AI can free up staff to focus on other aspects of the business, insurers should keep them as a critical component of the equation. Limited insurers have inclusive design or human-centric design principles in place to facilitate human–machine collaboration. To get the maximum pay-off from AI, insurers need to involve humans in many facets of their business. Customer-experience staff will need to be well-versed in the AI technology that is available to assist them in their jobs, as well as which procedures can be automated.

4.10 The State of Insurance in 2030

The insurance industry will be profoundly affected by AI and related technologies in all areas, including distribution, underwriting, pricing, and claims. With policies being quoted, bought, and bonded in almost real time, data and advanced technology are already having an impact on distribution and underwriting. It seems extremely fascinating to take a closer look at what insurance would look like in 2030.

a. Distribution:
Cycle times for obtaining a business, auto, life policy, or non-life insurance policy will be shortened to minutes, if not seconds, with adequate data on individual actions and AI algorithms creating risk profiles. Auto and house insurance providers will continue to hone their ability to quickly provide coverage to a wider range of customers as in-home Internet of Things (IoT) devices and telematics spread and pricing algorithms develop.

Simplified issue plans are being experimented with by many life insurance firms, although they are typically more expensive and available to only the healthiest candidates. As AI permeates life insurance and carriers are able to analyse risk in a much more nuanced and sophisticated manner, the future will see the emergence of a new class of mass-market quick-issue policies.

Smart contracts powered by blockchain instantaneously approve payments from a customer's bank account. The costs associated with acquiring new customers for insurers are decreased by the removal or simplification of contract formalities and payment verification. Commercial insurance purchases are also expedited since a combination of drones, IoT, and other data sources provides enough details for cognitive AI models to issue a binding price in advance.

Plans that are very dynamic and tailored to the behaviour of specific clients are known as usage-based insurance (UBI). Insurance evolves away from a "purchase and annual renewal" model and toward a continuous cycle as product alternatives adjust to a person's behavioural habits. Additionally, users can tailor products to their particular needs by breaking them down into micro coverage components (such as airline delay insurance and phone battery insurance), and they can compare pricing from several carriers in real time for their tailored baskets of insurance goods. To address the evolving nature of living situations and travel, new products are being produced. As physical assets are shared among numerous parties, UBI becomes the standard, with a pay-per-mile or pay-per-ride model for automobile sharing and pay-per-stay insurance for home-sharing services like Airbnb.

The position of insurance salespeople will significantly change by 2030. The number of active agents declines as active agents retire, and the remaining agents will rely heavily on technology to increase productivity. Agents now play the role of product instructors rather than process facilitators. Almost any type of insurance can be sold by the agent of the future, who adds value by helping customers manage their insurance portfolios for experiences, life, health, mobility, homes, and personal property. Agents use AI-enabled bots to find potential transactions for clients and intelligent personal assistants to make their work more efficient. These solutions allow agents to handle a much larger client base while shortening customer interactions because each engagement will be tailored to the precise current and prospective needs of each client.

b. Underwriting and pricing:
By 2030, underwriting as it exists today will be obsolete for the majority of personal and small-business products in the estate, life, and liability insurance sectors. The underwriting process is largely automated and supported by a combination of machine learning and deep learning models that are integrated into the technological stack, taking only a few seconds. These models are powered by internal data as well as a wide variety of external data gathered through interfaces for application programming and third-party analytics and information providers. Numerous data warehouses collect information from devices provided by reinsurers, mainline carriers, product producers, and distributors. These data sources enable insurers to make ex-ante underwriting and pricing choices, thus facilitating a proactive outreach with a bindable quote for a product bundle customised to the buyer's risk profile and coverage requirements.

A clear method of evaluating a score's traceability is required since regulators review machine learning and AI-enabled models. To determine whether data usage is acceptable for marketing and underwriting, regulators examine a range of model inputs. To ensure that the outputs of the algorithm are within acceptable bounds, they also develop test standards that service providers must adhere to when establishing pricing for online plans. Public legislation restricts access to some critical and prognostic data (such as genetic and health information), which limits underwriting and pricing flexibility and raises the anti-selection risk in some market sectors.

c. Pricing:
Consumers' decision-making is mostly based on price, although carriers are innovating to reduce price competition. Customers and insurers are connected through sophisticated proprietary systems that provide customers with unique experiences, features, and value. Price competition is increasing, and razor-thin margins are the norm in some areas, while unique insurance offers provide margin expansion and differentiation in others. The speed of pricing innovation is quick in jurisdictions that accept change. Pricing is offered in real time and is based on consumption and a dynamic, data-rich risk assessment, giving consumers control over how their activities affect coverage, insurability, and pricing.

d. Claims:
Carriers' major duty in 2030 will still be claims processing, although automation will have replaced more than half of claims activities. Initial

claim routing is handled by advanced algorithms, which improves efficiency and accuracy.

IoT sensors and various data-capture tools, like drones, have largely supplanted traditional, human means of first notice of loss. Repair and claims triage services are usually started immediately when a loss occurs. A policyholder might record a streaming video of the damage after an automobile accident, which is later translated into loss descriptions and estimates. When an autonomous vehicle sustains minor damage, it will automatically steer itself toward a repair facility for maintenance while simultaneously dispatching another autonomous vehicle. IoT devices will be used more frequently in homes to proactively monitor temperature, water levels, and other important risk factors, warning tenants and insurers of dangers before they materialise.

The majority of interactions with policyholders are handled by speech- and text-based automated customer support apps that use self-learning scripts to interact with the claims, service, medical, repair, policy, and fraud systems. Instead of taking days or weeks to resolve many claims, it only takes minutes. Complex and unusual claims, random manual reviews of claims to ensure adequate oversight of algorithmic decision-making, contested claims where human interaction and negotiation are assisted by analytics and data-driven insights, and claims linked to systemic issues and risks posed by new technology (for instance, hackers infiltrating important IoT systems) are all areas where human claims management focuses on.

For claims companies, risk monitoring, mitigation, and prevention are becoming more crucial. IoT and new information sources are used to assess risk and initiate responses when parameters exceed AI-defined thresholds. To prevent future loss, customers are encouraged to participate in insurance claims groups. People who can be connected to automated inspection, repair, and maintenance activities receive real-time notifications. Provided that cell phone service and power have not been affected in the area, insurers use telematics, integrated IoT, and mobile phone data to monitor homes and autos in real time during large-scale catastrophe claims. Data aggregators combine information from observatories, weather services, networked drones, and policyholder data in real time when the lights go off, enabling insurers to profile claims. The largest carriers have pretested this technology across several catastrophe scenarios, ensuring that considerably precise loss estimates are reliably reported in an actual event. For speedier reinsurance fund flow, detailed reports are automatically sent to reinsurers.

4.11 Conclusion

The insurance sector has long been a pioneer in the use of data and statistics, and it is now on the verge of the widespread adoption of cutting-edge AI. In light of the current situation, AI-based goods include insurance coverage for smart sensors, self-driving cars, and cybercrime damages. Important procedures such as prevention, risk assessment, asset management, and claims processing will also be aided by AI. The way AI implications are increasing in the insurance field, the idea of artificial intelligence evaluating data, forecasting outcomes, and assisting with decision-making in the insurance industry is not so far-fetched.

Given the current situation, insurers must adapt according to the changing business scenario as AI becomes more entrenched in the sector. Insurance executives must explore and comprehend the underlying factors that will influence this shift, as well as how the advent of artificial intelligence will influence claims, underwriting, distribution, and pricing. This knowledge could help them in developing the relevant skills and talent, assist in embracing emerging technology, and in creating the cultural vision required to survive and succeed in the insurance industry of the future [11].

While AI can improve insurance, it can also have unintended harmful consequences for insurers. Regulation of AI usage will increase responsibility and prevent misuse, removing legal barriers and increasing customer confidence. This should not be viewed as a barrier to AI adoption by insurers, as the same rule requires AI use in road safety and healthcare. In the insurance sector, AI progress is still in its infancy and is now an excellent time to nurture its development for the benefit of both society and humanity. Insurers who will quickly and continuously adapt artificial intelligence will stand out and rise above the competition.

References

[1] Bohnert, A., Fritzsche, A. and Gregor, S., Digital agendas in the insurance industry: The importance of comprehensive approaches. *The Geneva Papers on Risk and Insurance- Issues and Practice,* 44, 1, 1–19, 2019.

[2] https://earlymetrics.com/infographic-artificial-intelligence-in-the-insurance-sector/

[3] Kelley, K.H., Fontanetta, L.M., Heintzman, M. and Pereira, N., Artificial intelligence: Implications for social inflation and insurance. *Risk Management and Insurance Review,* 21, 3, 373–387, 2018.

[4] https://www.datamation.com/artificial-intelligence/ai-in-insurance/

[5] https://www.propertycasualty360.com/2021/05/10/three-ways-the-insurance-industry-can-adopt-ai-responsibly/

[6] https://marutitech.com/artificial-intelligence-in-insurance/

[7] Zarifis, A., Holland, C. and Milne, A. Evaluating the impact of AI on insurance: The four emerging AI- and data-driven business models. *Emerald Open Research,* 1, 15, 1–12, 2019.

[8] Eling, M., Nuessle, D. and Staubli, J., The impact of artificial intelligence along the insurance value chain and on the insurability of risks. *The Geneva Papers on Risk and Insurance- Issues and Practice*, 47, 205–241, 2022.

[9] Riikkinen, M., Saarijärvi, H., and Sarlin, P., Using artificial intelligence to create value in insurance. *International Journal of Bank Marketing*, 36, 6, 1145–1168, 2018

[10] https://www.oecd.org/finance/Impact-Big-Data-AI-in-the-Insurance-Sector.pdf

[11] https://www.mckinsey.com/industries/financial-services/our-insights/insurance-2030-the-impact-of-ai-on-the-future-of-insurance

[12] Hall, S., How Artificial Intelligence is changing the insurance industry. *The Center for Insurance Policy & Research*, 22, 1–8, 2017.

[13] Akhusama, P.M. and Moturi, C., Cloud computing adoption in insurance companies in Kenya. *American Journal of Information Systems,* 4, 1, 11–16, 2016.

[14] Kumar, N., Srivastava, J. and Bisht, H., Artificial Intelligence in Insurance Sector. *Journal of the Gujarat Research Society*, 21, 7, 79–91, 2019.

[15] Dale, R., The return of the chatbots. *Natural Language Engineering,* 22, 5, 811–817, 2016.

[16] Bologa, A.R., Bologa, R. and Florea. A., Big data and specific analysis methods for insurance fraud detection. *Database Systems Journal,* 4, 4, 30-39, 2013.

[17] www2.deloitte.com/content/dam/Deloitte/de/Documents/risk/article-risk-and-compliance-implications-of-ai.pdf

[18] https://www.mantralabsglobal.com/blog/challenges-in-ai-implementation-insurance/

[19] https://www.efma.com/study/5067-artificial-intelligence-challenges-and-opportunities-for-insurers

[20] Sood, K., Seth, N., & Grima, S. (2022). Portfolio Performance of Public Sector General Insurance Companies in India: A Comparative Analysis. In Simon Grima, Ercan Özen, Inna Romānova (Ed.), In *Managing Risk*

and Decision Making in Times of Economic Distress, Part B. (pp.215–229). Emerald Publishing Limited,UK.

[21] Grima, S., & Marano, P. (2021). Designing a Model for Testing the Effectiveness of a Regulation: The Case of DORA for Insurance Undertakings. *Risks*, *9*(11), 206.

[22] https://www.tcs.com/blogs/ai-adoption-insurance-bias-risks

5

Data Science in Insurance

Kuldeep Singh Kaswan[1], Sandeep Lal[2], Jagjit Singh Dhatterwal[3], Simon Grima[4], and Kiran Sood[5,6]

[1]School of Computing Science & Engineering, Galgotias University, India
[2]Punjab Institute of Technology, Rajpura, Punjab, India
[3]Department of Artificial Intelligence & Data Science (AI&DS), Koneru Lakshmaiah Education Foundation, Guntur, Andhra Pradesh, India
[4]Department of Insurance and Risk Management, Faculty of Economics, Management and Accountancy, University of Malta, Malta
[5]Postdoc Researcher, University of Usak, Faculty of Applied Sciences, Dept of Finance and Banking
[6]Chitkara Business School, Chitkara University, Punjab, India

Email: kaswankuldeep@gmail.com; Sandeep_mohar@rediffmail.com; Jagjits247@gmail.com; simon.grima@um.edu.mt; kiransood1982@gmail.com

Abstract

The globe is currently creating massive volumes of data, with data output in recent years increasing at an unprecedented rate. Most of this additional knowledge is being collected in novel ways, and technological developments allow it to be stored and analysed much more efficiently than usual. Against this background, there has been a lot of recent discussion about big data and data science. That is the capacity to analyse and derive valuable implications from increasing amounts of data from various sources quicker than ever. Data science is already transforming many facets of modern life, and it has enormous potential to foster innovations in the insurance sector. Insurers have traditionally collected information to understand premiums and risks better. Data science, coupled with increased computer capacity, provides a step-change in vulnerability assessment by allowing insurers to observe these risks in much greater depth continually. This can benefit insurers and policyholders, with room for innovations in how insurance is presented and priced and how claims are managed. As customers' aspirations of all sorts of information, ability to respond quickly, and ways of conducting business rise, so will their expectations of the insurance sector. This chapter discusses many machine learning algorithms for effectively analysing insurance claims and comparing their performance using various criteria.

5.1 Introduction

The globe is creating massive volumes of data, with data output increasing exponentially in recent years. Every day, no less than 2.5 quadrillion bytes of data (i.e., 2500 million billion bytes) are produced, according to estimates in 2015. To demonstrate the rate of increase, it was estimated that 90% of all data in existence at the time had been produced in the preceding three years alone [1]. We are not just witnessing increased data quantities but also data being collected in novel ways:

- massive amounts of customer information being gathered via internet search engines such as Google;
- the increase in data created via social networking sites such as Facebook and Twitter;
- data acquired by cell phones, tablets, and smartwatches, and telecommunications and accessories.

We are rapidly living in what is referred to as the Internet of Things, or IoT. This is the increase in the number of household devices that are all connected to each other via networking and are all creating data and interacting with us. In the United Kingdom, the Internet of Things is gaining popularity in home heating systems, allowing for more economical energy consumption in smart houses. With the growing prominence of wearable gadgets that monitor workout routines and give well-being statistics, technology is also revolutionising fitness regimens. Insurers have already seen potential due to these specific examples being applied to home and insurance coverage. Thanks to technological advancements and processing power, data can now be stored, manipulated, and analysed considerably more rapidly and inexpensively than in the past [2].

5.2 Data Science and Machine Learning in Insurance

Insurers work in a more data-rich and encryption method environment, where increased processing power enables computers to gather, convert, and analyse data more effectively. Data science and machine learning allow actuaries

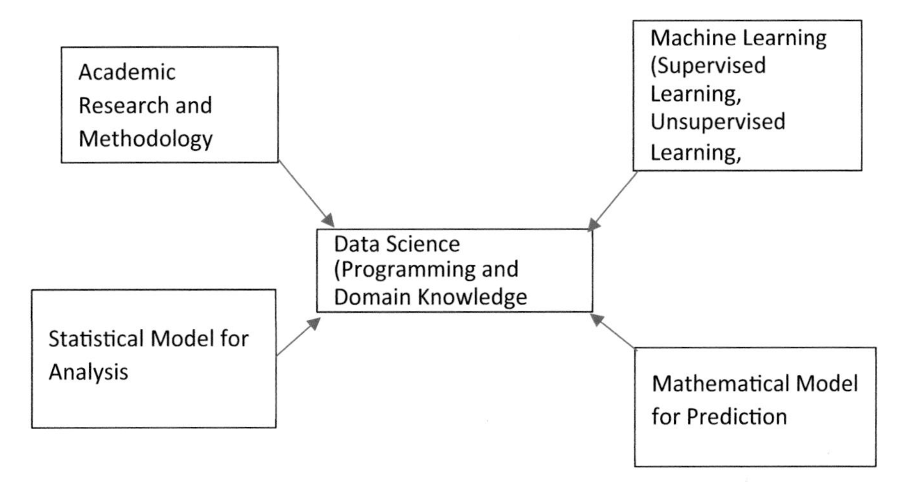

Figure 5.1 Data science framework.

to develop established actuarial domains while embracing new approaches to improve various business activities, governance procedures, and customer happiness. To compete in this fast-changing and demanding business, insurers must invest in business intelligence [3, 17, 18].

5.2.1 Data scientists and actuaries

Insurance businesses are increasingly demanding their employees to have data science skills. Machine learning often possesses three components: coding, computer studies, and domain expertise. While computing enables data translation and algorithms development, mathematical principles enable it to utilise data to construct models and predict future events. Furthermore, machine learning must be able to comprehend actual events and regulations to tackle genuine problems. As a result, data science spans the whole spectrum of data management, not only machine learning and scientific techniques. These qualities, together with professional actuarial competence and regulatory understanding, are becoming increasingly desirable in the employment of accountants, as shown in Figure 5.1.

Although computer science and parametric modelling have many similarities, putting the actuaries' professions in a good position to take advantage of emerging data analytics approaches, they differ in how computer scientists and actuaries happen in reality. The primary distinction comes when creating and executing sensible solutions. Actuaries often utilise their domain expertise to pick relevant models before focusing on adjusting characteristics that are appropriate for achieving the goal. On the other hand, data engineers

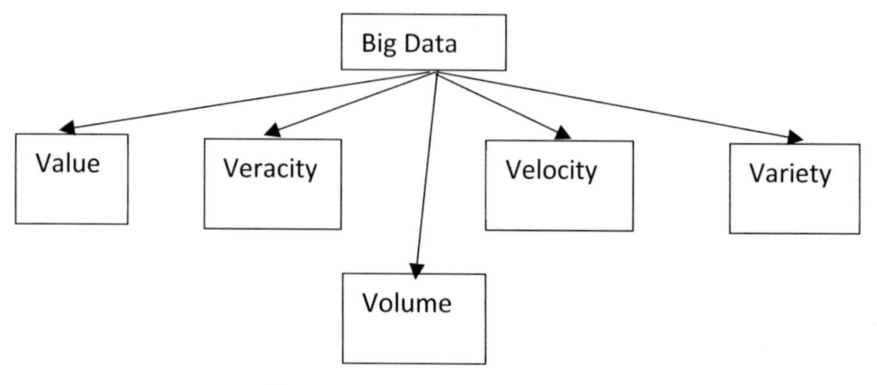

Figure 5.2 Framework of big data.

spend time and effort testing several algorithms before calculating suitable model parameters. Furthermore, these fields differ because actuaries create economic models, but computer scientists frequently rely on outside specialists to grasp knowledge components. As a result, there are variances in how hypotheses are validated, characteristics are chosen, and the model fit is assessed [4].

5.3 Big Data: Challenges and Opportunities

The insurance industry is no stranger to discovering patterns in enormous amounts of data. Nonetheless, this remains a significant difficulty since disorganised and fluctuating data is difficult to analyse using typical technologies. Big data demands creative technology and ways to capture, store, and interpret massive volumes of data in addition to obtaining significant business information, as shown in Figure 5.2.

The development of big data is prompting insurers and governments to develop best practices for its usage. It is typical in big data literature with respect to the "5 Vs" to explain the primary issues that organisations face [5]:

- Volume – Data volumes have expanded considerably in recent decades and continue to do so. IFRS 17 standards, for example, will raise information requirements owing to the need to aggregate regulations at more granularity degrees and improve data technologies to run directly relevant.
- Variety – Digital databases can be created from various sources with varying data architectures. Historically, insurance companies depended on heavily organising the data housed in relational database systems,

but with the proliferation of telematics gadgets like smartphones, unsupervised learning has developed, giving insurers extra information about products, services, and consumers.

- Velocity – Data must not only be obtained rapidly but it must also be analysed and interpreted swiftly. Understanding and analysing data quickly can assist organisations in speeding up decision-making operations and maintaining their position in a competitive marketplace.
- Veracity – With a rise in the volume and diversity of data, it is vital to guarantee that the data can be trusted and depended on. Data must be dependable and correctly understood in order for database administrators and actuarial to provide relevant analyses and, as a result, make high-quality judgements.
- Value – Big data analytics-driven management decisions should contribute to economic gains and product differentiation [6].

The quantity of information (and the rising storage capacity) necessitates using cloud computing technology for exterior information management. Cloud computing provides insurers with mobile innovations to manage steadily growing data volumes and improve data storage. Cloud computing clearly places new responsibilities on IT department, database administrators, and the business processes they enable, but they may also provide enormous value across an insurance organisation:

- Underwriting – Using bigger datasets to ensure risks more efficiently than rivals and optimise marketing strategies through more precise predictive modelling approaches [7].
- Fraud detection involves determining policyholders who are more likely than others to commit fraud during the health insurance marketplace and monitoring the concealment or falsification of application documentation. Big data may also be used to track claims in insurance firms. Social media for indications of potentially fraudulent behaviour.
- Claims management is putting in place systems that rely on experiences and data from the internet to effectively screen suspect claims, accelerate the claim cycle, and cut costs.
- Explicit knowledge from social networking sites can assist insurers in gathering policyholders' thoughts on goods and, as a result, establish a strategy to improve customer engagement [8].

5.3.1 From classical statistics to machine learning

Statisticians have used sophisticated statistical approaches for many generations, but in a modernising business, they are becoming obsolete. GLMs, for example, was previously utilised in the non-life insurance business for pricing and preserving to identify how critical variables (e.g., regularity and seriousness of claims) varied with rating criteria. GLMs have also found traction in the life insurance industry, where actuarial frequently use them to represent the most important risk factors and influence the calibrating of depreciation assumptions [9].

However, GLMs have their own set of restrictions. They are parametric models that rely on a predefined Weibull distribution and connection function. Furthermore, they are unsuitable for identifying relationships between variables and intricate connections. Such restrictions can result in poor goodness-of-fit and incorrect projections of future data.

To overcome these constraints, and as a result of rapid technological improvements, machine learning (ML) is becoming more used in the insurance industry. Without explicitly programmed, ML can construct algorithms that recognise complicated patterns, make sound decisions, and generate educated predictions based on data inputs. Essentially, machine learning can learn from past experienced information and make recommendations without the need for human interaction.

This enables the formulation of more complicated links between attributes and consequences than standard models allow. A detailed analysis to improve in contentment to reduce efforts of anomaly detection in the business. A Rapid reaction to changes in system architecture and real business conditions. ML algorithms are often classified into three types based on the type of issue to which they are implemented.

The purpose of supervised methods is to predict the future values of an output measure using a large number of input measurements. Because of the presence of an endogenous construct that controls the educational process, the learning process is supervised. Examples include regression, machine learning, and tree-based approaches such as randomised forests.

There is no assessment instrument in unsupervised classification; the purpose is just to explain the correlations and patterns among a collection of inputs. Cluster analysis and principal component analysis (PCA) are two examples.

Reinforcement learning incorporates anticipated results into the algorithm to enhance the next predictions. The algorithm's forecasts improve over time as it understands something about the environments in which it functions, and the models it employs are regularly updated. It is not presently commonly employed in finance and accounting, although this may change as statistical methods and processing power improve.

Penalised regressions (for example, lasso, ridge, and elastic net), which try to reduce the number of variables, are particular types of supervised learning ML approaches that can overcome some of the shortcomings of GLMs. By restricting and reducing parameters, these approaches can minimise the variability of estimations at the expense of a negligible bias [19,20].

Other machine learning approaches, such as decision trees, random forests, and machine learning, make inroads into probability and statistics. Underwriting is one area where it may be used as a prediction method to classify new policyholders and decide whether to accept or reject standard conditions. Similarly, the same approaches may be used for marketing and preserving; for example, historical policyholder data, such as actual claim amounts and periodicity, can be used to optimise marketing strategies and anticipate future losses.

ML modelling is generally used to develop recommendations independently or in conjunction with other multivariate approaches, such as clustering and PCA, to optimise certain study elements. Clustering and PCA are standard exploratory analysis techniques used to decrease the computational burden and eliminate duplicate features. Lowering the dimensions of a training input set may increase training time, and datasets can be reduced to only a few parameters, making data visualisation straightforward [10].

Cluster analysis may also be used to improve the production of model points. Due to time and processing capacity limits, parametric models using grouped model variables rather than entire data runs are frequently required. This method creates a user-defined number of similar advantages without requiring manual interference by recognising groupings of rules with comparable distinctive and differentiating traits. Software firms have produced new packages to meet the increased need for more complex data analytics methodologies and a broader choice of ML methods.

Nonetheless, ML advancement in the insurance business is still in its early stages. There are several reasons why insurance is apprehensive about abandoning traditional statistical procedures in favour of machine learning technologies in Figure 5.3. To begin with, linear models are a straightforward and well-known statistical methodology, and standardisation software tools for implementing such approaches are readily accessible. Second, insurance firms have recently begun establishing business intelligence teams; so company-wide objectives and plans are still being developed. Because data science specialists are frequently dispersed throughout organisations, expertise

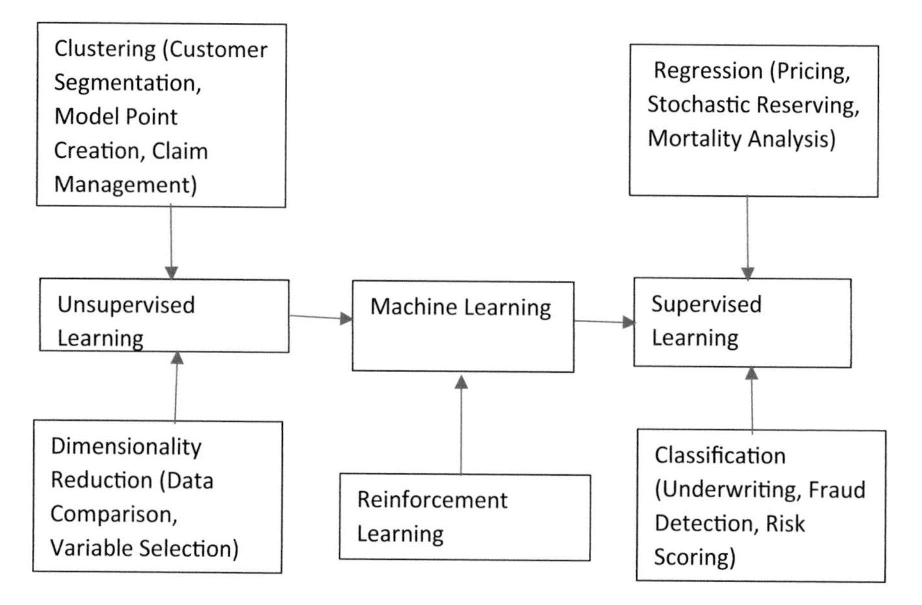

Figure 5.3 Machine learning-based insurance.

and operations are not adequately connected and organised. The advent of big data and sophisticated analytics necessitates investment in new technology, the provision of specialised training, and the implementation of extra management.

5.3.2 Reaping the benefits

Actuarial competence paired with development skills and an understanding of visualisation tools can assist insurers in realising the advantages of system improvement and improved client involvement.

Data analytics and machine learning can significantly improve information retrieval and comprehension of client risk profiles. Restructuring systems to promote effective manner identification can provide a clear signal of future unfavourable events and possible expenses [11].

Implementing cutting-edge machine learning solutions will allow an insurance company to increase productivity and acquire a comparative benefit. The possible applications involve many parts of actuarial modelling, including model point creation, pricing, sparing, claims administration, and reporting automation. Furthermore, by converting systems to understand through data trends, find new situations, and react properly to analyse each scenario autonomously, these technologies can help to underwrite correctly calculated risks and detect probable false accusations. This enables insurers to save time and money in areas where procedures are mostly manual while also providing a quick and efficient customer experience. Predictive modelling may also eliminate speculation and aid in the rapid identification of critical market categories and client groups, as well as forecast customer behaviour, allowing insurers to tailor goods and marketing to the specific demands of their customers.

Choosing a supplier that provides sensible solutions will provide actual value to the business, particularly: in-house technology and development – the application of approaches to real-life situations is constantly studied. Make it clear to all stakeholders – no naive faith in technology or the simulation model.

Approach comparison – increase complexity only when necessary. Multiple validations are required to have confidence in both accuracy and interpretation. Flexibility in development and deployment – make the solution commercially feasible.

5.4 Data Science in Insurance: Benefiting from the Digital Experience

Based on the risk, a suitable payment would be determined. On the other hand, data science enables insurers to see their applicants' risk profiles in far greater depth. This more detailed hazard identification should result in better-informed underwriter choices and allow insurers to set rates that are more properly aligned with the relevant degree of risk. In certain circumstances, more precise and thorough risk management might enhance or decrease medical insurance, such as when a risk would be refused under a broad assessment but found acceptable in light of more details. For example, better clarity on the underlying degree of risk might result in younger drivers being assigned to maximum and minimum risk groups, with the latter perhaps able to drive more safely. To get a better deal on car insurance, some customers with impairments may also benefit from a more thorough risk assessment. Data science may also minimise the time and effort necessary to receive an insurance quotation if most of the information for underwriters is acquired immediately, lowering the number of direct questions the customer must answer. More precise underwriting can also imply less inaccurate disclosure, allowing insurers to lessen the chance that applicants would take advantage of the knowledge that the company does not yet have [12].

5.4.1 Stronger engagement with consumers

Consumer participation may also improve during the life of an insurance term, which benefits both the customer and the provider. Lower rates may be offered to subscribers if their risk profile increases, for example, through a telematics device in the case of vehicle insurance or a fitness tracker gadget in the case of health coverage. Furthermore, the incentive of lower premiums may motivate policyholders to lessen the "riskiness" of their way of life. Data science provides demonstrable societal advantages by rewarding insurance subscribers in this manner. Consumer interaction is also being pushed to the next level with "on demand" insurance: at least one car insurance company is already giving the ability to turn on and off coverage via a cell phone [13].

5.4.2 Data science: potential for innovation in insurance

Data science is already transforming many sectors of modern life, including healthcare insurance, research and technology, politics, and even sports. Furthermore, data science can promote significant innovation in the insurance business about how and to whom policies are sold. Risk management is fundamental to assurance, and underwriters collect data to analyse the threats and risks they are underwriting. Data science provides insurers with a step-change in vulnerability assessment by allowing them to understand the hazards of the items they insure – such as automobiles, people, or property – in far more detail than normally available. It also enables the discovery of connections that would otherwise be concealed inside data and the creation of new connections. Predictions that are more accurate as a result of a continual updating of much larger datasets. Actuaries are already using advanced analytics in a variety of insurance technologies, such as telematics devices in auto insurance, wearable fitness devices in health and care insurance, and improved corporate governance (more extensive analysis of experienced analytics) in life insurance [14].

Data science benefits customers at every stage of the liability insurance lifecycle, including better:

- consumer segmentation and represent tasks;
- vulnerability assessments, underwriter, and pricing;
- consumer interaction throughout coverage;
- claims administration, including the detection and prevention of fraud.

5.4.3 Better consumer targeting/product design

Insurers can use new data sources to target targeted clients to specific, effective, and perhaps more appropriate products. Investigation of information retrieval patterns or social media information aids in predicting customer preferences and behaviours. These studies can then be used to connect people with specific items. A similar moving average of preferences and expectations increases an insurer's awareness of changes in customer demands, which may be valuable in the creation of creative new solutions and the implementation of related features. The increased usage of smart apps by the community as a whole, as well as insurers' study of their use, may make it simpler for various consumer groups to get insurance, including fulfilling untapped areas of demands for healthcare coverage. A more comprehensive perspective of customers and their requirements may imply that they are not skipping out on vital healthcare insurance or doubling up on coverage someplace [15].

5.4.4 More accurate risk assessment, underwriting, and pricing

Much of data science's promise in the insurance business is related to the increased understanding that can be obtained during hazard identification. It has long been accepted procedure for insurers to collect data on applicants' (or their homeowner's) attributes and use it to predict the likelihood and expense of claims. Following that, an underwriter decision was made: take the risk on regular terms, approve on amended terms, or refuse [16].

5.4.5 Better claims management

Data science can help insurers and policyholders by improving insurance settlement administration and related complaints procedures more efficiently. Data analytics may be used to prioritise claims, with simple instances prioritising speedy settlements and more complicated situations being highlighted for further investigation by claims evaluators. Analysis of social media interactions and relationships may also be used efficiently to detect false statements activity by groups of individuals collaborating to generate a succession based on unsubstantiated claims. As customers' expectations of all types of computers, ability to adapt quickly, and ways of doing business rise, so will their expectations of the insurance sector, both in terms of claims administration and more broadly. Finally, developing high-tech "Insure tech" enterprises have the potential to affect the insurance sector. Such Insure tech enterprises may be highly nimble in their marketing strategy, with data-driven thinking and a computerised perspective.

5.4.6 Data: property of insurer or individual?

Another dilemma is who maintains the data obtained on a policyholder. Is it owned by the appropriate insurer or by the insurance policy? If the insurer owned the data collected by a telematics or wearables device, it might limit the consumer's opportunity to get a better price elsewhere. If it were the consumer's possession, could they be forced to share what was before data with an insurance if they switched providers?

5.4.7 Transparency and judgement

There may be a lack of openness surrounding data science and related analytics, and comprehending what is underneath the analytical "black box" might be difficult. With the increased usage of data science, unanswered questions between insurers and customers may expand. There is also the risk of relying too much on analysing model outputs: no model can completely replace human knowledge and judgement. The human judgement raises additional public-interest considerations. With data science becoming more pervasive, there is a risk of prejudice (or bias) entering into any risk profile used. Because some sophisticated pricing models depend on algorithms rather than clear rating variables, it might be impossible to tell whether a pricing strategy is being used. On the other hand, increased visibility of the risk may minimise dependence on analysts' biases by making judgements more data-driven.

As data science penetrates more into the insurance business, concerns about ethics and the broader public interest arise. Insurers and customers may be influenced in novel and unanticipated ways, potentially leading to unforeseen effects and health insurance industry inefficiencies.

5.4.8 Insurance unavailable for some?

Certain segments of the ordinary population may discover that data science negatively influences the cost or accessibility of insurance. If insurers have a better grasp of an individual's characteristics, users in specific market sectors may find it more difficult or costly to get insurance. For example, young drivers with little racing dynamics may find it difficult to obtain auto insurance; people with increased morbidity and mortality may find that healthcare insurance becomes prohibitively expensive; and older travellers may require insurance policy, which means the method used to set premiums and deductibles are commonly broad-brush rather than detailed in the environment.

This, thus, raises what is fundamentally a public policy issue that will need to be resolved. Applying a premium based on the individual's risk may be considered an adequate treatment of consumers. On the other hand, individuals or their property may be in "greater danger" owing to intrinsic elements or changes in situations over which they have no control. If this is limited by the availability of cover at a fair price is regarded to constitute a market failure. Preventing it will require involvement from the administration, regulations, or the insurance sector itself.

5.4.9 Less pooling of risk

Data science might potentially influence the related idea of risk pooling. Risk pooling, or sharing risks among policyholders with roughly comparable risk profiles, is a long-standing component of insurance. The growth of data science is projected to gradually diminish the volume of each pool, lowering present levels of cross-subsidy across various policyholders. In the United Kingdom, there is now a tendency toward more precise risk assessment within Insurance, such as the introduction of damaged life and modified expectancies in the mid-1990s, which provide a larger income to smoking or those with symptomatic health concerns. This trend may develop in data science, with risk classification getting finer and less complex. Risk can be pooled into ever-smaller portions.

However, society may demand a degree of cross-subsidisation by keeping risk-pooling when this is regarded as "fair." Sexual identity insurance pricing was outlawed across the EU in 2012 due to sex inequality. In the United Kingdom, the establishment of Flood Re* acknowledges that when flood risk grows, both individuals at highest danger and the enterprise require time to respond; in this case, a transitional subsidy from policyholders with a lower risk of dehydration to those with heightened hazard.

5.4.10 Price discrimination

As the quantity of data acquired on individuals grows, the scope of pricing discrimination by insurers may also grow. Price discrimination refers to the practice of basing insurance premiums not just on an individual's level of risk and related expenses but also on broader criteria such as price sensitivity and brand loyalty. It is not a new concept, but expanding the breadth of such sensitivities raises the possibility that such actions would negatively impact vulnerable consumer groups. Insurers will have to assess whether such pricing discrimination raises any concerns about conduct risk.

As insurers collect greater amounts of personal data, they may be regarded as being unduly invasive and operating in a "Big Brother" manner. It is also critical that insurers be honest in their use of data to avoid undermining the degree of faith that policyholders have in them and, as a result, limiting the scope of data policyholders are willing to give the insurer access to. However, if insurers communicate the practical benefits of data collecting to policyholders, such as a prospective decrease in insurance rates through the use of a telematics device, they may conclude that this trumps any security issues.

5.4.11 Cyber risk

Finally, cyber risk is a major rising risk area, and when more data is generated, data science increases insurers' susceptibility to it. The dangers of data loss, corruption, or theft, in particular, are critical considerations for users of data application areas to address. Procedures with sufficient safeguards against hackers and other unauthorised people accessing this information will need to be created.

5.5 Regulation Data Protection

Data science is heavily reliant in accordance with existing and forthcoming data protection regulations. The General Data Protection Regulation (GDPR) will go into effect in the United Kingdom on May 25, 2018. Previous data protection law in the United Kingdom stretches back to the 1990s, and much has altered in the intervening years. We further highlight that the UK Government has declared that the UK's decision to exit the EU would not affect the GDPR's implementation in the UK. However, because insurers operate worldwide, they will face the added issue of conforming to numerous data privacy standards.

The notion of permission to data usage is critical, but users must be aware of the bounds of data that have been collected. For example, have policyholders, and social media outlets necessarily consented to the use of data obtained from social media activity for insurance assessment or promotional reasons? The information gathered should also be pertinent to the purpose for which it is being utilised. As previously stated, there is an additional concern about ownership and access. As a result, insurers will need to have strong data administration systems and appropriate controls in place to guarantee that relevant processing complies with existing and developing data privacy regulations. Data science also raises worries about the potential behavioural risk for insurers. They will need to assess whether product development and pricing, as well as broader uses of data science, prioritise customers' demands and are in the general public's best interests. Failure to do so may necessitate regulatory action to resolve the potential conflict of interest or enhance customer consequences.

5.5.1 Regulating actuaries in the public interest

Given the possible ethics and broader public benefit challenges raised by the growing use of data science, it is critical to evaluate the regulations of individuals working in this sector, whether they be actuarial, computer engineers, risk managers, or others. The IFoA governs our members to protect the public interest while promoting business and technology. Ensuring the interests of the public refers to the protection of the general public and society as a whole, but it also includes preserving public trust in the professional world by upholding its good reputation. Insurance companies who adhere to high-quality standards of work and the professional skills of the actuaries who deliver it serve the people.

5.6 Data Science in Insurance

By integrating machine learning and enormous additional datasets from autos, houses, and wearables, you can gain a more in-depth knowledge of present and potential threats. With this information, insurers may set pricing more accurately, even on an interpersonal basis, optimising both customer attractiveness and company performance. To make educated portfolio selections, get a clearer understanding of the company's overall potential risks.

5.6.1 Underwriting and claims processing

Automate regular transactions to enhance agility, cut costs, and raise customer happiness. Many insurance applications and complaint filings may be handled without a person's involvement. Algorithms powered by machine learning seek signals of irregularity or worry and either expedite uncomplicated situations or provide crucial contextual data when appropriate.

5.6.2 Fraud detection

Data science can be an effective deterrent to applications and claims of dishonesty. Machine learning can examine enormous datasets and interactions to discover and flag potentially erroneous and/or duplication claims more effectively than human assessment. After a claim is filed, new data sources, such as online and social channels, might be exploited to provide proof of fraudulent conduct.

5.7 Conclusion

Data science is already transforming the insurance sector, with demonstrable implications for insurers, the insured, and the general public. However, the vastly increased use of data science raises a number of potential public benefit issues for the insurance companies, regulators, the administration, and other related authorities to consider. For example, certain subsets of the common person may find that bioinformatics has a negative impact on the cost or availability of insurance. Current innovations are constantly making their way into numerous industries of business. In this regard, the insurance business does not lag behind the competition. Statistics have long been used in the insurance industry. As a result, the fact that insurance firms are actively employing data science analytics is not surprising. In essence, the goal of using data science analytics in insurance is the same as in other industries: to enhance advertising strategies, improve organisational performance, increase income, and decrease expenses.

References

[1] Albrecher, H., D. Bauer, P. Embrechts, D. Filipovic, P. Koch-Medina, R. Korn, S. Loisel, A. Pelsser, F. Schiller, H. Schmeiser, J. Wagner, 'Asset-Liability Management for Long-Term Insurance Business', European Actuarial Journal, Vol. 8, pp. 9–25, 2018.

[2] Albrecher, H., P. Embrechts, D. Filipovic, G. Harrison, P. Koch, S. Loisel, P. Vanini, and J. Wagner, 'Old-age provision: past, present and future', European Actuarial Journal, Vol. 6, pp. 287–306, 2016.

[3] Jagjit Singh Dhatterwal, Kuldeep Singh Kaswan, Vivek Jaglan, and Aanchal Vij, 'Machine Learning and Deep Learning Algorithms for IoD" in book entitled "Internet of Drones: Opportunities and Challenges' in "Apple Academic Press (AAP), Canada, Published, Hard ISBN: 9781774639856, 2022.

[4] Kuldeep Singh Kaswan, Jagjit Singh Dhatterwal, 'The Use of Machine Learning Sustainable and Resilient Buildings" in book entitled "Digital Cities Roadmap: IoT-based Architecture and Sustainable Buildings', Published by John Wiley & Sons, ISBN No. 978-1-119-79159-1, 2020

[5] Preety, Jagjit Singh Dhatterwal, Kuldeep Singh Kaswan, 'Securing Big Data Using Big Data Mining' in the book entitled "Data-Driven Decision Making using Analytics' Published in Taylor Francis, CRC Press, ISBN No. 9781003199403, 2021.

[6] Preety, Kuldeep Singh Kaswan and Jagjit Singh Dhatterwal, (2021) 'Fog or Edge-Based Multimedia Data Computing and Storage Policies', In the book entitled "Recent advances in Multimedia Computing System and Virtual Reality", Published by Routledge Taylor & Francis Group, ISBN No. 9781032048,239, 2021.

[7] Albrecht, P., M. Huggenberger, 'The Fundamental Theorem of Mutual Insurance', Insurance: Mathematics and Economics, Vol. 75, pp.180–118, 2017.

[8] Ayuso M., Guillen M., A.M. Pérez-Marín, 'Telematics and gender discrimination', some usage-based evidence on whether men's risk of accidents differs from women's, Risks, Vol. 4, 10, 2016.

[9] Kuldeep Singh Kaswan, Jagjit Singh Dhatterwal, Santar Pal Singh, 'Blockchain Technology for Health Care', book entitled "Healthcare and Knowledge Management for Society 5.0: Trends, Issues, and Innovations" Published in CRC Press 2021, ISBN 9781003168638, 2021.

[10] Eling, M., M. Lehmann, 'The Impact of Digitalization on the Insurance Value Chain and the Insurability of Risks', The Geneva Papers on Risk and Insurance - Issues and Practice, Vol. 43, pp.359–396, 2018.

[11] Karoui, N. El, S. Loisel, Y. Salhi, 2017, 'Minimax optimality in robust detection of a disorder time in doubly-stochastic Poisson processes', The Annals of Applied Probability, Vol. 27, pp. 2515–2538. 2017.

[12] Gatzert, N., H. Schmeiser, 'The Merits of Pooling Claims Revisited', Journal of Risk Finance, Vol. 13, pp.184–198, 2012.

[13] Lopez O., X. Milhaud, P. Thérond, 'Tree-based censored regression with applications in insurance', Electronic journal of statistics, Vol. 10, pp.2685–2716, 2016

[14] Naveed, M., Ayday, E., Clayton E., Fellay, J., Gunter, C., Hubaux, J-P., Malin, B., Wang, X., 'Privacy in the Genomic Era', ACM Computing Surveys, Vol. 48, pp.1–43, 2015

[15] Verbelen, R., K. Antonio, G. Claeskens, 'Unravelling the predictive power of telematics data in car insurance pricing', Journal of the Royal Statistical Society: Series C (Applied Statistics), Vol. 67, 1275–1304, 2018

[16] Wüthrich, M. V., 'Neural networks applied to chain-ladder reserving', European Actuarial Journal, Vol. 8, pp.407–436, 2018

[17] Sood, K., Seth, N., & Grima, S. (2022). Portfolio Performance of Public Sector General Insurance Companies in India: A Comparative Analysis. In Simon Grima, Ercan Özen, Inna Romānova (Ed.), In *Managing Risk and Decision Making in Times of Economic Distress, Part B*. (pp.215–229). Emerald Publishing Limited, UK.

[18] Grima, S., & Marano, P. (2021). Designing a Model for Testing the Effectiveness of a Regulation: The Case of DORA for Insurance Undertakings. *Risks*, *9*(11), 206.

[19] Bassi, P., & Kaur, J. (2022). Comparative Predictive Performance of BPNN and SVM for Indian Insurance Companies. In Big Data Analytics in the Insurance Market (pp. 21–30). Emerald Publishing Limited.

[20] Trivedi, S., & Malik, R. (2022). Role and Significance of Data Protection in Risk Management Practices in the Insurance Market. In Big Data Analytics in the Insurance Market (pp. 263–273). Emerald Publishing Limited.

6

The Application of Big Data in the Insurance Market

Kuldeep Singh Kaswan[1], Jagjit Singh Dhatterwal[2], Reenu Batra[3], Kiran Sood[4,5], and Simon Grima[6]

[1]School of Computing Science & Engineering Galgotias University, India
[2]Department of Artificial Intelligence & Data Science Koneru Lakshmaiah Education Foundation, India
[3]Computer Science and Engineering, SGT University, India
[4]Postdoc Researcher, University of Usak, Faculty of Applied Sciences, Dept of Finance and Banking
[5]Chitkara Business School, Chitkara University, Punjab, India
[6]Department of Insurance and Risk Management, Faculty of Economics, Management and Accountancy, The University of Malta, Malta

Email: kaswankuldeep@gmail.com; jagjits247@gmail.com; reenubatra88@gmail.com; kiransood1982@gmail.com; simon.grima@um.edu.mt

Abstract

The term "big data" was used to characterise the enormous quantities of data that conventional techniques for managing such data cannot handle. Big data is crucial in numerous fields, such as agriculture, banking, data mining, education, chemistry, finance, cloud technology, marketing, and healthcare stocks. Analysis of massive datasets for hidden insights and previously unexplained correlations that might guide strategic planning and decision-making is known as "big data analytics." Big data's rising popularity may be attributed to its fast growth and many applications, such as Apache Hadoop, an open-source technology written in Java and supported by the Linux platform. The primary objective of this chapter is to provide a suitable and free option for big data applications in a network system and to discuss its advantages and simplicity of usage.

6.1 Introduction

All facets of society have benefited from the exponential growth of informational innovations over the past 60 years (Samuel, 2015). The recording, processing, and restoration of enormous amounts of information previously processed on paper exemplified the use of digital technologies for back-office administration during 1960–1970. Although only a small percentage of the population had access to digital technology, those who did saw huge reductions in computing demands and cost savings (Kuchipudi, 2015).

In the 1970s and 1980s, new technologies, scripting languages, and programs allowed for the creation of adaptable management systems and increased the use of these tools. Consequently, HRM has become more reliant on IT in every facet, including accounting and management oversight, salary disbursement, and more. Aside from managing payroll and benefits, bookkeeping, and other human resource-related difficulties, as well as monitoring and assessing the business's operations. There has been a steady increase in the formation of internal IT services and a plethora of features within these goods due to the globalisation of internet technology usage inside enterprises. Note that the French market has its own set of inputs and constraints (especially those imposed by the government) (Mukherjee, 2016).

The decade between 1980 and 1990 saw the proliferation of programming languages, the improvement of integrated circuits and quality control of instrumentation, the development of system interoperability, and the implementation of master plans within businesses to ensure the justification of investment opportunities and the adoption of a long-term vision for the innovation of IT resources. IT was used for almost all present-day operations by firms of all sizes, but it lacked the necessary staff sophistication to develop into a distribution channel at the time (Misra, 2014).

Through 1990–2000, we saw a reduction in equipment prices due to the ongoing development of solutions for various forms of information and communication technology and the improvement of ergonomics. Because of these changes, it became more advocated that people arm themselves. Businesses have benefited from the proliferation of big data, thanks to advancements in storage and processing technology (Venkata, 2015; Sood, Seth & Grima, 2022; Grima & Marano, 2021).

The rapid growth of the Internet beginning in the second half of the decade greatly benefited efforts to improve the use of information technology in economic relationships between organisations, and between companies and individuals, notwithstanding the bursting of the Internet bubble in 2000. Also, the urbanisation of communications systems became more dispersed because of the proliferation of internet technologies and the dissemination of information from the years 2000–2010. Increased flow requirements and the emergence of a solutions-focused ecosystem (Kaswan, 2022) are direct outcomes of developing distributed computing methods and capabilities.

Since 2010, there has been a rise in both the need for available finances and the variety of uses of information equipment and computer services, thanks to the proliferation of online usage through smartphones and social networks. Public data has grown and diversified because of user-generated content and government support for open data. Value is created for stakeholders when their interests are considered (Ganjir, 2016).

One of the first industries to make extensive use of IT, it has shown to be a useful tool for carrying out simultaneous, sequential, and partitioned tasks, as seen in the insurance industry. This shift has had far-reaching effects on the dynamic between insurers and policyholders, with management enhancement giving way to a sufficient processing challenge and transforming insurance into a consumer product where stakeholder positional sense is already paramount in an increasingly competitive marketplace.

The next few years will likely see a meteoric rise in the prevalence of the Internet of Things (IoT) and cognitive computing. There is no doubt that certain insurers will gain an advantage by improving upon this technology since the most progressive companies are currently investing heavily in it, increasing the value of their business relationships, increasing revenue, and delivering ever-more-tailored products and services to their customers (Sun, 2013).

Increasing connectivity in several developing nations will also add to the annual rise in data generated and exploitable in various industries. Since they have almost no IT infrastructure at present, they will be able to adopt cutting-edge latest technology directly, especially as the costs associated with these technologies have significantly decreased and, given their legal and demographic surroundings, they see rapidly expanding businesses springing up (Naik, 2016).

6.2 Claim Management in Insurance

Due to its importance in determining premiums, insuring risks, and processing claims, data has become an increasingly valuable resource for insurance companies in the age of information technology.

The data exchange between subscribers and insurers is fundamental to building trust and mutual commitment. Insurance companies manage their operations and limit their exposure to risk by making educated guesses about the likelihood of future claims. For the insurance sector to thrive in the long run, insurers need to maintain a competitive stance and grow their market share, dependent on a firm grasp of risk and customer behaviour. Improved policyholder support and lower claims expenses (Agarwal, 2014) result from increased access to a rising amount of information and the ability to evaluate this data in real time.

Due to their ability to assess massive volumes of data in a data-driven manner and, subsequent industrialisation, in an automated manner, machine learning algorithms are potentially essential tools in terms of data value. They allow insurers to make the most of data gathered in the course of digitising their operations and other data sources (Agarwal, 2010) as well as data already in possession of insurers.

6.3 Regulatory Framework in Insurance

Directive 2009/138/EC (Solvency II), Article 82, includes data quality standards in calculating technical provisions. To ensure that the data used in this framework is relevant, complete, and accurate, insurance and reinsurance companies must implement internal procedures as outlined in this Article (Kaswan, 2022).

Compliance with the safety audit trail and enhanced entry requirements dedicated to the use of data sources, as outlined in Article 19 of the Delegation mentioned above of responsibility Legislation (Barron, 2003), the data performance requirements connected with Solvency II might impose significant restrictions on the use of big information for the adopted calculating system. One factor pushing the early use of big data for actuarial applications is the need to meet this limitation. Quality assurance and the launch of pan-European goods should result from the adoption of universal criteria at the European level and the health insurance requirements of particular groups of the European population. In the Table 6.1 scenario, thanks to data science's enabling characterisation, we may use a data-driven method to objectively measure new homogeneous risk groups' creation. A tailor-made home insurance plan from Axe Art covering appreciating assets, artwork, government recourse, cyber dangers, and assistance was released to the public at the end of December 2016. It was initiated in four European nations (France, Germany, Belgium, and the United Kingdom). Spain, Italy, Germany, and the Dutch will join the lineup in early 2017.

6.3.1 Main insurance assets

The fundamental source of confidence comes through encounters with threats involving people beings, substances, and, thus, power (Brown, 2009).

Table 6.1 Data quality in insurance.

Points	Descriptions
Data collection	The data provide adequate context for analysing the nature of the underlying threats and spotting patterns within. The appropriate ones can be found for all relevant groups of individuals with similar risks.
Exact data	The data give sufficient background for assessing the nature of the underlying dangers and identifying trends within. All essential groups with a similar risk profile may find the right ones.
Data consistent (valid data)	The accuracy of the data is commensurate with its use. Due to their size and composition, the estimations produced on their basis for the computation of technical provisions are very likely to be accurate (that is likely to influence the decision-making or judgment of users of the calculation result, including the supervisory authorities). The technical provisions calculated using actuarial and statistical methods are compatible with the fundamental assumptions. They represent the dangers the insurer or reinsurer faces because of its insurance or reinsurance contracts. They are gathered, analysed, and used according to a well-documented approach that ensures openness and consistency. When calculating the system adopted, insurance or reinsurance firms ensure their data is always utilised similarly.

Source: Authors' compilation.

> DEFINITION – "*To be insurable, a risk must be random, future, lawful, independent of the policyholder's will and sufficiently common to be subject to calculation of its probability of occurrence without being almost certain.*"

The insurance industry has undergone a digital revolution in communicating with policyholders. Machine learning techniques enable the proof of identity of the most discriminatory treatment on the basic variables in predicting behavioural responses or the incidence of a given risk, significantly affect the rate at which information is gathered, and change the configuration of existing facts through perfectly in sync data collected with contextual information. As a result, investors' views on risk tend to improve, and price swings may occur (Brown, 2007).

Variations that occur on a less frequent timescale than a year are caused by fluctuations in policyholders' spending patterns (for example, seasonality of automobile use or occupation of a secondary residence). Data from interconnected objects might be used to predict external factors such as cost and, by extension, insurance coverage for the act. The capacity to customise the protections provided within an arrangement and allow the owner of the policy to contribute to just a fraction of the assurances supplied (for example, house insurance plans may contain lump-sum assurances that are inappropriate for young consumers) (Chandler, 1992).

While digital information sources have already allowed for the customisation of current goods and assurances, they may also facilitate the development of assurances for hitherto uninsurable risks. At last, emerging risks like cyber risk may be more easily insured against.

6.3.2 Insurance principle

The concept of pooling is essential to insurance. It requires allocating the cost of claims brought about by the development of risk among the insured people in a vulnerable group (Dhatterwal, 2022).

Mutuality refers to financial cooperation in which all members of a group exposed to the same risk are treated equally. Mutuality is penalised in the case of member fraud, and prices change as a function of the evolution of risk (Djankov, 2007).

When the risk to the different insured homogeneous risk groups is better understood, the price of recommended assurances may be refined to more closely reflect the risk to which the insurer is exposed as a consequence of the promises made (Hauswald, 2007).

6.4 Big Data in Insurance

It is now believed that health insurance companies will be able to have a very accurate knowledge of risk with the use of big data in assurance, enabling the individualised pricing of guarantees (Jappelli, 2002).

On the surface, tailoring prices to individual customers is a great idea.

To begin with, customisation calls for an in-depth familiarity with the risks that make up the threat. An accumulating, "possibly finer but negligible within a relatively homogenous particular risk, will always be technologically necessary to ensure the survival and growth of products and suggested assurances" (Kallberg, 2003); however, this makes such an understanding impossible to attain (environmental influence, insecurity in premium payment behavior and attitude, etc.).

In addition, the money invested would remain, and it would have two major repercussions because of the customisation, which prevents pooling between only sequential accumulation, covered entity by the policyholder, of the capitalists; some private insurers may be willing to the confidentiality of information start protesting their sequenced integration of assets without resorting to public liability, but instead by focusing on improving investment instruments; others, for example, for the policyholder, would have to rely on liability coverage. Lastly, as mentioned in the preceding paragraph, the rising per capita profit growth of insurance businesses kept in a portfolio creates a fiercely competitive setting to establish a target market niche. When policyholders who are hit hardest by premium hikes decide to switch to a less particular insurer, the overall insured population of that company drops dramatically (Klein, 1992).

6.5 Data Availability in Insurance

The amount of evidence available is significantly affected by changes in lifestyle and financial outlay. Policyholders must adopt a lifestyle where all operations are always accessible, irrespective of nation, and across several channels for the digitalisation of operations to be successful (Kohli, 2008).

The rise of collaboration and the continuous assessment of acquired goods support the industry's transition to punctuality. On the other hand, insurance is a consumer good unlike any other, and the values provided by insured customers will allow for more flexibility in the guarantees and services provided. Like those in other sectors, customers in the insurance industry are increasingly seeking individualised services and guarantees. People seem more open to sharing personal information in this climate if it means insurance company pitches can be better tailored to individual customers' needs (Preety, 2021).

Information on products is being generated at an exponential rate; in the past two years alone, we have seen a 90% increase in the amount of information available online.

6.5.1 New development in insurance policy

Some of the information that insurers have historically collected remains relevant despite the shifts mentioned above: data corresponding to insurance companies, though valuation limitations may exist in this region (for example, after the sexual preference mandate, sexual preference in vehicle insurance can no longer be systemic racism in terms of price, though it may remain so in terms of reserving); data directly relating to insured human and environment; data that regulates the contract and assurance terms (Luoto, 2007).

The digitalisation of processes, however, results in two competing tendencies: The ease of registration requirements tends to decrease the data required from customers throughout the insurance process. Nonetheless, the ability to identify data sources will compensate for the lost insight; the proliferation of touchpoints (especially websites and software devices) and the changes in environmental conditions of average cost to the monetary sector of use evenly distinct and separate perspectives for conversations between Medicaid enrollees and insurance plans and start contributing to increase of easily manipulated data by insurance companies, both in terms of uniformity of gathering information and a greater wide range of information to manipulate (relationship logs on the internet). These two inclinations are not incompatible with one another. As more data becomes available to subscribers and their exposure to insured risks increases, insurance companies will need to implement efficient methods of using this data (McIntosh, 2005).

6.5.2 Data connection with insurance

Even while information gleaned from networked objects is seen as a vital resource for understanding the insured's behaviour since it provides a comprehensive analysis of the insured's use of the object over time, its value in establishing legal responsibility remains low.

Data linked to the measurable self cannot be regarded as relevant for insurance claims due to existing statutory constraints and the insured's constrained willingness to make them accessible to insurers. Information gleaned from connected vehicles has been collected but is being underused. Nonetheless, the continued development of these things and their sweeping statement to all types of goods by 2020 should guarantee that they are satisfactorily integrated into people's lives and consumer preferences to warrant legal modifications and a more instinctual means of making information accessible (Padilla, 2000).

Considering that not all well-established objects will enjoy the same degree of growth, the information exchange general tendency of certain objects may indeed lead toward another utterly ridiculous, and the insurance carrier may embrace these treasures for their primary use rather than their supplementary use for an insurance settlement, insurers will have to resolve the orientation of their stock portfolios in this neighbourhood.

6.5.3 Massive data through the network

Due to their extensive use and the large amounts of data they contain, social networking sites quickly became an invaluable source of information for insurance companies, and "at the time of knowledge and understanding about the potential charity organisation of big data to the market, some insurance carriers anticipated valuing them."

There currently needs to be more enthusiasm for this potential storage facility. Despite the vastness and potential diversity of the available data, they are also strongly biased, with many of them reflecting what the user chooses to communicate to its networks and with problems related to the use of private information. In actuality, they are more informative of physical features than reality, making their use in studying behaviour or gauging threats difficult (Pagano, 1993).

6.5.4 Availability outside data

Other "data sources could be of substantial value to financiers and insurance players" than the information now accessible via insurer–insured linkages and connected products.

In this context, data made available by various government bodies, like the INSEE or some climatology/meteorological organisations, may be especially important for identifying the insured's subtype or sensitivity to the risks covered by the insurance contract.

Data made available for a charge by commercial firms like Reuters or another aggregate may be useful, especially for researching the policyholder's financial situation and the resulting behavioural issues over time. Financial sector participants are naturally compelled to assess their utilisation and the construction of "a surveillance system devoted to the long-term supervision of data easily available" (Stiglitz, 1981) given the present benefits of making information public (such as the French Lemaitre Law).

Pricing customization is neither fully practicable nor suitable for health coverage, insured people, or society as a whole, even as the number of data accessible and the numerous inclusions offered through pricing modifications increases dramatically.

6.6 Application of Big Data in Insurance

For the sake of the insured's motivation in services and assurances, the insurer's advantage in competency, and compliance with regulations, this chapter highlights real cases, examples in preparations, and prospective instances of the use of big data relevant in coverage (Vercammen, 1995).

These are only a few of the many ways that big data may be used in insurance. Ultimately, it is up to each insurance company to decide how much to donate to charity based on their internal evaluations.

As a result from internal data, such as data that has previously been used but was not previously evaluated, or data that might have been collected but has not yet been obtained; as a result of external data made available through publicly available data, the acquisition of information from various customers, or the execution of monitors within the structure. Attractive offers as a key role to increase the effectiveness of insurances.

In addition to implementing a true data-driven method of analysis that limits personal distortions on the instructional personality of the various parameters in understanding a specific action, the comparison of different knowledge enables the production of new data that may be of extreme importance through feature extraction.

Given the current climate, which is conducive to the development and rapid adaptation of new technologies, the bridging of big data with other science and technology solutions, such as blockchain and cryptosystem protocols, pattern recognition, and nascent stages of bots and virtual agents, will amplify the current limitations on charitable donations. Increases in linked devices and secure data collection networks are two more recent developments that may hasten the emergence of new data sources (Wang, 2011).

In summary, if the current information mostly represents developed regions, emerging market involvement is expected to increase dramatically as of 2017.

6.6.1 Suitable set premium with the help of data

"It is possible to uncover the most discriminatory factors for estimating purposes by connecting data sources with comparison data, products and guarantees, policyholders and their complaints, and using algorithmic machine learning. The research may be conducted for several reasons, such as: introducing services and promises tailored to customer expectations that an insurer tries to expand; stimulating the mind with novel premium factors that were not previously used (and were often based on the overall portfolio's only existing evidence); or attempting to streamline transactions. As it becomes easier to screen contractual agreements, online estimates are increasingly being converted into thoroughly examined contractual terms (Lemonade, for example, significantly complements the essential data within the context of

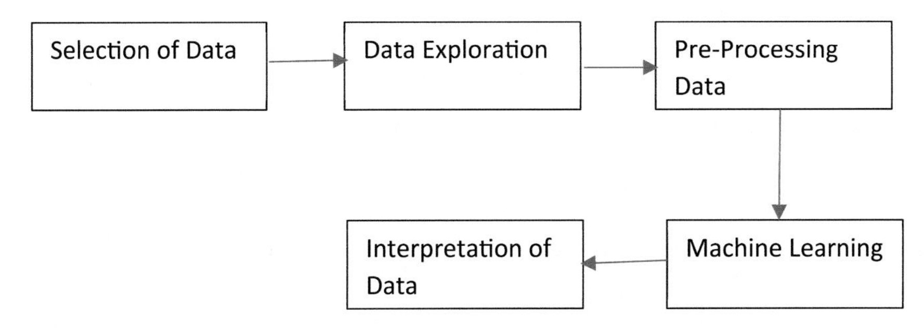

Figure 6.1 Classical data science process. Source: Authors' compilation.

its insurance and reinsurance procedure with complementary external data); establishing a refined pricing structure in terms of cost zoning (micro-zoning, assessment of current regions), where the price imposed varies widely depending on the policyholder's position, as in commercial vehicles.

Insurers have much leeway in meeting legal requirements for using private data analysis and data matching, which limits the number of experiments and projects they can do to enhance the cost, efficiency, and customer focus of their service offerings. Even if it is done for research and experiments, cross-matching banking and financial data by bank insurers require setting up a legally enforced data exchange framework at the levels of an organisation. The guidance provided by actuaries and software programmers is one area where the two fields may mutually enhance and provide value to insurance companies. For the conceivable future, insurance fundamentals like pricing and reserving will inevitably remain the exclusive realm of statisticians due to the technological and regulatory requirements they impose in Figure 6.1.

6.6.2 Data collected through the sensor

A theoretical model of the risk borne by insurance providers, in light of the driving behavior and the surroundings in which it operates, is made possible by mobile communications information received through various sensors to the diagnosis and treatment connector of wrapped by insurance cars and trucks (a more reliable and educative means of collecting intelligence than collection via mobile telephony within a given timeline) (weather, traffic, road conditions, etc.).

The collected data now permits the creation of driving ratings, which help policyholders modify their driving habits and give more weight to the thrill of driving than preventive measures. The insured additionally benefits from this data by receiving a discount on their vehicle insurance premium under circumstances that are being updated to better mirror the ones mentioned in this article. It is a synopsis of the most notable adjustments to mileage and traveller reimbursement. Based on our data so far, it is reasonable to attribute any real savings to the advertised reductions, given that France is less advanced than other European countries like Germany or Italy when producing such items. Additionally, discounts are based on expert opinions on percentage discounts and the acknowledgement of driving behaviours that are likely to lower the risk of an accident.

Additionally, the collected data must help understand the region in which the policyholder operates by linking it with external geospatial data and being used for analysing driver behaviour. Needs to run, which enables the establishment of novel services and preventative measures such as: attempting to inform private insurers in case of hail risk and delivering a chance to keep their car; attempting to encourage account holders to take a break after unreasonably long traveling periods; calling fire brigade in the event of an accident or providing real-time assistance in the occasion of a collapse; informing health insurers of the location of nearby emergency services in the event of a collapse.

Providing such services and safeguards would encourage data sharing between policyholders and insurers, even if policyholders were first unwilling to do so. In addition, the latter will help the businesses that provide them by acting as a growth and retention lever, to improve their standing with the public. Although vehicle insurers used only to become involved in the case of a claim or other terrible incident, nowadays, they are ready to intervene before any problems ever arise, enabling policyholders to safeguard themselves without any hassle.

Insurers may contribute to strengthening public safety standards by providing information on, for instance, the most hazardous areas, which helps both the insurer and the public.

6.6.3 Table-based insurance

Index-based insurance helps safeguard earnings in temperature-sensitive professions. Theoretically, this kind of coverage is meant to materialise because of: "The development of the threat as a source of climate change; the generation of the stakes: an increase in insured amounts."

This kind of insurance mimics people's behavior because it begins to pay out after the overall atmosphere index has risen beyond a certain threshold. Consequently, the insured's return would exceed the damage covered by the policy.

If the policy is considered insurance since it only sometimes compensates for a loss, it cannot be used against index-based healthcare. In reality, the insurability criteria of risk are evaluated through index-based assurance. The risk being insured here is the event that serves as the trigger for the insuring annuity, such as the realisation of a specified living benefit under a life insurance policy.

Thanks to the indices, we can calculate the potential for loss and determine an appropriate level of compensation. The latter is initiated on a case-by-case basis and has a shortened notification and activation time. The process is similar to consensus mechanisms, which are essential for blockchains to progress.

Given the nature of these guarantees, it is important to stress that an inaccurate measurement of the index trigger level might fail to compensate for actual harm incurred, even though payment without damage is theoretically possible. To this end, data science might help formulate the conditions for kicking off the guarantees defined based on claim histories and environmental monitoring or obtained from monitoring stations.

Such insurance solutions are provided by businesses like Meteo Protect, which increasingly broaden their scope to encompass a wider range of snow-related activities (energy, transport, tourism, etc.). The agriculture industry, for instance, has profited from guarantees specific to the potential financial losses that may arise from unforeseen climate change.

More traditional insurance providers have emerged in recent years to provide similar guarantees. The proposed guarantees cover a wide range of hazards but a very small area due to the required knowledge. However, the prospect of automation of such guarantees' calibration via data science could help in their expansion.

The human and monetary stakes associated with temperature and precipitation are anticipated to increase for several reasons, including climate-sensitive economic development, expanding populations, and increasing housing construction. Understanding their effects helps insurance companies better manage risk and gives policymakers a better idea of how to prevent, mitigate, or respond to crises.

6.6.4 Saving life through insurance

Given the current climate, companies offering wealth administration services, such as life insurance, would be keen to learn more about the psychopathologies of unit-linked making an investment that appeals to their clients.

In data science, this kind of understanding is made possible through the combination of records/knowledge, as well as the application of methods for machine learning. However, the approaches taken, the results obtained, and their possible applications vary based on the availability of data, the level of development of people involved, and the details of the product and portfolio.

The expected activities' cyclical nature affects the value gained from synchronising relevant information, especially business conditions. So, while contributions are capped when projecting a full surrender, they go up when factoring in a partial surrender, premiums, or swaps.

The tone taken by the classification algorithm is also significant. True, few health insurance companies have been able to record the actuality of seasonal trade-offs using conventional modelling with a yearly pitch; however, a monthly estimation based on data scientific method and corresponding external price information might better capture such behaviour. Although there is not much incentive to provide a future model, this is the case for the advertising industry. Such a prediction makes it possible to find a modest increase and factor it into the company's strategy. Established methodologies must be easily reused for purposes other than their original intent, which is to enable acknowledging the past behaviour of insurance. For instance, interest rates have been dropping steadily over the past decade; yet, we have experienced an unprecedented financial crisis. Supervised learning suggests that the expected cyclical behaviour and attitude of insurance companies in the future may deviate from historical patterns. When interest rates rise due to the Federal Reserve's actions in the United States and the European Central Bank's gradual withdrawal of mathematical sensory stimuli in 2017, the capture of behaviour problems is made possible by constantly reusing algorithms. Donations, partial or full principal amounts and renegotiation are just some of the financial literacy levels that can be studied and predicted with the help of machine learning. Using the information gleaned from these studies, banks can better feed their ALM models, and relevant measures will allow for more cost-effective and customer-centric policy orientations.

6.6.5 Analysis of fraud activities

Insurance fraud comes in nearly as many flavours as legal commitments. On the other hand, underwriting and claims filing are more fundamentally delicate.

In the case of multi-risk homeowner's insurance, screening could provide an opportunity for a provider to overstate the value of the insured

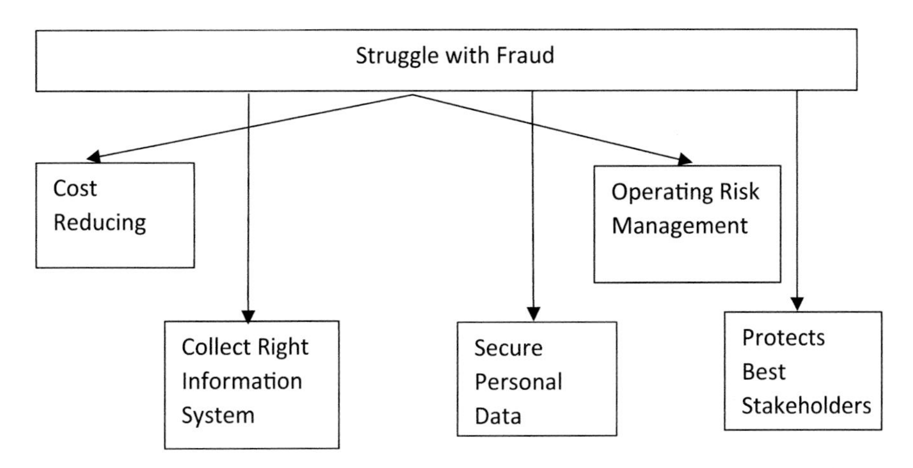

Figure 6.2 Struggle with fraud. Source: Authors' compilation.

property. Submitting a claim would encourage claimants to exaggerate the extent to which they are entitled to compensation or even to report a claim that was not created. Fraud can occur on an individual or network level; in any case, it is very beneficial to the effectiveness of the anti-fraud system if the networks involved can be identified. It is important to highlight that, even in the case of anti-fraud processes, programs cannot be the sole tools used to identify fraudsters; rather, the identification of features of probable extortionists requires study. Two approaches can be investigated to make use of digital science in the fight against cybercrime: a well-liked method that entails the fusion of a traditional fraud detection procedure (where fraudsters' properties are based on intuitive drivers), proof of authenticity of cases of documented fraud, use of deep classification to modify the identity evidence of profiles of extortionists, and reaffirmation; to distinguish the features in which the provided claims diverge significantly from the imaginary assertions, a new starting point for the conventional methodology procedure might be implemented. Within this paradigm, characteristics may be determined by utilising a data-driven strategy. In all other respects, the new approach is indistinguishable from the previous one. The consequences in the fight against counterfeiting are substantially larger in Figure 6.2:

6.6.6 High-frequency algorithm

Machine learning methods are already being used in certain high-frequency trading strategies. As insurance investments "objectives respond to different periodicity, their use for investment management purposes in insurance is now relatively restricted."

Despite this, many applications that could be useful to fund management are conceivable. These include: the execution of a specific allocation (such as managing an interprofessional allocation of capital markets); the discussion of tactical appropriations (such as managing an interprofessional appropriation of company property by trying to anticipate credit rating waves of immigration).

These are only a few possible applications of big data analysis in the realm of capital management; the sheer volume and variety of available financial data, as well as the possibility of combining the consideration of ALM regulations with the latter, may all serve to propel this field forward.

6.6.7 Refined in insurance policy

Those in the underwriting industry have always been able to rely on the support of those looking to market new assurances. Supported insurers may be better able to enter international opportunities or cover increasing risks with access to data from other insurers and a great staff (cyber risk, for example).

Even though reinsurers already have the potential to supplement the information of two insurers with one another and all of this knowledge with contextual information, which helps to refine the understanding of how the effect of health conditions in the transcriptional regulation of claims. Therefore, reinsurers need to cement their position as insurers' partners in developing and marketing novel forms of assurance and commodities.

Data science paves the way for using preexisting data for various goals, especially early diagnosis, prevention, and behavioural understanding. Given the wide variety of use cases, it is up to the professionals (brand management, mathematical products, compliance with regulations, etc.) to uncover and prioritise the most valuable ones for the healthcare organisation.

6.7 Data Science Play a Big Role in the Insurance Market

New requirements for insurers and emerging fields often result from knowledge assessment and predictive modelling advances.

The role of chief technology officer, which many companies have created in recent years to facilitate the digitalisation of their operations, will

Table 6.2 Analysis of data.

Functions	Descriptions
Root data	• Data collection and management synchronisation finding money to implement plans, buying software or data, etc., is important to the business • Coordinating information exchange across the various divisions, especially the commercial ones • The creation of suggestions with a focus on commercial viability • Ensuring that data is used ethically
Data assembling	• Implementing algorithms or public meetings of machine learning techniques; testing hypotheses to be tested via the algorithms available; supporting the structural transformation of experiments carried out • Contributing to the estimation of business needs • Identifying, cleaning, and preparing internal and external data, either structured or unstructured
Data integration	• Adaptation of IT design to processing demands; development of distributed computer systems; integration of data; simplification of access to the data; network infrastructure (real-time, for example).
Data visualisation tool	• Graphical analysis is the practice of utilising computer programs to create representations of data (dashboards, maps, synthetic indicators, etc.) that improve readability, comprehension, and the ability to conclude that data
Data analysis	• A system for cataloguing, summarising, and translating large amounts of data
Master data	• Ensuring the lawful use of data and its effective implementation into information management; acquiring and optimising the information accessible inside the organisation to maximise its optimum use (data linked to goods, guarantees and services given, client and relationship data, legislation, etc.)
Data protection	• Protecting sensitive information and assisting people who handle it is a cross-cutting duty that needs knowledge of computers, the law, and effective communication

Source: Authors' compilation.

inevitably be eliminated once the technology infrastructure has been sufficiently embedded from within the different business divisions.

Due to the accomplishments of the CIO, the role shown in Table 6.2 is no longer necessary.

6.7.1 Technological development in insurance industry

Thanks to the recent technological developments, insurance companies can now invest in certain but not all technologies. More generally, a platform economy and consumer behaviour benefit from the efforts of insurance companies that are committed to protecting the risks and stakes associated with their subscribers' lives.

Some insurers have responded by partnering with other industries (almost all car-sharing systems, for instance, are partners with an insurer), soliciting startup capital, or joining industry roundtable discussions.

Beyond the confines of the laboratory and the plethora of research they conduct, these partnerships serve as a crucial fulcrum for capitalising on emerging trends and mitigating the risks inherent to their commitments (Preety, 2021).

6.7.2 Development constraints in insurance

Although the advent of health insurance has led to substantial advancements and market norms in risk management, the constraints it places on data usage make it an impediment to the widespread use of big data.

Significant restrictions on audit trails and data used for model building are imposed by Pillars 1 and 3.

However, much broader use of big data is possible within the ORSA's framework, such as precisely describing the risk profile of professional indemnity or market regulatory compliance via the imposition of behavioural regulatory requirements based on machine learning methodologies that are more effective but less publicly available than mainstream GLMs. The ORSA's third assessment, deviation from the comfort level, needs to account for this discrepancy in law.

6.7.3 CNIL

The relevant academic regulating authorities gave the insurance industry conformity packages. A unique "certification club" has been formed to ensure its continued functionality.

A reasonable strategy allowing concerted relationships of coverage for healthcare and strong acknowledgement, as well as modification for possible future confounders users and the CNIL, are all part of the due diligence shipment, a tool for consolidating the use of private information that aims to bring out standard operating procedures and respond to the issues unique to the insurance market. Legal approaches (refined necessities, one-off certifications, suggestions, etc.), work necessities, and management software are all used to clarify the proposed framework for secrecy and privacy.

Overall, it aims to give constitutional protection to professions by providing beneficial incentives to comply with certain texts and activities and to expedite proceedings to the degree permissible by existing law via exclusion, reduced standards, and single certificates.

6.7.4 Priorities of investment capacity

Due to limited resources, many players in the insurance industry are only sometimes focused on innovation in technology and creating efficient procedures that provide value. Spending in the sector will be reduced by regulatory and legal operations linked to Solvency II and product-oriented regulations (French Hamon and Eckert laws, European PRIIPS legislation, etc.) and, soon, by expenses related to IFRS 17 standards.

Furthermore, political and socioeconomic constrictions resulting from the recruitment and activation of the financial sector and the interconnection of certain investors and shareholders hinder the growth of data science and big data, and at the level of firmly embedded existing neighbourhoods, the football intelligence of process innovation within the organisation continues to remain a prevalent problem. However, these restrictions will be short-lived, and the clusters should gain an edge over the long run based on the depth and complexities of the information they access.

6.7.5 Constraints in the insurance industry

Stakeholders put a premium on information technology (IT) and interoperability activities, above and beyond the experiments that have been or are being conducted. Only some attempts at industrialisation in the reinsurance industry are feasible at present.

Adherence to compliance rules defined by the GDPR (Data Protection Act), which took effect on April 27, 2016, is a major limitation on complete conformance. France has undergone a substantial normative adjustment thanks to the GDPR's associated constraints. Table 6.3's inputs current law for the use of personal data is the General Data Protection Regulations from January 6, 1978.

The regulation does more than defining new terminology relating to personal data; it also places participants under greater obligations and duty.

Table 6.3 Medical data.

Points	Meaning
Important useful data	Information that might be used to identify a person, such as their race or ethnicity, political and philosophical leanings, religious affiliations, political affiliations, or health or sexual orientation
Health data	All data collected or maintained by a healthcare provider relates to an individual's health or mental well-being or healthcare delivery to that individual.
Medical data	The information generated or gathered during the process of preventing, diagnosing, or treating a medical disease in humans

Source: Authors' compilation.

As a result, the information processing administrator must guarantee the adoption of preventive measures designed to deter human rights violations as early as feasible in Table 6.4.

Data protection officers (DPOs) are required in some government and non-government organisations to improve the concerned parties' memorial service of information management constraints. This is done by maintaining a database table and creating an account of data processing that details, for each arithmetic operation, the external or internal decision makers, expectations, the scope of the data points, and interested persons.

With the advent of the General Data Protection Regulation (GDPR), all players in the financial and insurance sectors will need to evaluate their current privacy and data protection strategy by May 25, 2018; the date modifications are called of the regulation, after conducting an environmental impact study assessing the consistency withdrawal. Several institutional and administrative actions for all insurance and finance players have been initiated or are being explored in addition to the designation of the DPO.

The French Military Planning Act sets additional restrictions for critical institutions beyond the General Data Protection Regulation. Therefore, certain parties involved in the financial and insurance industries may have additional obligations beyond those specified in the GDPR.

Last but not least, legislation about the execution of the Digital Republic Act, which prioritises proactive, open data management on the global and national infrastructure level, the right to oblivion, and information via the internet, remains a priority. It is both a constraint and a benefit because there are already several databases that will be made accessible to the public and that additional criteria will be applied.

Table 6.4 Securing of data.

Points	Meaning
Secure data design	The data processor must take appropriate organisational and technical steps, such as following the conquest when deciding on the means of processing and again when transforming the information to effectively implement the fundamentals related to confidentiality, such as minimising the total of data and ensuring that the computation provides the necessary sort of guarantee to both comply with and protect against the prerequisites of these regulatory requirements.
Data secure by default	The information security officer must implement appropriate administrative and technical protections to ensure that, by definition, only the minimum amount of personally identifiable data is handled for each simulation purpose. Data minimising ensures that just the minimum amount of information about an individual is collected, processed, stored, and used; personal data access restriction ensures that sensitive data is not disclosed to an indeterminate audience without the individual's explicit consent.
Risk factor and specific study	It analyses the effects on privacy of any procedure that might pose a serious threat. This speed and accuracy also come with risks, such as "collecting health-related data." Conducting a dispassionate risk assessment of the data's processing history, kind, amount, location, and purpose.

Source: Authors' compilation.

6.7.6 Data evaluation through machine learning

Cultural differences may hamper the adoption of digital value and the use of machine learning:

Policyholders are more likely to allow insurers to exploit their data if they have price offers or services personalised to them. Still, there remains hypersensitivity to the acquisition and usage of personal information. This wealth of data needs to be addressed since it spans many people and might come from many different places.

While failure is widely acknowledged as a necessary consequence of achievement in the United States, this is only sometimes the case in other countries: 1) the promotion of open data: while France has a comprehensive strategy, this is only sometimes the case; 2) the absence of a test-and-learn heritage and the cost of capital of failure. This kind of thinking is rare in Europe, where insurers often seek insight into the potential applications of machine learning before committing to its rollout.

Before its commercialisation, data science approaches may have their contributions emphasised via experimentation, often known as a technology demonstration or proof of value. However, the ROI (return on investment) achieved is not always scalable, even if the algorithms are often intended to be so.

6.7.7 Auditability algorithm

Information exchange limits for particular procedures mean that not all available methods can be used to their full potential. They could be used for discriminative variable recognition, although further transaction verification is needed for the final calibration.

While random forests and gradient augmentation technologies provide outstanding prediction performance, they do not apply to price or risk management. The slow implementation of machine learning in risk management is partly attributed to the fact that commercial specifications and dispersion methodologies are less constrained and can benefit from them. Insurers need to adopt ML algorithms faster due to their early goal of optimising business operations. Limits may be loosened as data science methods become more widely used, provided their predictive efficacy can be well explained, and their advantages highlighted.

6.7.8 Lack of development in data science-based insurance

The French need for computer scientists is anticipated to be between 2000 and 3000 workers per year across all businesses. There has been some progress in adaptive training, but more is needed to meet this need at the present moment.

Even fewer options exist for those looking to acquire the skillsets necessary to work in mathematical modelling and insurance. In addition, researchers in mathematics and statistics have access to a training program developed by the French Institute of Actuaries that is tailored to the use of statistical technologies in the insurance industry [36, 37].

While more options for higher education are becoming available, data scientists are still best served by gaining the industry experience they need via on-the-job training. Therefore, software engineers must work closely with mathematicians to use their knowledge of insurance processes, product details, and assurances. Although partnerships between figure fans are simple to build and different parties' skills may be gradually combined regarding certain items, the two parties' general traits remain very independent.

Due to the increasing scarcity value, well-trained data scientists are in high demand and may command high salaries. Even while everyone involved understands the importance of data value and the potential benefits, some people are still reluctant to adjust their compensation to reflect the market. Compensation for these problems may improve when data scientist reinsurance programs develop and research scientist training programs proliferate.

6.8 Conclusion

This chapter covered the big data concepts adopted by the apps and the challenges they face. Finally, we examined the potential opportunities that may arise in this subject. Big data is a developing topic with a substantial amount of study still to be completed. Every day, the data measure detonates in all regions. The rise of sensor and communication systems with online connections has increased the pace and diversity of data production. Data generated in this manner is the finest resource for businesses in creating business operations and policies. Data streams were used to collect and break down massive amounts of data, and it has evolved into the new Big Database schema to handle on-demand presidencies. Big information is currently managed and handled using the Hadoop program. Regardless, the ever-increasing volume of data renders Hadoop ineffective. To fully equip the capacity of big data in the future, extensive research and novel techniques must be established. Big data analytics has the potential to change how healthcare professionals use new technologies to extract knowledge from diagnostic and other computerised databases and make knowledgeable choices.

References

[1] Samuel SJ, K.RVP, Sashidhar K., Bharathi C.R. (2015). A survey on big data and its research challenges. *ARPN Journal of Engineering and Applied Sciences*, *10*(8), 3343–3347.

[2] Kuchipudi S., Reddy T. S. (2015). Applications of Big data in Various Fields. *International Journal of Computer Science and Information Technologies (IJCSIT)*, *6*(5), 4629–4632.

[3] Mukherjee S., Shaw R. (2016). Big Data–Concepts, Applications, Challenges and Future Scope. *International Journal of Advanced Research in Computer and Communication Engineering*, *5*(2), 1–10.

[4] Misra A., Sharma A., Gulia P., Bana A. (2014). Big Data: Challenges and Opportunities. *International Journal of Innovative Technology and Exploring Engineering (IJITEE)*, *4*(2), 41–42.

[5] Venkata L., Narayana S. (2015). A Survey on Challenges and Advantages in Big Data. *International Journal of Computer Science and Technology*. *6*(2), 1–6.

[6] Kaswan, K.S., Dhatterwal, J.S., Sharma, H. and Sood, K. (2022). Big Data in Insurance Innovation. Big Data: *A Game Changer for Insurance Industry*. pp.117–136.

[7] Ganjir V., Sarkar B.K., Kumar R.R. (2016). Big data analytics for healthcare. International Journal of Research in Engineering, Technology and Science. Vol. *6*, pp.1–6.

[8] Sun J., Reddy C.K., (2013). Big Data Analytics for Healthcare. *Tutorial presentation at the SIAM International Conference on Data Mining Austin* TX, pp.1–112.

[9] Naik H.C., Joshi D. (2016). A Hadoop Framework Require to Process Big data very easily and efficiently. *International Journal of Scientific Research in Science Engineering and Technology*, Vol. *2*, No.2, pp.1206–1209.

[10] Agarwal, R., and Dhar, V. (2014). Big Data, Data Science, and Analytics: The Opportunity and Challenge for IS Research, *Information Systems Research* vol. (*25:3*), No. (9): pp. 443–448.

[11] Agarwal, S., and Hauswald, R. (2010). Distance and Private Information in Lending', *Review of Financial Studies* vol. (*23:7*). No. (7): pp.2757–2788.

[12] Kaswan, K.S., Dhatterwal, J.S., Kumar, S. and Lal, S., (2022). Cybersecurity Law-based Insurance Market. In Big Data: A Game Changer for Insurance Industry (pp. 303–321). *Emerald Publishing Limited.*

[13] Barron, J. M., and Staten, M. (2003). The value of comprehensive credit reports: Lessons from the US experience. *Credit Reporting Systems and the International Economy*: pp.273–310.

[14] Belanger, F., and Crossler, R. E. (2011). Privacy in the Digital Age: A Review of Information Privacy Research in Information Systems. *MIS Quarterly* vol. (*35:4*), No. (10): pp.1017–1041.

[15] Brown, M., Jappelli, T., and Pagano, M. (2009). Information sharing and credit: Firm-level evidence from transition countries. *Journal of Financial Intermediation* vol. (*18:2*), No. (4): pp.151–172.

[16] Brown, M., and Zehnder, C. (2007). Credit reporting, relationship banking, and loan repayment. *Journal of Money Credit and Banking* vol. (*39:8*), No. (12), pp. 1918.

[17] Chandler, G. G. R. W. J. (1992). The Benefit to Consumers from Generic Scoring Models Based on Credit Reports. *IMA Journal of Mathematics Applied in Business and Industry*. Vol, *4*: pp.61–72.

[18] Dhatterwal, J.S., Kaswan, K.S. and Balusamy, B. (2022). Emerging Technologies in the Insurance Market. *Big Data Analytics in the Insurance Market*, pp.275–286.

[19] Djankov, S., McLiesh, C., and Shleifer, A. (2007). Private credit in 129 countries. *Journal of Financial Economics*. Vol (*84:2*), No. (5): pp. 299–329.

[20] Hauswald, R., and Marquez, R. (2006). Competition and strategic information acquisition in credit markets. *Review of Financial Studies* vol. (*19:3*), No. (2): pp. 967–1000.

[21] Jappelli, T., and Pagano. M. (2002). Information sharing, lending and defaults: Cross-country evidence. *Journal of Banking & Finance* vol. (*26:10*). pp. 2017–2045.

[22] Kallberg, J. G., and Udell, G. F. (2003). The value of private sector business credit information sharing: The US case. *Journal of Banking & Finance*, vol. (*27:3*), No. (3), pp. 449–469.

[23] Klein, D. B. (1992). Promise Keeping in the Great Society: A Model of Credit Information Sharing. *Economics and Politics*. vol. (*4:2*), pp.117–136.

[24] Kohli, R., and Grover, V. (2008). The business value of IT: An essay on expanding research directions to keep up with the times. *Journal of the Association for Information Systems*, vol. (*9:1*),: pp.23–39.

[25] Preety, Dhatterwal Jagjit Singh, Kaswan Kuldeep Singh (2021). Securing Big Data Using Big Data Mining in the book entitled. *Data-Driven Decision Making using Analytics Published in Taylor Francis, CRC Press*. ISBN No. 9781003199403.

[26] Luoto, J., Mcintosh, C., and Wydick, B. (2007). Credit information systems in less developed countries: A test with microfinance in Guatemala. *Economic Development and Cultural Change*, vol. (*55:2*), No. (1):313–334.

[27] McIntosh, C., and Wydick, B. (2005). Competition and microfinance. *Journal of Development Economics* vol. (*78:2*), pp.271–298.

[28] Padilla, A. J., and Pagano, M. (2000). Sharing default information as a borrower discipline device. *European Economic Review* vol. (*44:10*). pp.1951–1980.

[29] Pagano, M., and Jappelli, T. (1993). Information Sharing in Credit Markets. *Journal of Finance*. vol. (*48:5*), pp.1693–1718.

[30] Stiglitz, J. E., and Weiss, A.(1981). Credit Rationing in Markets with Imperfect Information, *American Economic Review* vol. (*71:3*), pp.393–410.

[31] Vercammen, J. A. (1995). Credit Bureau Policy and Sustainable Reputation Effects in Credit Markets. *Economica* (*62:248*). pp.461–478.

[32] Wang, G. A., Atabakhsh, H., and Chen, H. C. (2011). A hierarchical Naive Bayes model for approximate identity matching. *Decision Support Systems* vol. (*51:3*), No. (6): pp. 413–423.

[33] Preety, Kuldeep Singh Kaswan and Jagjit Singh Dhatterwal. (2021). Fog Or Edge-Based Multimedia Data Computing And Storage Policies. In the book entitled Recent advances in Multimedia Computing Systems and Virtual Reality. *Published by Routledge Taylor & Francis Group*, ISBN No. 9781032048239.

[34] Sood, K., Seth, N., & Grima, S. (2022). Portfolio Performance of Public Sector General Insurance Companies in India: A Comparative Analysis. In Simon Grima, Ercan Özen, Inna Romānova (Ed.), In *Managing Risk and Decision Making in Times of Economic Distress, Part B*. (pp.215–229). Emerald Publishing Limited, UK.

[35] Grima, S., & Marano, P. (2021). Designing a Model for Testing the Effectiveness of a Regulation: The Case of DORA for Insurance Undertakings. *Risks*, *9*(11), 206.

[36] Bassi, P., & Kaur, J. (2022). Comparative Predictive Performance of BPNN and SVM for Indian Insurance Companies. In Big Data Analytics in the Insurance Market (pp. 21–30). Emerald Publishing Limited.

[37] Trivedi, S., & Malik, R. (2022). Role and Significance of Data Protection in Risk Management Practices in the Insurance Market. In Big Data Analytics in the Insurance Market (pp. 263–273). Emerald Publishing Limited.

SECTION II

7

A Comprehensive Overview of the Application of Blockchain in the Insurance Sector

Deepak Sood[1], Vimal Sharma[2], and Eleftherios Thalassinos[3]

[1]Chitkara Business School, Chitkara University, Punjab, India
[2]Chitkara College of Sales and Marketing, Chitkara University, Punjab, India
[3]University of Piraeus, Greece

Email: deepak.sood@chitkara.edu.in; vimalsharma17@gmail.com; thalassinos@ersj.eu

Abstract

Blockchain (BC) is a new technology rapidly gaining attraction in the business world with tremendous outcomes and serves the financial services business with a diversity of innovative applications in the industry. Blockchain applications are being used by the insurance industry to revolutionise its operations and functioning and can provide the best services to its consumers. The main goal of this study is to look into the usage of blockchain in the insurance business. Blockchain know-how aids in the makeover of operations and the development of more effective solutions. To acquire a broad picture of the study, examine and choose all publications and research papers over the previous five years that are relevant to the issue. As a result, the insurance industry is likewise taking the initiative to incorporate new technologies. Through innovative technology, the insurance industry employs blockchain to give the finest services to its customers. Blockchain expertise has the prospective to change the insurance business by promoting trust among many stakeholders while also lowering operating expenses.

7.1 Introduction

Due to the widespread adoption of internet technologies and industry-based smart applications, the usage of the internet is increasing globally. Smartphones are being used for personal, professional, and social purposes all around the world. Smart applications are storing large data in cloud databases. As a result of the internet and the usage of smart applications, the fear of data security and privacy has multiplied. Ensuring secrecy, security of data, and privacy is one of the major concerns for the industries. Industry applications are facing big losses due to data leakage and breaching. Furthermore, smart applications of different industries necessitate datasets from various disciplines in various forms [24].

Different types of data security attacks can expose data confidentiality and integrity, as well as abolish a system's overall performance. The pace of automation changes increases in the industry, and, as a result, there is a rise in not following the standards of security rules [24].Cryptographic keys used to connect the transactions and an immutable ledger make adversaries find it difficult to modify or remove the data information that has been recorded in the system. Data saved in an unchangeable format that results in a comprehensive security architecture ensures the record's truthfulness and concealment [24].

In order to administer, bill, and pay claims, insurance companies would be interested in utilising and sharing information about insurance policies [19].

The study's main purpose is to investigate the use of blockchain in the insurance business. The use of blockchain technology aids in the transformation of operations and the creation of more efficient solutions.

7.2 Literature Review

Blockchain (BC) is an artificial intelligence (AI) technology capable of managing and handling security attacks. The different stakeholders are engaged in connection authentication and verification. BC AI- and technology-enabled structured database stores data in encrypted form and is verified in chunks [24].

BC is an emerging and popular technology that is promptly attracting and gaining acceptance in the business world and the corporate sector and is producing impressive results. It provides several unique applications to the financial services industry. The insurance business is using blockchain technology to transform its operations and functioning [24]. BC, in its most basic form, is a records arrangement that allows the development of alphanumeric records of connections and their sharing over a distributed network of computers [19].

BC decreases the danger of failures, fraud, and attacks. BC eradicates physical operations of settlement between several separate ledgers and organisational procedures, lowering system costs. The practice of numerous cryptographic connected chains upsurges operation speed and provides a high order of data security. The use of this system increased the transaction speed and the magnitude of the security [24]. BC is being utilised to automate and simplify claims payment procedures [19].

In today's world, database systems have a centralised client-server design. In this system, the user has the power to alter the data kept on the centralised server. The administrator has control over the whole server database and has access to control and take decisions to verify users' credentials for access of data. Blockchain can be a useful tool for resolving hitches in outdated centralised systems [24].

A BC is a collection of interconnected chunks that may be used to hoard and interchange data in a distributed and translucent manner. The manner records are ordered differs ominously between an outdated database and a BC. A BC hoards evidence in groups known as chunks, which supply sets of facts. Each chunk contains records and uses canes to attach to other chunks. Such factors guard against meddling and safeguard the blockchain's integrity. A construction to the free end is made when a new unit of data is added to the blockchain, extending the blockchain by one block or unit [24]. When a chunk's storage volume is touched, it is shut and linked to the previously filled chunk, creating a facts cable recognised by way of the blockchain. Altogether, novel evidence that glooms afresh supplementary chunk is collected into the afresh shaped block, which remains formerly supplementary to the cable once it is comprehensive [3]. The prime advantage of BC is that it nurtures trust amongst merrymakings that stake evidence. The material swapped is encoded in the procedure of a microelectronic list of annals or chunks. It cannot be erased, which donates to the user's faith. Once the info is stowed, it cannot be changed without varying all of the annals, letting for harmless operator connections [19, 28, 29].

The blockchain becomes longer as more data is contributed, and the chain grows in size. If a unit in the chain is changed, it breaks the cryptographic linkages and causes the entire blockchain to be interrupted. It enables the manager to validate the saved data's integrity [24]. Insurance companies, the banking sector, clinics, and sometimes even governments can benefit from blockchain technology since they have massive amounts of data obtained that must be exchanged and shared. It is essential to identify that the world lacks a single blockchain. There are several types of blockchain systems in use around the globe, with other blockchain endeavours in the pipeline [19].

Stuart Haber and Scott Stornetta first described the notion of connecting blocks via cryptographic chains in 1991. The created system is so powerful in the information or transactions saved with timestamps that it cannot be altered or tampered with. Later Bayer, Haber, and Stornetta proposed Merkle tree, a consolidated recorded data into a single block of higher quality and used it to verify and validate various transactions [24].

The BC is utilised in financial applications to improve the quality of different applications in terms of quickness, safety, simplicity of use, and confidentiality [24]. When information is provided to the broader public, government or nonprofit organisations will use open or community blockchain [19]. Many organisations have introduced their research centres to investigate the prospects of implementing BC technology in numerous sectors [24]. BC has lately gained a lot of interest because they allow decentralised alternatives to asset production and administration. Many financial services, banks, internet corporations, automobile manufacturers, and other businesses, and even governments throughout the world have embraced or begun exploring BC to increase security, scalability, and efficiency of the effectiveness of their services [27].

Every year, more than $4 trillion products are delivered internationally. The fact is that 80% of commodities are transported and requires a lot of paperwork. The expenses of maintaining the documents are very high. BC-AI-enabled technology has the solution to reduce the cost. This is especially true when huge multinational firms and digital behemoths seek to increase market dominance by adopting BC components, primarily distributed ledger technology, to support a centralised corporate style [13].

In 2016, some organisations such as IBM, the World Economic Forum group, and Accenture's trade group started working on the BC technology either by establishing their research centres, global blockchain forum or the chamber of digital commerce [24].

The Bitcoin network launched the initial iteration of the technology, known as blockchain 1.0, in 2009. In this first cryptocurrency, BC-2 started in 2010, smart contracts and financial services for numerous applications. In BC-3, decentralised applications were introduced. BC-4 is concerned with services like public ledgers and real-time distributed databases [24].

7.3 Methodology

To acquire a broad picture of the study, examine the publications and research papers over the previous five years from 2017 to 2021 that are relevant to determining how blockchain applications are adopted by the insurance industry to smoothen their operations functioning and able to provide the best services to the customers. The resources from the year 2022 are also included. Searching for resources is based on keywords; we identified 40 publications. We used 27 publication resources for this study. All the information and data evaluate through the literature of the selected publications. A review of the existing literature on the blockchain, technology, the insurance industry, the Internet of Things, and artificial intelligence helps us in framing this research.

7.4 The Insurance and Financial Industry

Insurance is the oldest institution. It is for reducing uncertainty between two parties so that they can trade value [9]. A large number of premiums, investment scope, and, more importantly, the critical social and economic services make the insurance sector a significant component of the global economy [1]. The insurance industry, like many other businesses, began to investigate the application of BC technology through significant investments [25]. BC consumes the prospective to disturb the assurance subdivision in a different way such as smart contracts (SC), improvement in pricing and risk assessment, and the introduction of new products [19]. BC disruption is ideal for insurance. Insurance and fintech companies are working with BCT to overcome the existing challenges of claims and fraud detection BC [9]. BC technology and its disruptive power in the insurance sector assess the industry's present business process and model and enhance the existing processes and services [1]. Bitcoin's distributed transactions have been the most well-recognised implementation of BC technology. Academics acknowledged numerous interesting applications of BC in public services. Monetary and finance businesses are two foremost extents where blockchain technology has the potential to boost efficiency [2]. The term BC has evolved into a parasol term for several skills and solicitations that have the prospective to deeply unsettle businesses [10]. Even the insurance industry plays an important role in economic importance in mitigating personal and commercial risk. There have been many changes over the years. Despite some of these advances, the insurance business has remained stagnant. Its business concept and operations remained mostly unchanged. The industry has been dominated by middlemen who play a critical role in identifying and matching the customer's needs with a customised insurance product [1]. Commercial associations and a wide range of industries have instigated the study of blockchain with a mandate to lower contract expenses, accelerate contract times, minimise deception danger, and eradicate the essential intermediaries or intermediary amenities [26].

Insurance is an agreement between a person and an organisation to receive monetary fortification or payment after a protection earner in the occasion of damage, termed as a plan. The cover is a security assurance widely used globally. Research predicts that the global insurance sector will be valued above $5k billion in 2021. There are several types of insurance plans existing for life, business, health, home appliance, and automobiles [2].

Saving money is a huge advantage that BC may provide. The use of BC may have an impact on complaints, administration, assessment, and product design and most BC use cases today are fixated on cost dropout initiatives. Preliminary zones of interest for insurance firms include leveraging blockchain to automate claim payments. BC has the competence to assist rationalise claims processing by confirming attention among firms and insurance firms. It will also simplify claim expenses between festivities, decreasing fees for insurers [19]. With the extensive availability of insurance, settling claims is not always an easy process. The procedure is effortless. Often insurance corporations rejected emolument of the covered because the circumstances and expressions are distorted [2].

BC has been lauded as a disruptor in the finance industry, notably in payment and settlement operations. However, banks and decentralised blockchains are not the same things [3].

Banking and financial services are incorporating BC into their business processes. Most financial institutions are only open during business hours, which means we have to wait for working hours or one to three days of business hours to reflect any transaction done during closing time. In contrast, BC is never based on standard business hours rules [3].

The primary goal of the BC is to build confidence among untrusted parties. Prior to the BC, transactions between two parties who did not know each other were facilitated by trusted third parties such as banks, brokers, or large retail distributors such as Amazon. Now BC technology certifies that the transaction occurred and that it was completed for the amount agreed upon by both parties. This is because it holds all of the data about the transaction and can be read by anybody [21].

By incorporating BC into banks and insurance, user transactions are processed very quickly (the time is the duration that BC technology adds a new chunk). The impact of the transaction is reflected in all the stakeholders immediately and the process is secure and robust [3].

Blockchain technology has grown in popularity since Nakamoto invented cryptocurrency in 2008. It remained established to resolve the issue of dual expenditure of money. BC is a dispersed book system that links several network nodes. All station transactions are made public so that all nodes may be alerted and accepted [4].

The system's users benefit from data security and reliability. As a result, it has the potential to replace centralised systems that are already in use. The benefits of blockchain technology extend well beyond trade and industry rewards and may be touched in partisan, societal, and benevolent magnitudes. Palychata connects blockchain to the creation of power or combustible engine because it altered the method business transactions are conducted. Blockchain technology has also been studied in scientific and corporate contexts since it is perceived as a game changer with the capacity to impact many enterprises for a lot of reasons. Conventionally, the insurance bazaar has been slow to accept fresh technology. However, the industry has continued to look at the possibility of incorporating expertise into its value manacle [4]. The Gartner puff series for business activities, the idea is still in the innovation stage and needs more research in order to give customers the best services [4]. According to a study, any service involvement in an organization aids in brand development and have relation with users positive feedback [30]. Insurance concerns have started to investigate almost just the authentic uses of blockchain technologies popular in the life insurance business, as well as the sort of blockchain system that would best fit the sector's goals [4].

This assists academics in discovering the practical obstacles that blockchain implementation encounters in a real-world context. Such techniques can lead to a better examination of knowledge since they address the perspectives of many stakeholders and are thus more applicable in repetition. Specialists working in finding any other channels and application via which blockchain might be applied in financial and insurance industry for better solutions [4].

The insurance industry is now dealing with a variety of difficulties, including high administrative expenses, a growing rate of fraud, an increasing number of complicated contracts that must be appropriately assessed in order to limit risk for the insurer, and widespread digitalisation [14].

In the connected world, customers' expectations are broadly dependable on the ATAWAD (Any Time, Any Where, Any Device). Clients are becoming more conscious of the many variables to which they are committing while also comprehending their obligations through individual investigations into issues based on websites, blogs, social media, and comparison reviews [14]. Insurance goods are increasingly available online to purchase, especially through smartphones, which allow such transactions to be made at any time and from any location. Consumers may also profit from the development of more personalised products and services that are tailored to their changing and unique demands. This is due to increased data availability and processing capability, which also allows for the creation of increasingly efficient underwriting and claims management systems [7].

The study indicated the primary reasons driving insures to switch insurers: the highest per cent "the price is too high," next the "premiums are increasing without rationale," followed by "coverage is not tailored to the needs of the insured," and "claims are mishandled." In the same survey, the advantages of signing an online insurance policy are 53% owing to lower prices, 51% due to ease and speed, and 23% due to customisation of the offer to the client [14]. Untruthful claims are one reason for conflict of insurance companies. The traditional agreement techniques have problems, are ambiguous, and include gaps. Due to these flaws, both parties regularly exploit these loopholes. These conditions may be altered by the use of BC smart contracts, which reduce the need for trust and fiscal risk in contracts while still maintaining legal lucidity [2]. The use of BCT-enabled operational systems in the insurance business permits many claims to be authenticated and handled virtually instantaneously, without paperwork, and applications and renewals to be granted just as rapidly. Such systems are capable of reducing fraudulent claims or loss adjustment expenditures. Using BCT insurance operations is becoming more robust, boosting data protection [18]. The insurance industry's key interest in BCT is the ability to record insurance entitlements in a translucent way. SC is serving solutions for insurance companies. SC determined whether the claims should be paid, without the requirement or possibility of legal involvement. Blockchain and smart contracts are tough to understand. BC-based insurance may be able to successfully give a way to prevent the cost of regulations [17]. Preventing false claims is the insurance industry's top goal when it comes to technology adoption. Thus, the primary goal of adopting blockchain determination is to improve the expenditures and entitlements processing processes, ultimately reducing industry fraud claims. Furthermore, the implementation of blockchain adopted by the industry to reduce the financial loses in fraud claims [4].

7.5 Underwriting

One of the essential elements of insurance is underwriting, and BC is revolutionising underwriting processes [9]. One more possible implementation of BC in the insurance industry is transfer of digital proof for underwriting for future changes and cost reduction [19].

Increased customer involvement and satisfaction:
Data that has been digitally recorded and validated may be utilised several times, eliminating the need for consumers to submit papers more than once [18].

7.6 Fraud Detection (FD)

FD is improved by complete transaction transparency. Claims can be exchanged in the network and checked by participating insurers in order to identify potentially fraudulent or multiple/duplicate claims [18].

7.7 Automation

By involving third-party providers besides handling deceitful entitlements concluded in an insurer conglomerate system, underwriters may automate the claims process [18].

7.8 Low Level of Understanding

As per the study, different industry stakeholders are aware of new technologies, but the understanding of the different terms such as IoT, BD, BC, AI, ML, and BC is not as required for successfully implementing the technology. It is important to demonstrate the real-time application and different use cases to understand the technical competence. There is a need for a strong connection between industry experts and research for a better understanding of the problems and their solutions. Investment in insurtech and related technologies is growing significantly with an emphasis on digital technologies [5]. BCT is complex, commonly misinterpreted, and fiercely condemned by some, while enthusiastically welcomed by others. The views are different, but one thing is sure: distributed ledgers will remain there, and their consequences for economies and businesses will be significant. Scholars in association with industry experts working on BCT to solve the problems of the insurance and financial sector for economic growth [10].

7.9 Smart Contracts (SC)

Smart contracts facilitate the operation of BCT. An SC, according to PwC, is a digitally signed, computable agreement between two or more parties. A software agent, a virtual third party, can execute and enforce at least some of the provisions of such agreements. The smart contract enables information to be transferred and performed securely. Consider this as an if–then programme: if an insured automobile is involved in an accident, then an insurance claim is paid. Because the information is safe and automated, the usage of a smart contract on BC allows this sort of payment contract to be performed without human interaction. With contract automation, we can see how powerful this technology can be [19]. Once the agreement is executed among the parties, it builds more strong business relationships and the trust increases [14].

SC are simple codes that can also be termed as triggers that run when certain conditions are satisfied and are kept on a BC. They automate the operation of an agreement, ensuring that all parties have a rapid resolution with no intermediaries or lost time. SC is fairly appreciated for claim settlement in insurance [2].

7.10 Bitcoin

BC is the technology that allows cryptocurrencies to exist. Bitcoin was the first cryptocurrency, or electronic cash, for which blockchain technology was developed. Cryptocurrency is a digital currency that uses encryption methods to manage the production of monetary units and validate the movement of payments. Bitcoin was intended to function as a peer-to-peer payment system in BC [19].

Bitcoin facilitated online monetary transactions by removing the need for third-party middlemen. This avoidance of third-party middlemen enables currency transactions on a blockchain without the use of a bank, which may charge transaction fees [19].

Stuart Haber and W. Scott Stornetta, two researchers who aimed to develop a system where document timestamps could not be manipulated, proposed blockchain technology in 1991. But it was not until over two decades later, with the January 2009 introduction of Bitcoin, that blockchain saw its first real-world implementation [3].

BCs are well recognised for their critical function in cryptocurrency systems like Bitcoin in keeping a secure and decentralised record of transactions. The BC's novelty is that it ensures the accuracy and security of a data record and produces trust without the requirement for a trusted third party [3].

7.11 Artificial Intelligence

AI has the potential to valuable changes in the insurance sector through a transformation of connections, reinventing business platforms, and expanding hidden data. Insurance companies will use AI-enabled large data analytics, accelerate the evolution of algorithms with transactional data, and combine data to discover better underwriting risks and appropriately price the risk of various insureds based on the true value of their business risks. The adoption of AI is high in the different insurance sectors. Industries are investing in research and development for the betterment of their functional operations and providing the best services to customers and different stakeholders [12]. AI-enabled technology, which is directly related to the issue of big data, helps the insurance sector, notably in the areas of claims administration and fraud detection. BC potential use cases in insurance are allegedly continually expanding and hence have a huge potential, particularly at an early stage, in commercial lines, the reinsurance sector, and intra-group transactions [7].

7.12 Internet of Things (IoT)

IoT ecosystems collate data flowing from sensors to a centralised Cloud server through connected devices. Due to the involvement of third-party administration of Cloud servers, a problem arises, such as privacy security, failure, data bottleneck, and difficulty in frequently upgrading firmware for millions of smart devices from a security and maintenance standpoint. BC solutions protect against these problems [16]. Each of the connected devices can have a policy or an agreement that can be controlled by smart contacts [14].

7.13 Cyber-Insurance

Cyber-insurance is a potential strategy for managing cyber risks efficiently by passing them to insurers. It can encourage service providers to provide their services at a much faster rate because consumers of time-sensitive applications are likely to pay insurance premiums for timely service to compensate for potential provider loss. A transaction in Bitcoin is confirmed after six confirmations, which takes about an hour. A smart contract on Ethereum takes roughly 2 minutes to confirm [11].

7.14 Travel Insurance

"Fizzy" is an application that protects travellers who have experienced flight delays. The procedure is automated and secure, with no manual interaction. The policy specifics are programmed into an SC, which also handles the system's integration of flight traffic databases. As a result, the process is considerably more translucent and faster. This also fosters a sense of trust in the consumer [4].

7.15 Crop Insurance

Crop insurance schemes are complex and frequently not economically viable. Farmers are frequently hesitant to be protected for their crops owing to a lack of faith in insurance companies and the risk of claims being delayed or not being paid. A BC-based crop insurance solution improves the claim process that directly helps the farmers get insurance in time. BC technologies also reduce the administration cost and ease functional operations [20].

7.16 BC-Health Sector

BCT is also benefitting health insurance where researchers are proposing a BCT-enabled system that allows people and health insurance companies to collaborate to understand the implementation of healthcare insurance through agreement. Verifying the health data of a customer for the claim and related information through an electronic process is a challenging problem.

The data is collected from different domains. Data encoding and interchange is a relatively tough problem in the health sector as the data is used among different stakeholders in the health insurance sector. Several standards have been set globally for the interoperable representation and exchange of health data in healthcare and research [6].

Treatment of some uncommon diseases is very expensive, which is a big problem for the healthcare system globally. The current system is not in favour to cut drug pricing. Researchers are working on massive group insurance (MGI), a type of mechanism that helps in reducing the cost of drugs by contributing intellectual-property-right (IPR) remuneration to the drug company from an MGI agency, which is responsible for collecting insurance premiums from the country on each orphan drug that requires the access for the citizens. The system will help in managing drug pricing [15].

With the significant rise in pharmaceutical expenditures, controlling spending has become a crucial responsibility for the Health Insurance Department. Insurance settlements are paid on a per-service basis, resulting in a slew of irrational charges. The single-disease-payment (SDP) system has been frequently used to address this issue, but there is a possibility of fraud. Researchers are working on presenting a methodology for detecting medical insurance fraud based on consortium BC and deep learning, which can automatically recognise suspicious medical records to assure the correct implementation of SDP [8].

BC technology is a mechanism for generating an absolute, secure, and distributed transaction database. However, the technical accomplishment has been extended to legal transactions, medical data, insurance billing, and smart contracts. BC technology is crucial to healthcare, which transforms medical database interoperability and can aid in improving access to medical records, imaging archives, and prescription databases. Given the importance of a patient's medicinal history in effective medicine, BC technology has the potential to significantly enhance medical treatment [22].

BC technology benefits biological and healthcare areas also. Well-organised data and operations help in providing the best services in the health industry [23].

7.17 Findings

As a result, the insurance industry is likewise taking the initiative to incorporate new technologies. Financial institutions are working on research and development of new applications using blockchain and artificial intelligence. Through innovative technology, the insurance industry employs blockchain to give the finest services to its customers. The main focus is to make the process simpler and provide the best services to the customers. The trust of the consumers increases and builds a strong relationship among their patrons. The use of BCT is increasing in the insurance industry to make it more powerful and remove the hurdle for better results.

7.18 Practical Implications

Blockchain technology has the potential to alter the insurance industry by fostering trust among many stakeholders while also lowering operating expenses. BC has the power to disrupt the insurance industry's challenges.

References

[1] A. Akande, "Disruptive power of blockchain on the insurance industry," *Master's Thesis, Univ. of Tartu, Inst. of Comput. Sci.*, Tartu, 2018

[2] A. Hassan, M. I. Ali, R. Ahammed, M. M. Khan, N. Alsufyani, and A. Alsufyani, "Secured Insurance Framework Using Blockchain and Smart Contract," *Scientific Program. 2021*, pp. 1–11, 2021, doi: 10.1155/2021/6787406

[3] A. Hayes, "Blockchain Explained," *Investopedia*, March 2022, https://www.investopedia.com/terms/b/blockchain.asp

[4] A. K. Kar and L. Navin, "Diffusion of blockchain in insurance industry: An analysis through the review of academic and trade literature," *Telematics and Inform.*, vol. 58, pp. 101532, 2021, doi: 10.1016/j.tele.2020.101532

[5] D. Popovic et al., "Understanding blockchain for insurance use cases," *Brit. Actuarial J.*, vol. 25, 2020, doi: 10.1017/s1357321720000148

[6] E. Chondrogiannis, V. Andronikou, E. Karanastasis, A. Litke, and T. Varvarigou, "Using blockchain and semantic web technologies for

the implementation of smart contracts between individuals and health insurance organizations, " *Blockchain: Res. and Appl.*, vol. 3, no. 2, pp. 100049, 2022, doi: 10.1016/j.bcra.2021.100049

[7] E. I. Roundtable, "How technology and data are reshaping the insurance landscape," *Summary from the Roundtable Organised by EIOPA*, vol. 28, pp. 17–165, 2017.

[8] G. Zhang, X. Zhang, M. Bilal, W. Dou, X. Xu and J. J. Rodrigues, "Identifying fraud in medical insurance based on blockchin and deep learning," *Future Gener. Comput. Systems*, vol. 130, pp. 140–154, 2022, doi: 10.1016/j.future.2021.12.006

[9] H. Kim and M. Mehar, "Blockchain in Commercial Insurance: Achieving and Learning Towards Insurance That Keeps Pace in a Digitally Transformed Business Landscape," *SSRN Electronic J.*, 2019, doi: 10.2139/ssrn.3423382

[10] H. Treiblmaier and A. Tumasjan, "Editorial: Economic and Business Implications of Blockchain Technology," *Frontiers in Blockchain*, vol. 5, 2022, doi: 10.3389/fbloc.2022.857247

[11] J. Xu, Y. Wu, X. Luo, and D. Yang, "Improving the efficiency of blockchain applications with smart contract based cyber-insurance," *In ICC 2020-2020 IEEE International Conf. on Communications (ICC),* pp. 1–7, Jun. 2020.

[12] K. H. Kelley, L. M. Fontanetta, M. Heintzman, and N. Pereira, "Artificial Intelligence: Implications for Social Inflation and Insurance," *Risk Management and Insurance Review*, vol. 21, no. 3, pp. 373–387, 2018, doi: 10.1111/rmir.12111

[13] K. Panetta, "The CIO's Guide to Blockchain" *Gartner*, Sep. 2019.

[14] K. Sayegh, "Blockchain Application in Insurance and Reinsurance," *Skema Business School*, vol. 1, pp. 38, 2018.

[15] L. S. Ho, T. Zhang, T. C. T. Kwok, K. P. Wat, F. T. T. Lai, and S. Li, "Financing Orphan Drugs Through a Blockchain-Supported Insurance Model," *Frontiers in Blockchain*, vol. 5, 2022, doi: 10.3389/fbloc.2022.818807

[16] M. A. Uddin, A. Stranieri, I. Gondal, and V. Balasubramanian, "A survey on the adoption of blockchain in IoT: challenges and solutions," *Blockchain: Research and Applications*, vol. 2, no. 2, pp. 100006, 2021, doi: 10.1016/j.bcra.2021.100006

[17] M. B. Abramowicz, "Blockchain-Based Insurance," *SSRN Electronic J.*, 2019, doi: 10.2139/ssrn.3366603.

[18] M. Crawford, "The insurance implications of blockchain," *Risk Management*, vol. 64, no.2, pp. 24, 2017.

[19] M. Lounds, "Blockchain and its Implications for the Insurance Industry," *Munich Re Life US*, Jul. 2020, https://www.munichre.com/us-life/en/perspectives/underwriting/blockchain-implications-insurance-industry.html
[20] N. Jha, D. Prashar, O. I. Khalaf, Y. Alotaibi, A. Alsufyani, and S. Alghamdi, "Blockchain Based Crop Insurance: A Decentralized Insurance System for Modernization of Indian Farmers," *Sustainability*, vol. 13, no. 16, pp. 8921, 2021, doi: 10.3390/su13168921
[21] S. K. Ghosh., "Adoption of Blockchain in Financial Services," *International Journal of Scientific Research in Science, Engineering and Technology*, pp. 161–165, 2022, doi: 10.32628/ijsrset229143
[22] T. Heston, "Why Blockchain Technology Is Important for Healthcare Professionals," *SSRN 3006389*, 2017.
[23] T. T. Kuo, H. E. Kim, and L. Ohno-Machado, "Blockchain distributed ledger technologies for biomedical and health care applications," *J. of the American Medical Informatics Association*, vol. 24, no. 6, pp. 1211–1220, 2017.
[24] U. Bodkhe, S. Tanwar, K. Parekh, P. Khanpara, S. Tyagi, N. Kumar, and M. Alazab, "Blockchain for industry 4.0: A comprehensive review," *IEEE Access*, vol. 8, pp. 79764-79800, 2020, doi: 10.1109/access.2020.2988579
[25] V. Gatteschi, F. Lamberti, C. Demartini, C. Pranteda, and V. Santamaría, "Blockchain and Smart Contracts for Insurance: Is the Technology Mature Enough?," *Future Internet*, vol. 10, no. 2, pp. 20, 2018, doi: 10.3390/fi10020020
[26] V. Langaliya and J. A. G. Gohil, "A Comparative and Comprehensive Analysis of Smart Contract Enabled Blockchain Applications," *International Journal on Recent and Innovation Trends in Computing and Communication*, vol. 9, no. 9, pp. 16–26, 2021, doi: 10.17762/ijritcc.v9i9.5489
[27] W. Chen, Z. Xu, S. Shi, Y. Zhao, and J. Zhao, "A survey of blockchain applications in different domains," In Proceedings of the 2018 *International Conference on Blockchain Technology and Application*, pp. 17–21, Dec. 2018.
[28] Sood, K., Seth, N., & Grima, S. (2022). Portfolio Performance of Public Sector General Insurance Companies in India: A Comparative Analysis. In Simon Grima, Ercan Özen, Inna Romānova (Ed.), In *Managing Risk and Decision Making in Times of Economic Distress, Part B*. (pp.215–229). Emerald Publishing Limited,UK.
[29] Grima, S., & Marano, P. (2021). Designing a Model for Testing the Effectiveness of a Regulation: The Case of DORA for Insurance Undertakings. *Risks*, *9*(11), 206.

[30] Manohar, S., Mittal, A. and Marwah, S. (2020). Service innovation, corporate reputation and word-of-mouth in the banking sector: A test on multigroup-moderated mediation effect, Benchmarking: An International Journal, Vol. 27 No. 1, pp. 406–429. https://doi.org/10.1108/BIJ-05-2019-0217

8

A Systematic Review of Publications on Blockchain Application in the Insurance Industry

Cheenu Goel[1], Payal Bassi[1], Pankaj Kathuria[2], and Igor Todorović[3]

[1]Chitkara Business School, Chitkara University, Punjab, India
[2]University School of Business, Chandigarh University, Punjab, India
[3]Faculty of Economics, University of Banja Luka, Bosnia and Herzegovina

Email: cheenu.pcte@gmail.com; payalbassi@gmail.com; pankajmca18@gmail.com; igor.todorovic@ef.unibl.org

Abstract

The service industry has lately experienced tremendous changes in the recent past, especially in the insurance sector. Blockchain has emerged as one of the most influential emerging technologies, providing lasting solutions to the issues faced in the insurance sector. Since the insurance industry is generating enormous data from different domains of insurance parameters, blockchain technologies are extending competent solutions for day-to-day as well as long-term concerns in the real world. The present chapter aims to conduct a thorough systematic review of papers published in the related field. Relevant inclusion and exclusion criteria were levied and finally a total of 413 research papers were retained from ProQuest and Scopus published during 2012–2022. The network association diagrams were created in VOS viewer to explore the most frequently occurring indexed keywords so that the latest concerns of research practitioners, scholars, academicians, and policymakers could be examined in the field of blockchain applications in the insurance sector. Thereafter, top 10 journals, year-wise publications, document type, co-occurrence of all keywords, and co-authorship with countries were done with the help of visual networks. Finally, future research could be developed by examining the various issues raised during the systematic review of the existing literature.

8.1 Introduction

Insurance industry has always been the best avenue for mitigating risk and safeguarding the lives of our near and dear ones. Initially, insurance was just considered as a possibility of financially securing a person's life and people were in dilemma whether a service industry can assure the possible solution for securing their future. But over the period of time, the awareness spread by government and various private players in the insurance sector brought a toll in the thinking pattern of investors and thus pushing them to explore new products in the insurance service industry such as ULIP, vehicle insurance, house insurance, motor insurance, health insurance, travel insurance, and many more. However, the amount of risk and volatility associated in insurance cannot be denied and thus companies are putting lot of efforts not only in creating awareness and attracting customers but also in retaining them in this competitive environment.

Therefore, the companies involved in the insurance sector have become more cautious towards the large amount of data available on cloud, so that the clients could be served effectively. For this, companies are making appropriate use and accurate assimilation of technology so that it is possible to devise a solution that can subdue the risk involved from the perspective of both the parties. One such attempt was cryptographically secured chain of blocks, which was started by Stuart Haber and W. Scott Stornetta in the year 1991. But the term blockchain grabbed attention when Satoshi Nakamoto designed a framework for a blockchain in his white paper. Further, in the year 2008, the concept became famous when a public transaction ledger was created for the cryptocurrency Bitcoin (a virtual trading currency) [1]. With the spur of digitalisation, blockchain technology became quite prominent in almost every field of economy especially healthcare, agriculture, pharmaceutical industry, financial sector, financial services, and insurance. Blockchain is a chain of blocks that store relevant information such as source of generator and receiver, thus ensuring mitigation of risk and creating ledger in the cloud.

The insurance industry being highly volatile has always wanted a one-point solution where the risk aversion could happen, and clients could gain confidence while availing any insurance policy for human or non-human assets. One such solution was provided by blockchain as it took the business on all together a new and escalated platform of customer connect, trust, security, data integrity, and maximum profits [2]. Each block in the blockchain has a special code retaining the information of source of generator and retriever, hash as a unique code and a previous block hash. The initial block is often known as genesis block, which does not contain any previous data as it is the first block in the chain. Any tampering done in the blockchain is not possible as the change will be immediately triggered to all the persons associated with the chain and change is allowed only once the people connected give their consent to make the changes. Thus, the chance of data manipulation is reduced to a minimum, and the system as a whole becomes completely proof. Moreover, a digital transaction ledger is created in the cloud, thus making all the transactions secured. Therefore, the application of blockchain is more viable in the insurance industry especially when claims are to be made and payments can be released, and for the purpose of maintaining records, it acts as the mode of communication between the insurers and the insured [24, 25].

8.2 Review of Literature

Blockchain has emerged as the latest technology that is immensely benefitting insurance industry, but still there are a lot of opportunities that are totally unexplored, and the insurance industry is still facing teething issues in success. One of the main reasons is the lack of proper awareness followed by the risk of online fraud, which has been quite prevalent in the insurance and banking industries [3]. The blockchain technology has been very functional in the creation of digital currencies, and it has eliminated the role of third parties giving financial help [4]. Blockchain can enhance the financial performance of any company and at the same time facilitate the stakeholders by ensuring security, risk minimisation, scalability, and addressing privacy issues [5].

Blockchain is very effective in tracing records of transactions by creating cloud ledgers, minimising risk, and enhancing performance of financial institutions [6–8]. Blockchain has emerged as a disruptive technology that spurred positivity in the economy and has aligned directly with the sustainable development goals and motivating almost all the economies of countries to apply blockchain for the stability of its economy [9]. However, the government is enforcing ethical and legal rules and regulations so that the stakeholders of every business can gain trust and adopt blockchain technologies in their business [10].

Blockchain, smart contracts, big data, metaverse, etc., have intensively changed the way insurance industry functioned, thus reducing inconvenience and making transactions smooth [11]. The insurance industry is directly related to majority of individuals and tycoons of business world as it ensures financial cover not only for their lives but also for their goods and assets. Traditionally, the entire process of getting insured and claiming the insurance was a tedious task, but with the help of technologies like blockchain and Ethereum, the transactions could be recorded in a digital form and a track could be kept of every transaction online in the form of Hyperledger. This not only reduces the time involved in verifying and confirming the records but also stops theft, intentional financial gain, and tampering of records [12].

Various challenges faced by developing countries to adopt blockchain and endorse the role of blockchain have become a catalyst in the world of digitalisation. Technology, governance, corporate itself, and the environment in which it operates along with knowledge are the major areas where companies are finding difficulties while practicing blockchain technology in their working [13].

The financial services industry has a tremendous amount of potential that can be unlocked and transformed, thanks to blockchain technology, which is based on five principles: computational logic, peer-to-peer transmission, irreversibility of records, distributed databases, and transparency with anonymity [14]. Blockchain technology is rapidly developing and attracting substantial contributions from academics around the world as decentralised banking, insurance, trade finance, financial markets, and the cryptocurrency market use it more and more.

Indian service industry is highly volatile, and to keep pace with the competitive world, one needs to adopt disruptive technologies to survive, especially the banking and insurance sectors [15]. Blockchain-enabled companies create digital database and ensure safety and strengthen the trust of people in online transactions, thus enhancing transparency in the transactions. It further reduces interference of third parties, providing security of transactions in the form of blocks and creating peer-to-peer networks [16]. The use of "smart contracts," a type of legal agreement in which some contracts between two parties are involved are recorded digitally and secured by a special code, has benefited service providers generally and the insurance industry in particular [17]. Blockchain-enabled smart contracts store immutable data record in the electronic database of both the insurer and the insured, thus making it retrievable without wasting any time or incurring any additional cost [18].

Distributed ledger technology is used in blockchain applications to reduce waste, high transaction costs, and lengthy claim processing times. As a result of the data and payments being securely recorded, there is less risk and the insurance is easier to obtain. The main blockchain applications currently being used in the insurance sector are "self-purchase insurance," "automated claim settlement," "fraud detection," and "money flow record monitoring" [19].

For the sake of the current study's goals, the following questions have been developed after analysing other research papers:

RQ 1: Which journals and countries have published the most articles, especially in the subject of blockchain applications in the insurance industry?

RQ 2: What are the most influential keywords where research in the said domain is being performed?

8.3 Research Methodology

In the current study, the applications of blockchain technologies in the field of insurance sector were intensively examined in the existing literature. In the light of relevant keywords that might satisfy the study's criteria, the research articles were found in the databases of ProQuest and Scopus. Both databases are trustworthy sources for data retrieval since they contain articles from indexed journals from prestigious publishers including Taylor & Francis, Wiley, Elsevier, Science Direct, Emerald, Springer, and many more [20]. The research scientists place a high value on the Scopus database's publication record and see it as the most trustworthy data source [21–23]. Additionally, all these journals are frequently regarded as extremely reliable sources of rigorous and authentic research parameters. The final papers were chosen based on the pertinent research prevailing in the field of blockchain and insurance industry especially in terms of minimising the risk and lessening the tendency of data tampering. To accomplish the said results, a framework was designed comprising keyword research, data analysis, and data selection.

8.3.1 Keywords

The initial database search used pertinent terms including "Blockchain Technologies," "Blockchain Applications," "Smart Contracts," "Hyperledger," "Risk Management," and "Insurance Industry" (ProQuest and Scopus). The following string was decided upon after conducting several searches while maintaining the specified research questions. Initially, 2508 documents were found using the string mentioned in Table 8.1.

Table 8.1 Search process.

Search expressions	Database	Period	String	No. of records
A systematic review of blockchain applications in the insurance industry	ProQuest	2012–2022	(“Blockchain Technologies” OR “Blockchain Applications” AND “Insurance Industry”)	1855
	Scopus	2016–2022	(“Blockchain Applications” OR “Blockchain Technology” OR “Blockchain” AND “Insurance” OR “Hyperledger”)	653

Source: Author’s compilation.

8.3.2 Data analysis

The inclusion of full-text, peer-reviewed scientific journals, articles, books, book chapters, conference papers, and conference proceedings was finalised; however, conference proceedings were omitted when the search domain was expanded. In ProQuest and Scopus, the research time frames were restricted to the most recent 10 years (2012–2022). Despite being the industry’s buzzword since 2009, blockchain only started to get the attention of corporates in 2014. As a result, in 2016, the academicians began to draw attention to their study on reputable forums like Scopus, and the first publication was published in Scopus in the year 2016. Therefore, 2498 papers were included in the study.

8.3.3 Data selection

The inclusion and exclusion criteria were used to envision the study objectives as shown in Table 8.2. Initial inclusion of topics such as blockchain, cryptography, digital currencies, algorithms, artificial intelligence, and many more in ProQuest resulted in a reduction of the data to 1672, whereas inclusion of topics such as computer science, engineering, decision science, social sciences, economics, and health professions in Scopus resulted in a reduction of the sample size to 609. Thus, a total of 2281 papers were retained. Then English was selected as the final language for all the published articles reaching a total of 2185. Further, the field of publication stage was selected for the finally published papers in both the databases and the inclusion of several journals, namely *Sensors*, *Wireless Communications & Mobile Computing* (Online), *Cryptography*, *PeerJ Computer Science*, *Blockchain in Healthcare Today*, and *Algorithms*, and other journals related to selected keywords were included and 909 papers were retrieved. After extensively evaluating all the articles based on their titles and abstracts, a final review was conducted, and duplicates were then eliminated. The final sample size was reduced to 413 as shown in Prisma Diagram Figure 8.1.

Table 8.2 Inclusion criteria.

Inclusion	**Duration**	10 years (2012–2022)
	Language	English
	Subject	Blockchain, Cryptography, Digital Currencies, Peer-to-Peer Computing, Big Data, Artificial Intelligence, Technology, Cloud Computing, Data Integrity, Systematic Review, Wireless Networks, Financial Services, Bibliometrics, and Neural Networks Databases.
	Publication title	*Security and Communication Networks*, *Wireless Communications & Mobile Computing* (Online), *Cryptography*, *PeerJ Computer Science*, *Blockchain in Healthcare Today*, *Journal of Risk and Financial Management*, *Journal of Legal*, *Journal of Sensors*, and many more.

Source: Author's compilation.

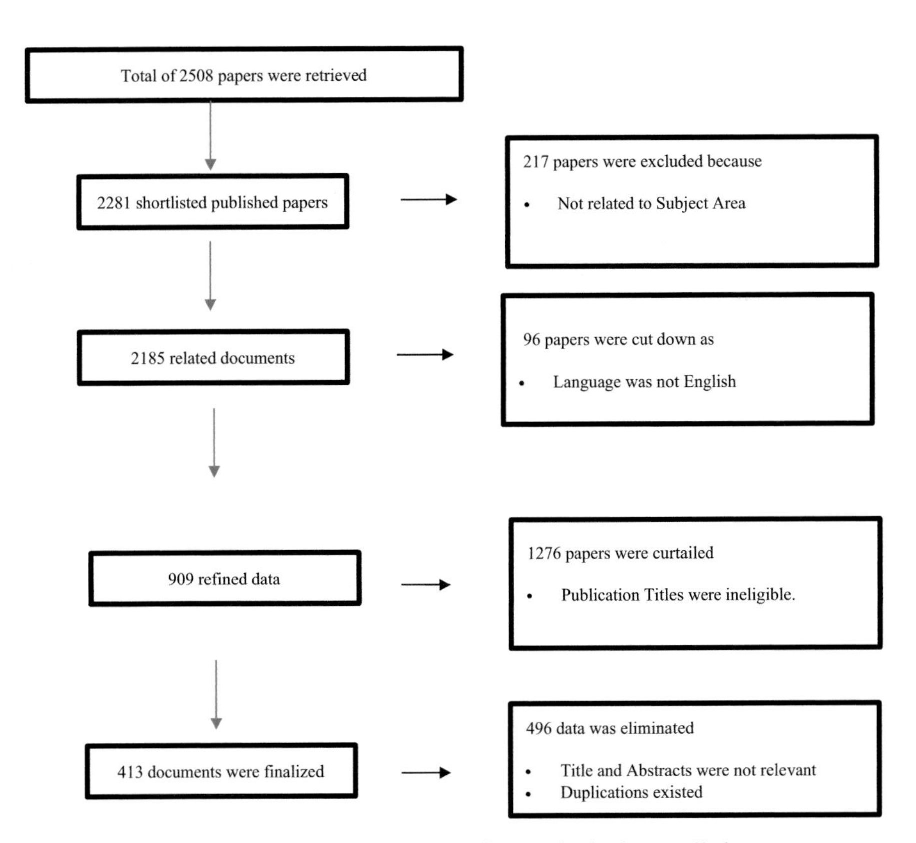

Figure 8.1 Prisma flow diagram. Source: Author's compilation.

Table 8.3 Publications by year.

Year	Total number of publications	Percentage
2022	83	20.09
2021	129	31.23
2020	88	21.30
2019	64	15.49
2018	41	9.92
2017	7	1.69
2016	1	0.24

Source: Author's compilation.

8.4 Data Analysis

8.4.1 Publications by year

For the aim of a systematic review, Table 8.3 provides information about publications on blockchain applications in the insurance industry for the period of 2016–2022. Although the chosen time period for the selected study was the most recent 10 years, from 2012 to 2022, no paper was able to make it through the databases of ProQuest and Scopus; as a result, Table 8.3 lists the year-by-year publication details for the period from 2016 to 2022, when articles first started getting published in the chosen databases. Except for 2022, when there was a very slight decline in publications for the associated industry, the data shows that there has been an exponential rise in publications connected to blockchain in the insurance industry from 2016 onwards. The maximum number of research articles to be accepted and published in ProQuest and Scopus databases throughout the study period was 129 in 2021.

8.4.2 Publication by document type

The document type considered in the current study for the aim of systematic literature is shown in Table 8.4. The sample showed that there are 187 conference papers, 118 articles, 63 academic journals, 29 book chapters, and 16 review papers in the whole sample.

8.4.3 Top 10 journals

The ProQuest and Scopus databases contain the most important journals in the disciplines of blockchain technologies, applications, and the insurance industry, which are listed in Table 8.5. Because it has published the most papers on

Table 8.4 Publication count as per the document type.

Document type	Total number
Article	118
Book chapter	29
Conference paper	187
Scholarly journals	63
Review papers	16

Source: Author's compilation.

Table 8.5 Influential journals on blockchain applications in insurance industry.

Source	Documents	Impact factor
Sensors	18	3.576
IEEE journals	12	3.476
International Journal of Advanced Computer Science and Applications	10	1.092
ACM International Conference Proceeding Series	9	0.55
Advances in Intelligent Systems and Computing	9	0.626
Security and Communication Networks	9	1.791
Communications in Computer and Information Science	8	0.476
Wireless Communications & Mobile Computing (Online)	8	2.336
EAI/Springer Innovations in Communication and Computing	7	0.93
Journal of Open Innovation: Technology, Market, and Complexity	6	3.61

Source: Author's compilation.

the topic, the table clearly shows that *Sensors*, with an impact factor of 3.576, is the best journal in the sample. IEEE journals, with 12 papers published, are ranked second in the current study's category of the most significant publications, followed by the *International Journal of Advanced Computer Science and Applications*, with a total of 10 articles. Despite having an impact factor of 2.336, *Wireless Communications & Mobile Computing* (Online) was only able to publish 8 papers throughout the year.

8.4.4 Keyword and title clouds

Wordart.com, an open-source word cloud generator, is used to produce a word cloud of the terms used by authors. According to the search results in Figure 8.2(a), the author frequently uses the keywords "blockchain,"

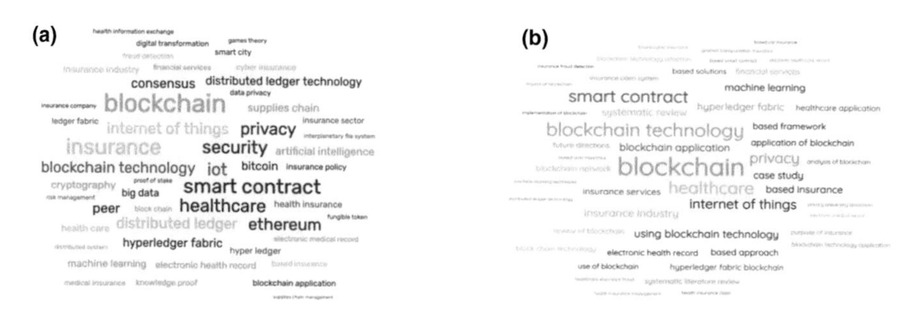

Figure 8.2 (a) World cloud of indexed keywords. (b) World cloud of title. Source: Author's compilation.

"blockchain technology," "privacy," "insurance," "blockchain network," and "Hyperledger fabric." While content written in small fonts has less representation in the chosen articles taken from both databases, the most prominent keywords with large font sizes are the most often searched key terms. Figure 8.2(b) shows a word cloud of the title in greater detail, with the most frequent occurrences of the phrases "artificial intelligence," "insurance," "Internet of Things," "smart contracts," "healthcare," "distributed ledger," "machine learning," "Hyperledger fabric," and "consensus."

8.4.5 Co-authorship with countries

According to Figure 8.3, out of all 42 nations, authors from the United States published the most articles on blockchain applications and technology with a focus on the insurance sector. While authors from all over the world are collaborating to produce ground-breaking research that will have an impact on the government, business leaders, academics, and policymakers, none of the authors of the papers that were chosen for this study have acknowledged the work of the others because they are all working on different facets of blockchain. There were 413 publications in the sample without any citations among the papers that were retrieved from both databases; however, the United States acquired 33 links overall.

8.4.6 Co-occurrence of all keywords

Figure 8.4 displays the findings based on the co-occurrence of all keywords. According to the statistics, 279 out of 2346 terms satisfy the requirement of having at least three occurrences, under certain conditions. In the chosen sample of papers, it was found that 48 clusters and 48 items emerged, with

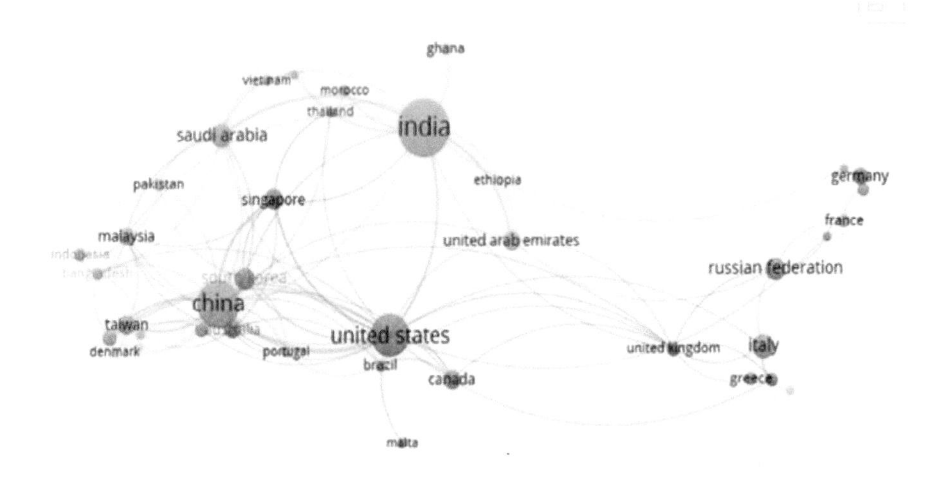

Figure 8.3 Network association of co-authorship with countries. Source: Author's compilation.

total connection strengths of 2051 and 987, respectively, with the keyword "Blockchain" appearing 300 times in total. VOS viewer software assisted in obtaining the results. Blockchain appeared 68 times in the search results, with a total link strength of 671, and was followed by smart contracts, healthcare, and insurance firms, which appeared 110, 48, and 53 times each. As shown in Figure 8.4, there are numerous additional related terms, such as cloud computing, cyber insurance, security, privacy, and the Internet of Things.

8.4.7 Network association of co-occurrence of index keywords

The network linkage of indexed keywords among the chosen papers under examination from the ProQuest and Scopus databases that were created in September 2022 is shown in Figure 8.5. The network uses a comprehensive counting method to reveal information about co-occurrence and indexed keywords. The co-occurrence of the indexed keywords blockchain, insurance, healthcare, insurance firms, Internet of Things, health insurance, network security, Ethereum, decentralised, insurance claims, etc., is shown in the figure.

8.5 Conclusion

In order to fulfil the goals of the current study, the chapter explored the most significant journals and the nations where the majority of work on blockchain and insurance has been done. The aforementioned goals have been attained

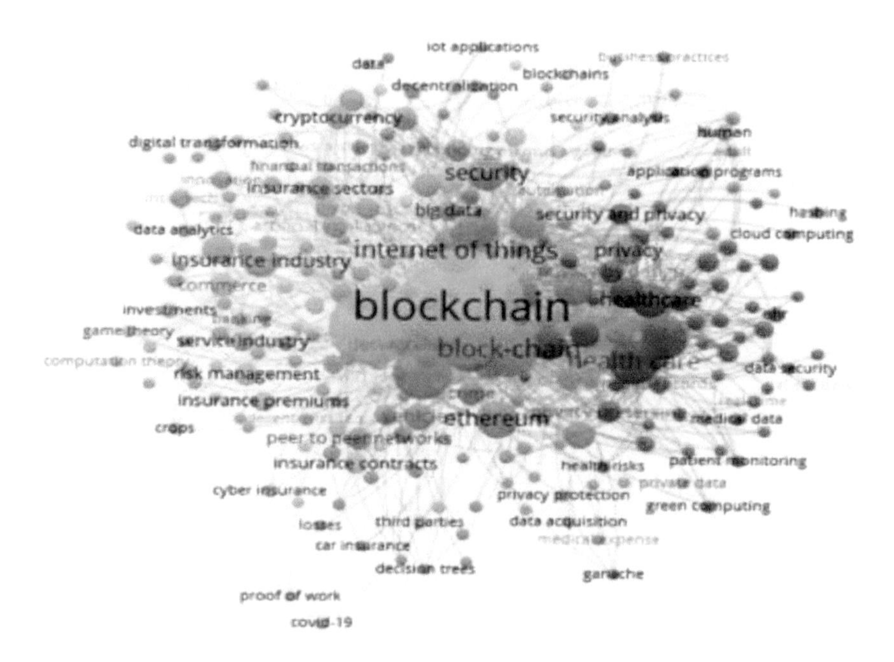

Figure 8.4 Network diagram of co-occurrence of all keywords. Source: Author's compilation.

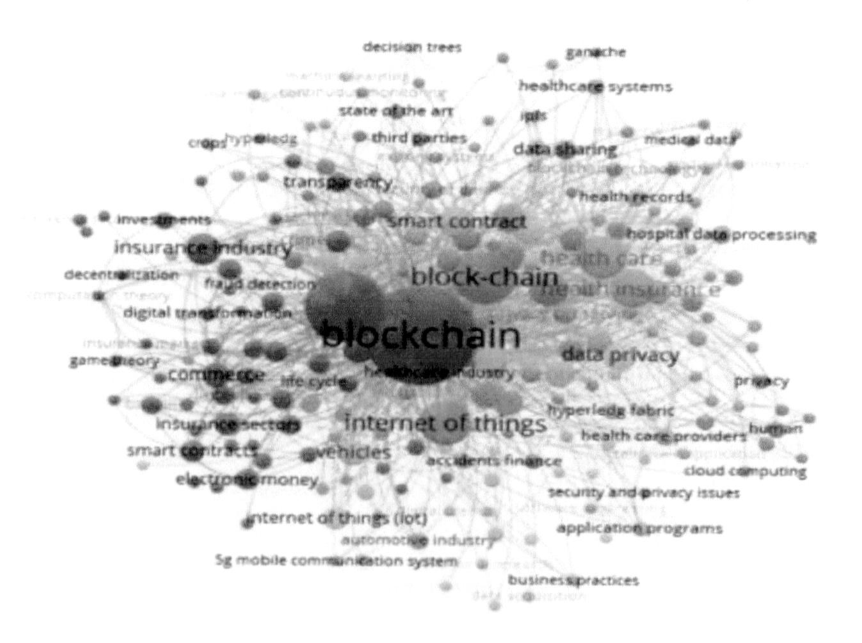

Figure 8.5 Network analysis of co-occurrence of index keywords. Source: Author's compilation.

with the aid of a thorough analysis of the blockchain and insurance fields in terms of insurance claim, privacy, security, digital storage, Ethereum, distributed Hyperledger, smart contracts, etc. The investigation has shown that IEEE journals, *International Journal of Advanced Computer Science and Applications*, and *Sensors* have made important contributions. Blockchain technologies and applications have changed the insurance sector with the rise of the internet, changing how records are kept and how transactions are protected against data tampering and risk. Each nation is doing their best efforts to conduct research in the area of interest, but the data shows that the United States has published the most documents in the field of blockchain with regard to a variety of insurance industry-related topics. Additionally, the terms "blockchain," "insurance," "healthcare," "insurance firms," "Internet of Things," "health insurance," "network security," "Ethereum," and "insurance claims" have been the most frequently co-occurring terms. With the incorporation of technology and blockchain applications, the sector has expanded and taken on a new direction. Blockchain applications have also decreased the time required for claim verification and given risk aversion, among other benefits. However, a lot of study has already been done in the area of blockchain, and rather than merely focusing on the opportunities and challenges present in the blockchain insurance industry, more attention has to be paid to blockchain applications in the technology and implications.

References

[1] Iansiti, M., Lakhani, K.R., & Mohamed, H., It will Take Years to Transform Business, but the Journey Begins Now. *Harvard Business Review*, 95,1, 118–128, 2017

[2] Wagire, A.A., Rathore, A.P.S., & Jain, R., Analysis and Synthesis of Industry 4.0 Research Landscape: using latent semantic analysis approach. *Journal of Manufacturing Technology Management*, 31,1,2019, doi: https://doi.org/10.1108/JMTM-10-2018-0349.

[3] Shetty, A., Shetty, A. D., Pai, R. Y., Rao, R. R., Bhandary, R., Shetty, J., ... & Dsouza, K. J., Block Chain Application in Insurance Services: A Systematic Review of the Evidence. *SAGE Open*, 12,1,2022, 21582440221079877.

[4] Alcazar, C. V., Data you can trust. *Air and Space Power Journal*, 31,2, 91–101,2017

[5] Chang, V., Baudier, P., Zhang, H., Xu, Q., Zhang, J., & Arami, M., How Blockchain can Impact Financial Services -The overview, challenges and recommendations from expert interviewees. *Technological*

forecasting and social change, 2020,158:120166. doi: https://doi.org/10.1016/j.techfore.2020.120166.

[6] Hassani, H., Huang, X., & Silva, E., Banking with Blockchain-ed Big Data. *Journal of Management Analytics*, 5,4,256-275,2018, doi: https://doi.org/10.1080/23270012.2018.1528900.

[7] Frizzo-Barker, J., Chow-White, P.A., Adams, P.R., Mentanko, J., Ha, D., & Green, S., Blockchain as a Disruptive Technology for Business: A systematic review. *International Journal of Information Management,* 51,102029 ,2020doi: https://doi.org/10.1016/j.ijinfomgt.2019.10.014.

[8] Palmié, M., Wincent, J., Parida, V., & Caglar, U., The Evolution of the Financial Technology Ecosystem: An introduction and agenda for future research on disruptive innovations in ecosystems. *Technological Forecasting and Social Change*, 151,119779,2020, doi: https://doi.org/10.1016/j.techfore.2019.119779.

[9] Parmentola, A., Petrillo, A., Tutore, I., & De Felice, F., Is Blockchain Able to Enhance Environmental Sustainability? A systematic review and research agenda from the perspective of Sustainable Development Goals (SDGs). *Business Strategy and the Environment,* 2021, doi: https://doi.org/10.1002/bse.2882.

[10] Kshetri, N., & Voas, J., Blockchain in developing countries. IT Professional, 20,2, 11–14, 2018, doi: https://doi.org/10.1109/mitp.2018.021921645.

[11] Yu, J., & Yen, B. P., Basic risk information component (BRIC) and insurance [Conference session]. Cross strait conference of information management development and strategy, Hong Kong, 2018.

[12] Gatteschi, V., Lamberti, F., Demartini, C., Pranteda, C., & Santamaría, V., Blockchain and smart contracts for insurance: Is the technology mature enough? Future Internet, 10(2), 20, 2018.

[13] Saif, A. N. M., Islam, K. A., Haque, A., Akhter, H., Rahman, S. M., Jafrin, N., ... & Mostafa, R., Blockchain Implementation Challenges in Developing Countries: An evidence-based systematic review and bibliometric analysis. *Technology Innovation Management Review,* 12(1/2),2022, doi: http://doi.org/10.22215/timreview/1479.

[14] Pal, A., Tiwari, C.K. and Behl, A., Blockchain technology in financial services: a comprehensive review of the literature, *Journal of Global Operations and Strategic Sourcing,* 14 , 1, 61–80, 2021, https://doi.org/10.1108/JGOSS-07-2020-0039.

[15] Bassi P., Mittal N., A comparative study of service quality dimensions in Indian Banking Industry," *IOSR Journal of Business and Management (IOSR-JBM).* 3,3, 26–31,2018,doi: 10.9790/487X-2003032631.

[16] Jindal T., Bassi P., Security and Privacy Issues of Block Chain Technologies: A Reviel.w Paper. Emerging Issues in Business and Finance Post Covid-19,34,2021, https://www.journalpressindia.com/website/ndim-intlconference-icgfbe2021.

[17] Christopher, C. M., The bridging model: Exploring the roles of trust and enforcement in banking, bitcoin, and the blockchain. *Nevada Law Journal*, 17, 139,2016.

[18] Nam, S., How much are insurance consumers willing to pay for blockchain and smart contracts? A contingent valuation study. Sustainability, 10(11), 4332,2018.

[19] Zhao, L., The analysis of application, key issues and the future development trend of blockchain technology in the insurance industry. *American Journal of Industrial and Business Management,* 10(02), 305–314, 2020, https://doi.org/10.4236/ajibm .2020.102019.

[20] Tseng, M. L., Islam, M. S., Karia, N., Fauzi, F. A., & Afrin, S., A literature review on green supply chain management: Trends and future challenges. *Resources, Conservation and Recycling,* 141, 145-162, 2019, doi: 10.1016/j.resconrec.2018.10.009.

[21] Fahimnia, B., Sarkis, J., & Davarzani, H.., Green supply chain management: A review and bibliometric analysis. *International Journal of Production Economics*, 162, 101114, 2015, doi: 10.1016/j.ijpe.2015.01.003.

[22] Malviya, R. K., & Kant, R., Green Supply Chain Management (GSCM): a structured literature review and research implications. *Benchmarking Int. J.* 22 (7), 1360–1394,2015.

[23] Seuring, S., & Müller, M., From a literature review to a conceptual framework for sustainable supply chain management. 16(15), 1699–1710, 2008, doi: 10.1016/j.jclepro.2008.04.020.

[24] Sood, K., Seth, N., & Grima, S. (2022). Portfolio Performance of Public Sector General Insurance Companies in India: A Comparative Analysis. In Simon Grima, Ercan Özen, Inna Romānova (Ed.), In *Managing Risk and Decision Making in Times of Economic Distress, Part B*. (pp.215–229). Emerald Publishing Limited,UK.

[25] Grima, S., & Marano, P. (2021). Designing a Model for Testing the Effectiveness of a Regulation: The Case of DORA for Insurance Undertakings. *Risks*, *9*(11), 206.

9

An Evaluation of the Operational and Regulatory Aspects of using Blockchain in the Insurance Industry

Sanjay Taneja[1], Aarti Dangwal[2], Simranjeet Kaur[3], Ercan Ozen[4], and Mohit Kukreti[5]

[1]Department of Management Studies, Graphic Era Deemed to be University, Dehradun, India
[2]Chitkara Business School, Chitkara University, Punjab, India
[3]USB, Chandigarh University, Mohali, India
[4]University of Usak, Faculty of Applied Sciences, 64200 Usak- Türkiye
[5]University of Technology and Applied Sciences, Ibri, Oman

Email: drsanjaytaneja1@gmail.com; aartidangwal13@gmail.com; kangsimrank@gmail.com; ercan.ozen@usak.edu.tr; mohit.ibr@cas.edu.om

Abstract

While insurance professionals are well-versed in their field, there is a distinct lack of a framework for implementing distributed ledger technology, including blockchain, in the industry. This chapter evaluates the operational and regulatory aspects of using blockchain in the insurance industry. Even though blockchain technology has piqued the involvement of several stakeholders, it is necessary to predict how this technology will affect financial services, including insurance. From 2017 to 2021, this script analysed 212 Scopus-indexed research papers. The graphical maps created via VOS shows the historical commitments of authors and countries, as well as trends through cooperation and journal. All selected articles assessed concepts of research, and shared study areas as well as potential future research endeavours.

9.1 Introduction

In December 2017, The DWWP (Digital World Working Party) surveyed the business world to learn more about the views and activities of insurance, as well as the practice and capability in assessing risks emerging from technology. Blockchain emerged as the most difficult topic for respondents to explain to a co-worker about digital advancement in the insurance market. This revealed a lack of knowledge about blockchain and its associated benefits and dangers. This research aims to figure out how to use blockchain to solve problems with insurance.

This chapter is split into three parts. First, the chapter gives some background information about the technology and what makes it different from other solutions. This is followed by a review of the literature to show how the technology can be used in the insurance field. Lastly, a bibliometric analysis is conducted, and the findings are presented using visualizations, specifically employing the VOS technique to create a graphical map of the study field concerning blockchain and insurance. Financial literacy and investment are tied to the popularity of insurance for both life and non-life insurance. In countries where new digital techniques are being actively developed, the dynamics of insurers are growing at a high rate. The financial sector in countries like China and India is experiencing rapid advancements in innovative digital technologies. Studies [17,19,20,21,22,23] indicate that within the financial industry, insurance innovations play a crucial role. This observation is supported by empirical evidence, highlighting the significance of these advancements within the broader financial landscape. It also highlights the development matrices for the development of an innovative co-concept in the insurance industry, artificial intelligence (AI), blockchain technology, data acquisition, and consumer protection devices in the information space. The study's findings may prove useful in future research. Interest in blockchain has gone up and down because the technology has not caught on quickly. But if considering the numbers of intellectual capital and unified power pervade into blockchain development, you cannot avoid a tipping point. It is referred to as Amara's law, which "we tend to overestimate the short-term impact of a technology while underestimating the long-term impact." As the internet did for information exchange, blockchain is an infrastructure technology that will enable an instantaneous peer-to-peer exchange of value. For example, end users (individuals, societies, and corporate entities) could pool/aggregate and transfer risk without the involvement of (re)insurance/broker companies. On a blockchain, based on distributed ledger technology, the current state (a snapshot of the data) is confirmed and checked without a trusted central authority. It is a database that is shared by multiple users and data would then be authenticated by many entities rather than one. Each participant spreads and stores the data. This article focuses on how this network can be used to resolve issues in the insurance industry rather than its technical definition.

How does it work? In layman's terms, coding, cryptography, distributed systems, and peer-to-peer networking all come together in a single technology known as the blockchain. One must have a firm foundation in such fields to comprehend the inner workings of this system, which is not what this paper is about. Instead, it gives a high-level summary that focuses on the trade-offs between technology and design choices. Here is an outline of designing and choosing the best solution for a certain use case. This section gives an overview of the most important components of blockchain/DLT technology. The software relies on a working knowledge of these concepts. It is not necessary to have a detailed understanding of the underlying protocols, however, as with web application development.

9.2 Literature Review

In-depth discussions with insurance industry experts who may be familiar with blockchain were used to consider the level of the technology and its potential for success in the region [1].To meet these security requirements, the authors in [2] came up with a blockchain-based system that combines mutual authentication of identities, pro between the two parties, and key security features. An arbitration mechanism was proposed in the event of a dispute to divide responsibilities. Both benefits and drawbacks of using modern digital disruptive technologies in insurance operations [3] sought to address in an article the comparative study of six specific insurance company digitalization implementation approaches on theoretical and empirical evidence. Carbone and Cavalcanti [4] presented a usage-based insurance (UBI) platform based on IoT and blockchain techniques, discussing potential partners, revenue streams, and interaction modes. Even the existing UBI products often use mileage, driving period, or driving province data to calculate insurance. Wang et al. [5] experimented with smart contracts relying on Hyperledger Fabric & Ethereum, which are often distributed accounts techniques, in comparing the use of private and community blockchains for insurers. Xenakis and Farao [6] discovered that numerous fields, such as cyber-insurance, have improved their apps and created a whole new structure for insuring a botnet whilst also incorporating secured bidding using the foregoing technology. Menges et al. [7] have put into making a viable solution for processing insurance-related transactions in which Hyperledger Fabric, an open source permissioned blockchain model used to inculcate an unconventional prototype as well as to discuss the key design criteria, and configuration initiatives, and to encode insurance processes as smart contracts. Several tests were conducted to examine the framework's performance and the proposed design's security. Wang et al. [8] used Ethereum smart contracts in the insurance market as a focal point, which proceeded with a description of how this technology works and what makes it unique. It was also found that under blockchain, insurance industry initiative companies have shown their interest by defining and regulating the method of streamlining their execution. Insurance premiums must be transparent to benefit from Blockchain based smart contracts. Omar et al. [9] considered the option that uses the concept of the Internet of Things and blockchain technology to make a self-organizing insurance framework that would do ahead with major drawbacks of insurers, and also the proposed project's virtues and shortfalls were evaluated. Ultimately, it has huge advantages and could be useful in the future, as per the findings. Analysis of omics data with other recent data forms, transmitting the implications of this tech to life sciences along with healthcare, has been discussed [10] in this review. There are both advantages and disadvantages to using BC technology in Romania, according to [11]. The dissolution of brokers that cause issues, forgery, and rising rates in the medical insurance mechanism will be attained via blockchain technology. Guo and Liang [12] confirmed a faster process of claims and treatment while managing the security and privacy of all medical records such as EHRs and EMRs. One thing that all successful uses of blockchain have in common is that business partners are willing to work together. The authors in [13] commented that companies need to be ready to work together and compete on the same network so that everyone can benefit [14, 15].

9.3 Applications of Blockchain in the Insurance Industry

Insurance is not unique in our lives. It has been running for millennia. Currently, the insurance industry is heavily reliant on brokers. Brokers commonly call people to sell them a policy. Making a paper contract allows for human errors in the document or when claiming insurance. Overall, it complicates things for everyone involved, including insurers, brokers, and buyers. We are now an industry-changing technology. It is not as impactful yet, but it looks promising. To better understand how blockchain affects or changes the insurance industry, we must first understand the risk associated with the entire process. We also need to know which use cases will have a long-term impact. Blockchain insurance is the latest phrase for insurance companies. It will lead to blockchain insurers that benefit the world of blockchain. Also, insurers must rebuild their entire business model. It will take time and testing

Table 9.1 Study criteria.

Sr. No	Research Elements	Category	Content
1	Elements	Articles, investigations, Book Reviews	No. of Publications & their types
2	Contribution	No. of Authors citations	Dominance of the writer in the field
3	Top countries	Name	Nations publishing more no. of papers
4	Most acclaimed publication	Title	journals with more work on this field

Source: Author's compilation through VOSviewer software.

to see substantial changes in the market. However, early adopters will have the edge of being first.

The four blockchain use cases and applications include the following:

1. Fraud detection and risk prevention
2. Claims prevention and management
3. Health insurance
4. Reinsurance
5. Property and casualty(P&C) insurance

9.4 Methodology

An aspect of the report called the bibliometric trial investigates authors and collaborations. The creation of VOS involved the use of authorship, geographic distribution, word frequency, bibliographic couplings, and mapping.

This source code evaluated 212 Scopus-indexed publications' bibliographies from 2017 to 2021 that were found using the keywords "insurance" and "Blockchain technology." With the help of a visual map, you can see how writers and countries have contributed over time, as well as market dynamics in collaboration and publication. As a result of this article, major research concepts and shared areas of interest and potential research projects were explored.

The themes of the current article are insurance and blockchain technology. An analysis of the above-mentioned areas was also done using quantitative research output measurement. Table 9.1 highlights the parameters included in the investigation.

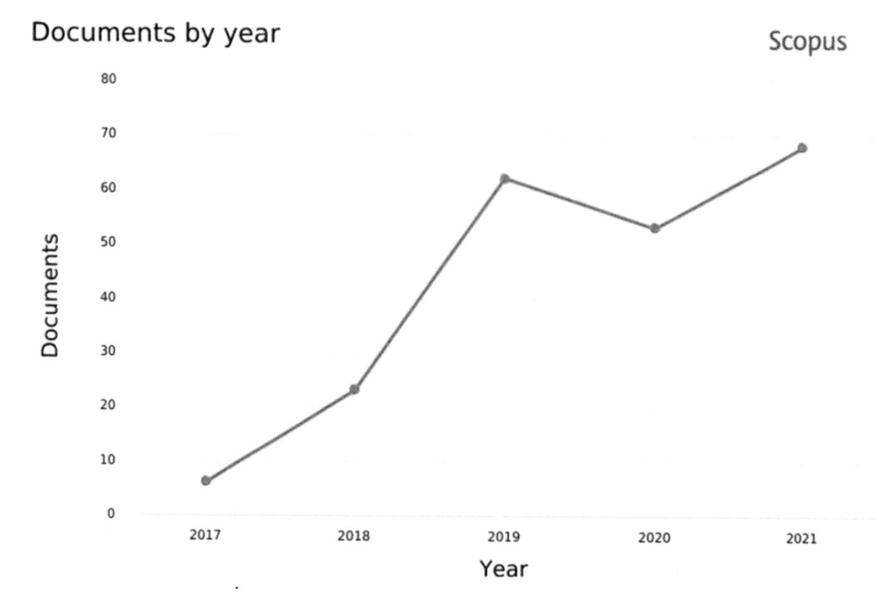

Figure 9.1 Publications between 2017 and 2021 through VOSviewer software.

9.4.1 Data extraction

The research paper incorporated multiple components, including keywords, authors, journals, citations, and publications, in its study. To gather relevant literature on the intersection of blockchain technology and insurance, the Scopus database was utilized as a valuable resource. This database facilitated the collection of pertinent information to support the research objectives. The main database of publications, i.e., Scopus, contains abstracts and citations to peer-reviewed literature from Springer, Elsevier, Emerald, T&F, and Interscience. The author accessed the insurance and blockchain studies that are referenced herein. To get to a manageable final count after removing papers from conferences, reviews of conferences, and peer-reviewed papers in multiple languages, the initial 222 results had to be downsized. Nevertheless, even after limited articles were removed because they did not fit the research's overall theme, this includes journals, book chapters, and other publications.

Figure 9.1 showcases the evolution of research papers from 2017 to 2021, revealing a significant surge in interest and attention towards the subject. However, it is noteworthy that in 2018 and 2019, there was a substantial increase in the number of researchers and concepts gaining traction. Out of the total 212 papers, only 138 were published during these two years. This observation suggests that blockchain and insurance as research domains are

Table 9.2 Authors in the area of blockchain technology in insurance (Scopus citations).

Ranking	Author	Documents
1	Aleksieva, V.	3
2	Chakravaram, V.	3
3	Huliyan, A.	3
4	Ratnakaram, S.	3
5	Valchanov, H.	3
6	Vihari, N.S.	3
7	Choo, K.K.R.	2
8	Demartini, C.	2
9	Demir, M.	2
10	Ferworn, A.	2

Source: Author's compilation through the VOSviewer software.

still relatively new and rapidly developing. The upward trend depicted in the graph demonstrates the growing body of literature and the increasing significance of these ideas within the academic community.

The next part of the research was to trace out whose journals, authors, and papers were mentioned the most. Articles about insurance and blockchain technology that were published in the top 10 journals gave more details. Table 9.2 highlights the top 10 authors based on the number of documents in each journal. This appeared in the fair number of publications in academic journals. Also, Table 9.2 depicts the citations and number of papers by a single author in his/her field based on Scopus citations.

Figure 9.3 illustrates the global distribution of documents, showcasing the number of documents originating from different countries. The data reveals that India has the highest representation with 50 documents, followed by China with 38 documents and the United States with 32 documents. The Russian Federation accounts for 11 documents, while Germany, Taiwan, and the United Arab Emirates each contribute 8 documents. Italy and Saudi Arabia both have 7 documents, while Canada and South Korea have 6 each. Australia and Singapore have 5 documents each, and Greece and Malaysia each have 4. Brazil, Bulgaria, France, Jordan, Latvia, the Netherlands, Poland, and Ukraine are all represented by 3 documents each. The United Kingdom, Hong Kong, Japan, and Malta have 2 documents each. These numbers provide an overview of the distribution of documents worldwide, showcasing the varying levels of document production across different countries.

9.5 Findings and Conclusion

In this review, it is shown that, however, it is true that blockchain's hype has not yet had a significant impact on the insurance sector. Acceptance of

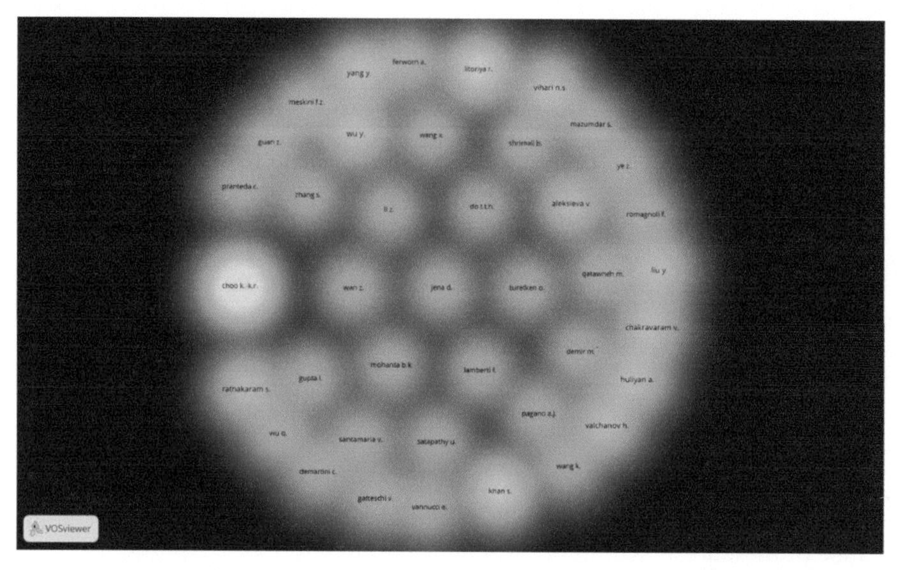

Figure 9.2 Density of authors with the highest citations through the VOSviewer software.

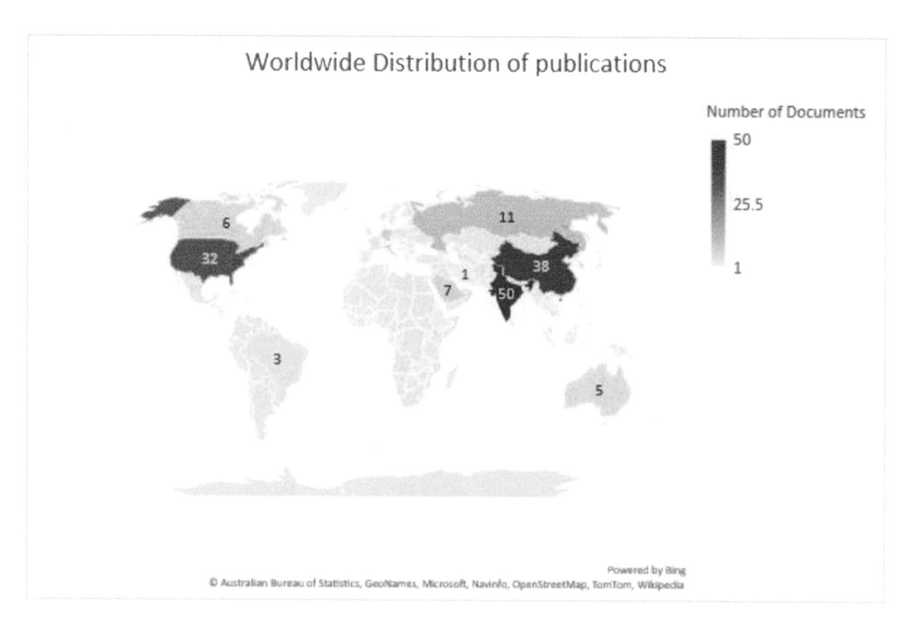

Figure 9.3 Worldwide distribution of publications through the VOSviewer software.

blockchain applications from a commercial perspective takes time, as does the adoption of most emerging innovations. Also, there is a practical guide for insurance industry practitioners to understand, assess, and adopt blockchain with bibliometric analysis, covering the distinctive characteristics of blockchain, real-world use cases developed, risks and challenges associated with adoption, and a guide to successfully adopt blockchain based on the framework suggested in the insurance industries. One of the two types of operation is taking place, while others are forming contact points for advice on how to implement this combination and a few are yet to begin the process at all. Adoption of IoT is a group sport that requires the cooperation of many stakeholders. Councils and policymakers, working with the organizations, need to clarify regulations so that further innovation can take place. Additionally, insurers must collaborate to develop blockchain standards and make sure that different solutions work together. Paper-based processes have to be changed with digitals, and there requires more trust in the blockchain where adoption has happened.

9.6 Research Limitations/Implications

This review illustrates the current state of play of blockchain in insurance from a commercial and regulatory point of view. It focuses the study has 212 Scopus-indexed publications is a key constraint and also the article explored only the concept of insurance and blockchain tech. Despite these constraints, the review effectively consolidates information from diverse sources to present a comprehensive understanding of the regulatory landscape for assessing insurance and blockchain technologies. Its intended audience includes academicians, regulatory authorities, and industry experts seeking insights into this dynamic field.

References

[1] Yu, T. -., Wang, A. -., Tseng, L. -., & Tsao, W. -. (2021). A preliminary study of the impact of blockchain technology on the application level of the insurance industry. Paper presented at the 2021 IEEE International Conference on Social Sciences and Intelligent Management, SSIM 2021, doi:10.1109/SSIM49526.2021.9555203 Retrieved from www.scopus.com

[2] G. Cheng, Y. Chen, S. Deng, H. Gao & J. Yin, "A Blockchain-Based Mutual Authentication Scheme for Collaborative Edge Computing," in IEEE Transactions on Computational Social Systems, vol. 9, no. 1, pp. 146–158, Feb. 2022, DOI: 10.1109/TCSS.2021.3056540.

[3] Ullah F, Sepasgozar SME, Wang C. A Systematic Review of Smart Real Estate Technology: Drivers of, and Barriers to, the Use of Digital Disruptive Technologies and Online Platforms. *Sustainability*. 2018; 10(9):3142. https://doi.org/10.3390/su10093142

[4] Carbone, A. F., & Cavalcanti, L. R. A Business Model for Vehicle Insurance Based on Blockchain Smart-Contracts

[5] S. Wang, L. Ouyang, Y. Yuan, X. Ni, X. Han & F. -Y. Wang, "Blockchain-Enabled Smart Contracts: Architecture, Applications, and Future Trends," in IEEE Transactions on Systems, Man, and Cybernetics: Systems, vol. 49, no. 11, pp. 2266–2277, Nov. 2019, DOI: 10.1109/TSMC.2019.2895123.

[6] Xenakis, C., & Farao, A. a Security ECONomics service platform for smart security investments and cyber insurance pricing in the beyond 2020 networking era.

[7] Menges, F., Putz, B. & Pernul, G. DEALER: decentralized incentives for threat intelligence reporting and exchange. *Int. J. Inf. Secure.* 20, 741–761 (2021).

[8] S. Wang, L. Ouyang, Y. Yuan, X. Ni, X. Han & F. -Y. Wang, "Blockchain-Enabled Smart Contracts: Architecture, Applications, and Future Trends," in IEEE Transactions on Systems, Man, and Cybernetics: Systems, vol. 49, no. 11, pp. 2266–2277, Nov. 2019, DOI: 10.1109/TSMC.2019.2895123.

[9] I. A. Omar, R. Jayaraman, M. S. Debe, K. Salah, I. Yaqoob, & M. Omar, "Automating Procurement Contracts in the Healthcare Supply Chain Using Blockchain Smart Contracts," in IEEE Access, vol. 9, pp. 37397–37409, 2021, DOI: 10.1109/ACCESS.2021.3062471.

[10] Dedeturk, B. A., Soran, A., & Bakir-Gungor, B. (2021). Blockchain for genomics and healthcare: a literature review, current status, classification, and open issues. *PeerJ*, *9*, e12130.

[11] Türkeş, M. C., Oncioiu, I., Aslam, H. D., Marin-Pantelescu, A., Topor, D. I., & Căpuşneanu, S. (2019). Drivers and barriers in using industry 4.0: a perspective of SMEs in Romania. *Processes*, *7*(3), 153.

[12] Guo, Y., Liang, C. Blockchain application and outlook in the banking industry. *FinancInnov* 2, 24 (2016)

[13] A. S. S Mishra, "Study on Blockchain-Based Healthcare Insurance Claim System,"2021AsianConferenceonInnovationinTechnology(ASIANCON), 2021, pp. 1–4, DOI: 10.1109/ASIANCON51346.2021.9544892.

[14] Dangwal A., Kaur S., Taneja S., Ozen E., "A Bibliometric Analysis of Green Tourism Based on the Scopus Platform," 2022 Developing Relationships, Personalization, and Data Herald in Marketing 5.0, pp. 242–255, DOI: 10.4018/978-1-6684-4496-2.ch015

[15] Sood, K., Seth, N., & Grima, S. (2022). Portfolio Performance of Public Sector General Insurance Companies in India: A Comparative Analysis.In Simon Grima, Ercan Özen, Inna Romānova (Ed.), In

Managing Risk and Decision Making in Times of Economic Distress, Part B. (pp.215–229). Emerald Publishing Limited,UK.

[16] Grima, S., & Marano, P. (2021). Designing a Model for Testing the Effectiveness of a Regulation: The Case of DORA for Insurance Undertakings. *Risks*, *9*(11), 206.

[17] Bhatnagar, M., Taneja, S. Kumar, P., & Özen, E., (2023a). Does Financial Education Act as a Catalyst for SME Competitiveness. International Journal of Education Economics and Development, 1(1), https://doi.org/10.1504/ijeed.2023.10053629

[18] Bhatnagar, M., Özen, E., Taneja, S., Grima, S., & Rupeika-Apoga, R. (2022a). The Dynamic Connectedness between Risk and Return in the Fintech Market of India: Evidence Using the GARCH-M Approach. *Risks, 10*(11), 209. https://doi.org/10.3390/risks10110209

[19] Bhatnagar, M., Taneja, S., & Özen, E. (2022b). A wave of green start-ups in India—The study of green finance as a support system for sustainable entrepreneurship. *Green Finance, 4*(2), 253–273. https://doi.org/10.3934/gf.2022012

[20] Bhatnagar, M., Taneja, S., & Rupeika-Apoga, R. (2023b). Demystifying the Effect of the News (Shocks) on Crypto Market Volatility. Journal of Risk and Financial Management, 16(2), 136. https://doi.org/10.3390/jrfm16020136

[21] Taneja, S. Kaur, S. & Özen, E., (2022a). Using green finance to promote global growth in a sustainable way. International Journal of Green Economics, 16(3), 246–257. https://doi.org/10.1504/ijge.2022.10052887

[22] Taneja, S., & Özen, E. (2023a). To analyse the relationship between bank's green financing and environmental performance. International Journal of Electronic Finance, 12(2), 163–175. doi:10.1504/IJEF.2023.129919

[23] Taneja, S., Bhatnagar, M., Kumar, P., & Rupeika-apoga, R. (2023b). India ' s Total Natural Resource Rents (NRR) and GDP : An Augmented Autoregressive Distributed Lag (ARDL) Bound Test. Journal of Risk and Financial Management, 16(2), 91. https://doi.org/doi.org/10.3390/jrfm16020091

10

The Role Played by Blockchain Computing Premium Payments and Creating New Markets

Kuldeep Singh Kaswan[1], Gurpreet Singh Sidhu[2], Jagjit Singh Dhatterwal[3], and Naresh Kumar[4]

[1]School of Computing Science & Engineering, Galgotias University, India
[2]Department of Management, Punjab Institute of Technology, Rajpura, District Patiala, Punjab, India
[3]Department of Artificial Intelligence & Data Science (AI&DS), Koneru Lakshmaiah Education Foundation, Andhra Pradesh, India
[4]Dean Applied Computational Science and engineering, G L Bajaj institute of technology and management , Greater Noida, India

Email: kaswankuldeep@gmail.com; 0083guri@gmail.com; Jagjits247@gmail.com; naresh.dhull@gmail.com

Abstract

Blockchain is an innovative technology that offers numerous advantages over centralised systems. For example, the Ethereum blockchain does not suffer from a single-machine breakdown. It eliminates the need for a trusted intermediary since the consensus process ensures confidence, saving both cost and money. Combining digital currencies with blockchain allows for the faster processing of transactions. Data integrity and encryption, once again, safeguard blockchain. As a result, it is often regarded as the most significant breakthrough since the internet. Many corporate apps are attempting to leverage blockchain to benefit from its benefits. One of them is the insurance industry, as the current system is plagued by fraud, the difficulty of managing paperwork, and the time-consuming nature of provides a systematic way. This chapter also reveals measure criteria to compute premium payments, build new health insurance, and create a new market for used car sales.

10.1 Introduction

The primary goal of distributed ledger technology is to solve the phenomenon like double expenditures in online transactions. Blockchain technology was initially utilised for cryptocurrencies, which is known as Blockchain1.0. After that, academics began exploring ways to use blockchain other than Bitcoin, such as providing automatic payment processing with security, known as Blockchain 2.0. Blockchain 3.0 refers to the use of blockchain for IoT devices [11].

Existing insurance sector labour in India is primarily done manually, from insurance purchase through systematic offers. People typically purchase insurance coverage through agents. They rely on agents to pay premiums, and agents process claims. Insurance clients must have all insurance documentation on file; otherwise, claims cannot be processed and possibly be unable to obtain healthcare insurance. For a claim to be processed, the insurance customer must provide the necessary documentation. And then, when all of those things have been verified, the claim will be approved, and they will get compensation. This procedure can take a long time, and beneficiaries may not get their benefits on time. This laborious procedure takes a long time, incurs additional costs in agent commissions, and has the highest risk of fraud [12].

Blockchain solutions for the insurance business will significantly alter current operations by allowing for faster administration, keeping data electronically safe and reducing costs. Blockchain technology will be employed in several insurance use cases such as fraudulent activities, assurance and underwriting, asset monitoring, and KYC. The use of blockchain in healthcare would drastically alter the old financial institutions, creating a new, quicker, simpler, and more secure market [13].

10.2 Literature Review

Ankitha Shetty et al. [1] discussed that the proportion of life insurance policies sold directly is more significant than 98%. By establishing smartphone apps, insurers gradually shift toward digital payments and claim administration. Insurers are emphasising the use of blockchain solutions to expedite claim management.

Insurers are implementing AI systems to assist agents in marketing the appropriate coverage, building up virtual locations, and handling commercial vehicle claims based on images.

In general, they are transforming IT systems by utilising innovative technologies like intelligent machines, extensive data analysis, and blockchain.

Amith Kumar et al. [2] explained that the central issue in the healthcare market is a loss for either of the participants, depending on the type of inaccuracy caused by the human entry. Blockchain is the ideal answer for these issues.

Anokye Acheampong et al. [3] observed that digital agreements differ from conventional commitments, making it difficult to deal with situations of breach of such arrangements because they lack evidential value in courts. Concentrate even more on the obstacles to blockchain deployment. Finally, advise that a current regulatory structure be established for blockchain networks.

Mayank Raikar [4] suggested a cryptocurrency architecture for insurance to provide fine-grained identity management and experimented by expanding the connection to verify the system's resilience. Finally, it was discovered that network size is directly connected to the condition to makes; the bigger the number of connections, the longer the confirmation time, i.e., the reduced quickly the network's connectivity [35, 36].

Najmeddine Dhieb [5] detected a fraud solution for insurance companies based on a public blockchain and modelling techniques, which was created. According to his research, the XGBoost algorithm outperforms the prediction model, SVM, *k*-nearest neighbour, and XGBoost algorithms in identifying fraud cases and forecasting consumers' future behaviour and complaint amounts.

Lamberti [6] reported that venture capitalists in cryptocurrency for pay-per-use first should confirm the national requirements for pay-per-use insurance plans, such as if a signed agreement is necessary. If this is the case, a change in general legislation would be required. Furthermore, additional measures might confirm that a strategy has been signed if the organisation is well-known. This research examined several areas where blockchain technology might provide essential advantages, while conventional methods performed well in others.

Zengxiang Li [7] suggested the fine-grained mobility coverage proposal based on blockchain and IoT capabilities. Enormous automobile trip information is kept on Hyperledger Fabric, giving more attention to blockchain to use private and public network performance. The insurance and payments mechanism is developed using the Ethereum architecture.

Mehmet Demir [8] used case for automobile insurance has been presented for managing healthcare information and combating phony proof of payment and intelligent contract mechanisation. This report outlines how players may benefit from transparency in this circumstance and how blockchain technology can help with accountability. Finally, how cryptographic protocols might be utilised to communicate between untrustworthy people is demonstrated and explained.

Celia Jenkins et al. [9] reported that India's Ministry of Finance reportedly stated that the government was trying to "examine the application of blockchain technology proactively." The Karnataka Government has also inked a Memorandum of Understanding (MoU) with numerous technological businesses to establish a network from unauthorised access in the state. It has recommended moving all of the State's official data connected to ownership rights and car ownership records onto a blockchain-based system to create a unified register accessible only to authorised people. About 1 million land documents are currently saved on cryptocurrency in the state.

Sumaiya [10] discussed the lack of research on availability and performance issues, a lack of investigation on launching decentralised applications on different blockchains other than Ethereum, a limited percentage of intelligent system business. There is a lack of inquiry on criminal enterprises in payment systems and a lack of significant investigation on payment systems.

10.3 Digital Currency-based Insurance Industry

With the right corporate use reasons and the right degree of cooperation from companies and government, India can compete with other leading nations like the United States and China. New business use cases, accommodating strategy for dealing with digital money, and a favourable environment to connect these networks are critical requirements for implementing blockchain. India has begun exploring cryptocurrency through several new initiatives, although it is still early [14].

Most research proposes a technique that focuses on architectural or model blockchain-powered; however, there is a shortage of technical information regarding the blockchain features employed [15].

We may create applications that address binary requirements using current cryptocurrency frameworks to develop blockchain for the insurance sector. In the firms collaborating on blockchain applications for assurance, there is a shortage of particular expertise about the insurance sector's quirks, rules, and other aspects.

In India, the financial, healthcare, and card industries have formed a consortium to capitalise on the applications of blockchain technology at the industry level [16].

While blockchain has many positives, it also has certain limitations, such as performance difficulties like sustainability, capacity bottleneck, transaction latency, and so on, on which academics must work. Another significant disadvantage is that, because blockchain is unchangeable, if we discover faults in smart contracts and need to change or update them,

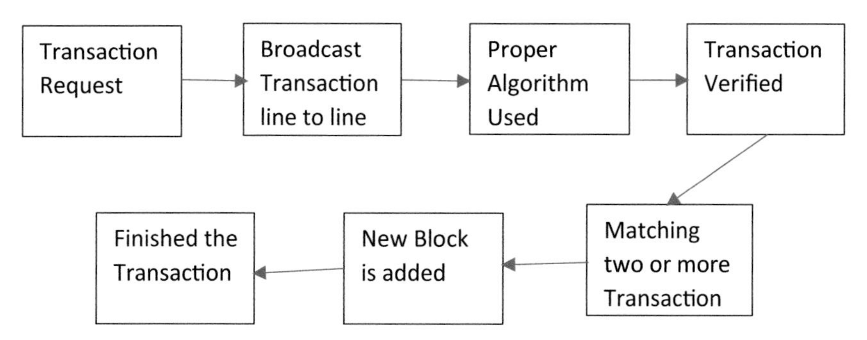

Figure 10.1 Structure of blockchain.

data saved in earlier intelligent agreements will not be immediately moved. Blockchain initiatives are still in the works. There is still a lack of industry benchmarks and a framework tailored to the sector. Despite widespread participation in consortia, B3i standards appear to be ways off, particularly in India [17].

10.4 Theories of Blockchain

Blockchain is an innovation in the computer science department with many applications. It is an open, decentralised cryptocurrency that efficiently, safely, and verifiably records interactions among both parties [18]. It is a growing collection of documents known as blocks connected and safeguarded using encryption. Each node in the network contains information about the car's state and transaction details. These types of information do not have to come from the same automobile owner. Because these blocks are built from a pool of previous transactions and vehicle status data, the information provided in each one might be tied to any of the cars [19]. It is encrypted with the car owner's shared secret key to identify data from different automobiles. We can ensure that the information corresponds to the specified individual because only the automobile owner has accessibility to his encryption key, as shown in Figure 10.1.

10.4.1 Intelligent contract

An intelligent contract is a two-party agreement maintained and recorded by the internet. It is run after all of the blocks in the sequence have been read [20]. Because the blocks are inherently unbreakable, the transaction's correctness can be ensured. Figure 10.2 depicts the operation.

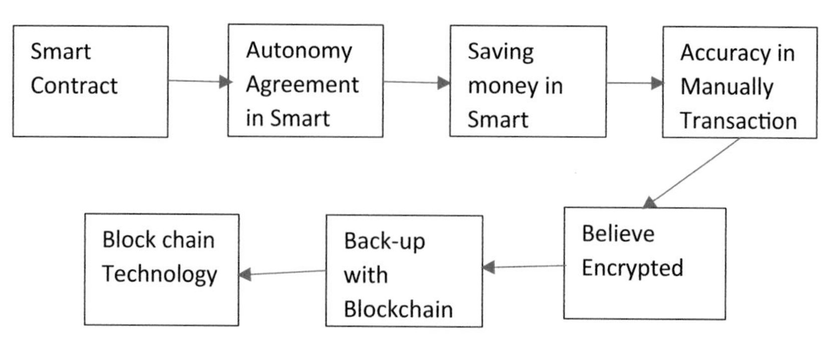

Figure 10.2 Structure of intelligent contract.

10.4.2 Issuing insurance license through IRDAI

To function, each insurance agency must get an IRDAI license. The IRDAI's issuance of this license and the contract terms to insurance firms are recorded and kept on the blockchain technology as a transaction. This indicates that the insurance market is permitted to provide auto insurance in India, and it is made available to all clients. The IRDAI engages in an intelligent contract with insurance firms for automatic license revocation or re-licensing at the end of the term [21].

Insurance firms sell automobile insurance to clients. The numerous insurance terms and conditions, such as insurance payment, damages compensation, claim policy, and so on, are kept and documented as transactions on the database. Clients and insurance providers enter into a permissioned blockchain for automated payments of healthcare premiums by the consumer and automatic payment of compensation by the insurance provider [22].

10.4.3 Insurance premiums paid by customers

Because the clients engaged in a blockchain network with the insurance industry, the higher insurance amount is usually charged every year (or at a more petite time frame), as all transactions on the database are continuously enforced. The number of healthcare defaulters is substantially decreasing. An intelligent contract also minimises the time to verify claims and speeds up clearing and settlement. Figure 10.3 depicts its application [23].

The usage of blockchain technology accelerates underwriters and claims to process. Because insurances, contractual relationships, and claims are all stored on the blockchain, a consensus mechanism may dynamically pick the application to be approved [24]. The related terms include costing, paying the premium, or settlement claims based on the existence of the circumstance.

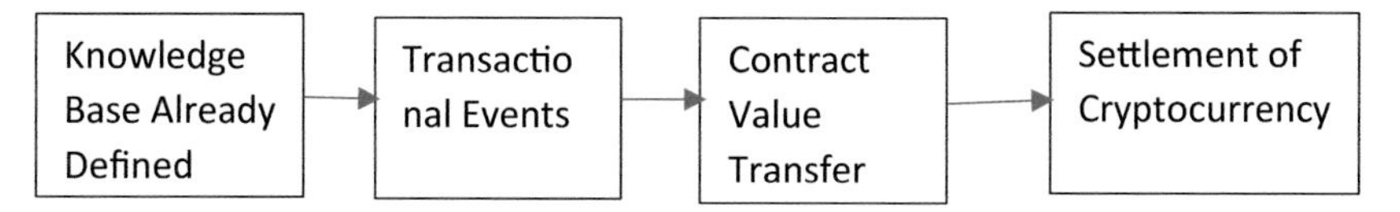

Figure 10.3 Smart contracts for vehicle insurance.

Set out in the arrangement and detect fraud. As a result, if the vehicle's damages can be detected using telemetry and added to the blockchain, an innovative insurance company may be issued using the smart contract. It saves the consumer time and effort in filing the claim. Because the data entered is genuine, the insurance claim requires no confirmation [25].

The settlement of claims is likewise stored in the blockchain as a transaction. This knowledge will be used in the future to calculate the annual premium in Figure 10.3.

10.4.4 Unique insurance number

Each insurance program is assigned a unique insurance number. A QR code is generated using this guarantee of compensation number and the cryptographic certificates of the insurance firm for the insurance policy. The QR code is distinctive to each insurance contract and legitimate since it bears the signature of the insuring business. As a result, the traditional financial service papers are being replaced with a simple, unique code [26].

10.4.5 Blockchain use in IoT and telematics for insurance policy

Vehicle data may be collected in real time, thanks to improvements in IoT and telematics. The acquired vehicle key is protected and saved on the blockchain regularly using the shared secret key decided upon in the insurance contract. The insurance agency encrypts and analyses the customer's vehicle data to determine driving habits. This is used to calculate the annual premium in PHYD (described later) [27].

Because the automobile data is stored on the database, no one can tamper with changing the annual premium. As a result, premium computation is precise and dependable. Furthermore, vehicle data storage in the database can compel claims resolution via a consensus mechanism. An android application may make the driving patterns and other metrics accessible to the associated consumer. Insurance carriers can give incentives to professional drivers to guarantee health and accident-free driving. This is referred to as MHYD, and it will be discussed further below [28].

10.4.6 Advantages of insurance system

This technique has several benefits and can close gaps in the present insurance system. The following are the benefits of employing this technology:

- No single point of failure
- A centrally controlled system usually accompanies with no singular entity influencing the blockchain
- No fraud insurance providers in the system
- Eradication of annual premium late payments due to the use of digital currencies
- Mechanisation of annual premium payments by vehicle owners
- Smart hypotheses instantly initiate the recuperation process, and to have, zero gently renewal

10.5 What Is Pay as You Drive?

Pay as You Drive (PAYD) is a common utilisation-based healthcare type. The insurance cost is calculated by the number of kilometres driven in the covered automobile during the protected duration. Individuals who go less are eligible for a lower insurance price. The insurance duration can be tailored to the requirements of the customers [29].

Pay How You Drive (PHYD) is a type of utilisation based health coverage that is quickly gaining traction in the market due to its several advantages. Instead of basing healthcare premiums on the brand and design of a vehicle, the driver's age, the driver's employment, and so on, PHYD bases prices on how the car is handled. This type of evaluation is more suitable because driving behaviour is a significant predictor of the user's propensity to submit a claim. A reckless driver, for example, is more likely to become involved in any accidents and, as a result, to make a claim [30].

10.5.1 Innovative technology used in blockchain

We have previously recommended using cryptocurrency to store vehicle data. Various vehicular data may be obtained from each car using different sensors such as milometers, inertial measurement units, accelerometer sensor monitors, generator excitation devices, GPS, torque sensors, engine and E-motor sensor, brake and clutch sensor, and so on. As a result, the car may be monitored in real time. The acquired data is added to the blockchain and is used by insurance firms to create models that are subsequently used to predict the user's premiums.

10.5.2 Insurance parametric format

Currently, the insurance coverage and premiums for a given car are determined primarily using the different indicators:

- IDV: This is the vehicle's insurance coverage amount. The coverage is determined based on the vehicle's value after devaluation.
- Age: As a person ages, he gets more conscientious and has less propensity to drive recklessly. As a result, a premium discount is available.
- Occupation: People in specific occupations are eligible for premium discounts. Medical physicians, finance professionals, and other professionals are among them.
- Claim history: If the motorist makes no claims over the insurance years, they may be eligible for a no-claim bonus.

In addition to these characteristics, the PHYD scheme includes the following:

- Average speed and explosion: The vehicle's acceleration and kinetic energy play a role in determining the insurance premium. A car driven at high speeds and accelerates rapidly is more likely to be dangerous to the vehicle.
- Braking and handling: If the driver often stops hard and turns at high rates of speed, the automobile wears out, and the operator may ultimately damage the car.
- Travel proximity: This is a typical factor in calculating premiums. A vehicle that travels vast distances is more vulnerable to degradation.
- Time of day and weekday: The car is more likely to be destroyed at night in the dark.
- Impact on the climate: If the car is running in rain, snowfall, or other inclement weather, the operator is more likely to damaging it.
- Manoeuvre type: Aggressive manoeuvres such as quick passes, approaching on a curve, breaking speed restrictions, and so on all enhance the likelihood of damage to vehicles.

- Good driving practices: Using indications, horns, and other warning devices is an excellent road behaviour, and its successful usage implies that the automobile is less likely to be damaged.
- Region: The province (district/city/town) where the automobile is most often used – rural regions or expressways. Vehicles utilised chiefly in the rural areas are less likely to make complaints.
- Service date: The length of time since the vehicle's previous business date is also considered. A car that has just been maintained is in a reasonable shape than one that has not been held in a long time. In addition, a vehicle that has been repaired several times in a short period may be driven by a reckless driver, who is more inclined to make an insurance claim.

The following criteria are analysed in real-time data obtained from the vehicle. Driving patterns are created to indicate whether the user is a safe or reckless driver. Each characteristic is assigned a percentage that shows its value or relevance in calculating the insurance premium. The basic based on the collection is the insured declared value (IDV). The relevant additions or deductions to the IDV are made based on the travel characteristics to compute the final premium amount [31].

Estimation techniques may be used to extrapolate the user's driving behaviours and forecast how the user will go in the perspective. This may be used to provide long-term coverage for secure and comfortable operators.

Customers can use a mobile phone app to retrieve the data observed, linear regression, and driving tendencies. This is to offer the user feedback on their driving directions. This gives birth to the phrase "manage how you drive" (MHYD). The software may also deliver user-specific ideas and guidance and support safe driving behaviours. By following the advice, the motorist can enhance driving [32].

10.5.3 Incentives provided for customers

Customers can be given the following incentives:

- Cashback: Offering monetary rewards to policyholders who meet specific predefined goals or milestones.
- Premium discount: If specified goals or milestones are met, policyholders will charge a portion of their agreed-upon competitive rate.
- Value-added benefits: Providing merchants discounts, automobile service notifications, and other supplementary services related to the policyholder's Pay How You Drive coverage.

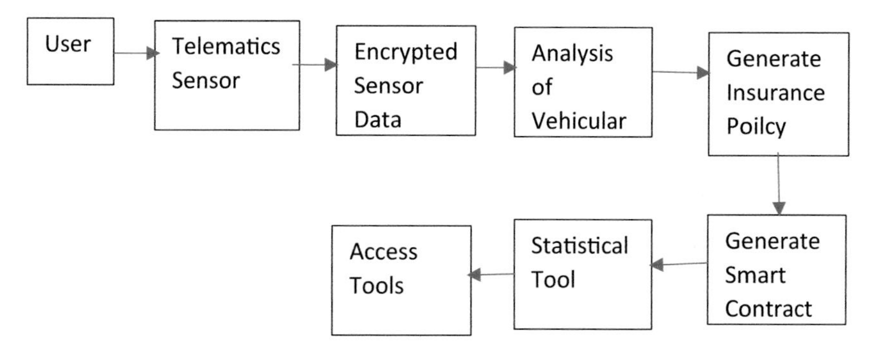

Figure 10.4 Structure of Pay How You Drive.

The abovementioned aims or milestones might be everything agreed upon by the driver and the firm when the insurance is issued. For example, if a client continuously follows excellent habits and only drives in amicable weather circumstances, it may be eligible for rewards points or a 10% discount on its premium amount [33].

The Pay How You Drive healthcare scheme's construction is depicted in Figure 10.4.

10.5.4 Benefits

The advantages of Pay How You Drive fall into three categories:

- Consumers can benefit: Capacity is the ability to decrease the premium by driving safely:
    - accountability in premium calculating owing to the usage of blockchain;
    - access to a variety of awards and significance programs based on driving habits;
    - companies' response times to accidents are slow.
- Advantages for the insurance carrier:
    - fight spurious claims;
    - improve pricing reliability;
    - identify, keep, and acquire lucrative client accounts;
    - reduce claim expenses;
    - allow customers to pay reduced premiums, enticing more consumers.

- Societal benefits:
  - reduce the frequency and severity of accidents;
  - minimise traffic on the roads;
  - promote safe driving behaviours.

10.5.5 Vehicle insurance facility

The inclusion of automotive data in the cryptocurrency also opens up the possibility of a dependable used-vehicle marketplace. Customers who want to sell their used automobiles can discuss the vehicle quality statistics that insurance firms perceive and analyse with other consumers. Customers looking to purchase used automobiles may utilise the vehicle's location to narrow down their options and haggle the price. The company car situation is genuine since it is kept on the blockchain and is continually monitored by the insurance carrier. Digital currencies may also be used to make money by selling old autos, making them legitimate, secured, and dependable. There is little need for a third-party distributor because the contracting parties may contact directly, as on the cryptocurrency [34].

10.5.6 Peer-to-peer insurance

Peer-to-peer assurance is an insurance plan in which a circle of friends, co-workers, or family and friends combine their funds to insure their automobiles. Insurance rates are cheaper than conventional insurance, and any remaining funds in the insurance programme are distributed to members at the end of the active subscription period. A mentoring reinsurance contract was formed and enforced using blockchain. Digital currencies may be used to construct insurance premium pools and resolve claims. As a result, a dependable and secure peer-to-peer insurance system may be established. The benefit of utilising blockchain is that this approach may be implemented safely and reliably, including between acquaintances.

10.6 Conclusion

Blockchain is a secure, trusted, and distributed database ledger. The use of blockchain in the Indian insurance sector will make dramatic changes. Blockchain will eliminate fraud corruption in the insurance sector. Smart contracts written on top of the blockchain will automatically make claim processing in a few seconds; the insurance client will not bother by keeping all documents and providing them at the time of claim processing.

This chapter explored blockchain in the insurance sector in the Indian scenario. Blockchain is a decentralised information ledger that is safe and reliable. The use of cryptocurrencies in the Indian insurance market will result in significant changes. Cryptocurrency will reduce fraud and corruption in the insurance industry. Intelligent contracts created on top of the blockchain will automate claim processing in a matter of seconds, eliminating the need for insurance clients to maintain all documents and provide them at the time of a systematic offer.

References

[1] Ankitha Shetty, Adithya D. Shetty, Rashmi Yogesh Pai, Rohini R. Rao, Rakshith Bhandary, Jyothi Shetty, Santosh Nayak, Tantri Keerthi Dinesh, and Komal Jenifer Dsouza1, 'Block Chain Application in Insurance Services: A Systematic Review of the Evidence' SAGE Open, pp. 1–15 DOI: 10.1177/21582440221079877, 2020

[2] Amith Kumar V, Abhiram Prasad, Rajashekara Murthy S., 'Application of blockchain in Usage Based Insurance', International Journal of Advance Research, Ideas and Innovations in Technology, Volume 5, Issue 2, pp.1574–1577, 2019

[3] Anokye Acheampong Amponsah Benjamin Asubam Weyori, Adebayo Felix Adekoya, 'Application of blockchain in Usage Based Insurance', International Journal of Advance Research, Ideas and Innovations in Technology, Volume 5, Issue 2, pp.445–466, 2019.

[4] M. Raikwar, S. Mazumdar, S. Ruj, S. Sen Gupta, A. Chattopadhyay and K. Lam, 'A Blockchain Framework for Insurance Processes,' 2018 9th IFIP International Conference on New Technologies, Mobility and Security (NTMS), Paris, pp. 1–4, doi:10.1109/NTMS.2018.8328731, 2018.

[5] N. Dhieb, H. Ghazzai, H. Besbes and Y. Massoud, 'A Secure AIDriven Architecture for Automated Insurance Systems: Fraud Detection and Risk Measurement', in IEEE Access, vol. 8, pp. 58546–58558, doi: 10.1109/ACCESS.2020.2983300, 2020.

[6] Lamberti, F., Gatteschi, V., Demartini, C., Pranteda, C., Santamaria, V. 'Blockchain or not blockchain, that is the question of the insurance and other sectors. IT Prof. [CrossRef], 2017.

[7] Z. Li, Z. Xiao, Q. Xu, E. Sotthiwat, R. S. Mong Goh and X. Liang, 'Blockchain and IoT Data Analytics for Fine-Grained Transportation

Insurance', IEEE 24th International Conference on Parallel and Distributed Systems (ICPADS), Singapore, Singapore, 2018, pp. 1022–1027, doi: 10.1109/PADSW.2018.8644599, 2018.

[8] M. Demir, O. Turetken and A. Ferworn, 'Blockchain Based Transparent Vehicle Insurance Management', Sixth International Conference on Software Defined Systems (SDS), pp. 213–220, DOI:10.1109/SDS.2019.8768669, 2019.

[9] Celia Jenkins, Anuj Bahukhandi and Swathi Ramakrishnan 'India: Blockchain Technology in the Indian Insurance Sector', 2018.

[10] Sumaiya and A. K. Bharti, 'A Study of Emerging Areas in Adoption of Blockchain Technology and it's Prospective Challenges in India', Women Institute of Technology Conference on Electrical and Computer Engineering (WITCON ECE), Dehradun Uttarakhand, India, pp. 146–153, doi: 10.1109/WITCONECE48374.2019.9092935, 2019.

[11] Zhang, X., 'Design and Implementation of Medical Insurance System Based on Blockchain Smart Contract Technology', Master's Thesis, Huazhong University of Science & Technology, Wuhan, China, May 2019.

[12] Esposito, C., De Santis, A., Tortora, G., Chang, H., Choo, K.K.R, 'Blockchain: A panacea for healthcare cloud-based data security and privacy?' IEEE Cloud Comput., 5, 31–37, 2018 [CrossRef]

[13] Novo, 'O. Blockchain meets IoT: An architecture for scalable access management in IoT', IEEE Internet Things J., 5, 1184–1195, 2018 [CrossRef]

[14] Wang, J., Li, M., He, Y., Li, H., Xiao, K., Wang, C., 'A blockchain based privacy-preserving incentive mechanism in crowdsensing applications.' IEEE Access, 6, 17545–17556, 2018. [CrossRef]

[15] Dorri, A., Steger, M., Kanhere, S.S., Jurdak, R., 'Blockchain: A distributed solution to automotive security and privacy', IEEE Commun. Mag., 55, 119–125, 2017 [CrossRef]

[16] Xia, Q., Sifah, E., Smahi, A., Amofa, S., Zhang, X., BBDS: Blockchain-Based data sharing for electronic medical records in cloud environments. Information, 8, 44, 2017 [CrossRef]

[17] Xu, J., Xue, K., Li, S., Tian, H., Hong, J., Hong, P., Yu, N. 'Healthchain: A blockchain-based privacy preserving scheme for large-scale health data', IEEE Internet Things J, 6, 8770–8781, 2019 [CrossRef]

[18] Liu, X., Wang, Z., Jin, C., Li, F., Li, G., 'A Blockchain-based medical data sharing and protection scheme', IEEE Access, 7, 118943–118953, 2019 [CrossRef]

[19] Chen, Z., Xu, W., Wang, B., Yu, H., 'A blockchain-based preserving and sharing system for medical data privacy', Future Gener. Comput. Syst., 124, 338–350, 2021 [CrossRef]

[20] Kuldeep Singh Kaswan, Jagjit Singh Dhatterwal, Santar Pal Singh, 'Blockchain Technology for Health Care" book entitled Healthcare and Knowledge Management for Society 5.0: Trends, Issues, and Innovations" Published in CRC Press, ISBN 9781003168638, 2021.

[21] Chiuchisan, I., Dimian, M., 'Internet of Things for e-Health: An approach to medical application', In Proceedings of the IEEE International Workshop on Computational Intelligence for Multimedia Understanding (IWCIM), Prague, Czech Republic, 29–30; pp. 1–5, 2015.

[22] Moosavi, S.R., Gia, T.N., Nigussie, E., Rahmani, A.M., Virtanen, S., Tenhunen, H., Isoaho, J. 'End-to-end security scheme for mobility enabled healthcare Internet of Things' Future Gener. Comput. Syst., 64, 108–124, 2016 [CrossRef]

[23] Azeez, N.A., Vyver, C.V.D. 'Security and privacy issues in e-health cloud-based system: A comprehensive content analysis', Egypt. Inform. J., 20, 97–108. 2019 [CrossRef]

[24] Li, C.T., Shih, D.H., Wang, C.C. 'Cloud-assisted mutual authentication and privacy preservation protocol for telecare medical information systems', Comput. Methods Programs Biomed., 157, 191–203, 2018. [CrossRef]

[25] Iribarren, S.J., Brown, W., Giguere, R., Stone, P., Schnall, R., Staggers, N., Carballo-Diéguez, A., 'Scoping review and evaluation of SMS/text messaging platforms for mHealth projects or clinical interventions', Int. J. Med. Inform., 101, 28–40, 2017. [CrossRef]

[26] Khemissa, H., Tandjaoui, D. 'A lightweight authentication scheme for e-health applications in the context of Internet of Things', In Proceedings of the 9th International Conference on Next Generation Mobile Applications, Services and Technologies, Paris, France, 29–30; pp. 90–95, 2015.

[27] Yang, Y., Ma, M. 'Conjunctive keyword search with designated tester and timing enabled proxy re-encryption function for e-health clouds', IEEE Trans. Inf. Forensics Secur., 11, 746–759., 2016 [CrossRef]

[28] Preety, Jagjit Singh Dhatterwal, Kuldeep Singh Kaswan, 'Securing Big Data Using Big Data Mining' in the book entitled "Data Driven Decision Making using Analytics" Published in Taylor Francis, CRC Press, ISBN No. 9781003199403, 2021.

[29] Chen, C.C., Deng, Y.Y., Weng, W., Sun, H., Zhou, M., 'A Blockchain-Based Secure Inter-Hospital EMR Sharing System', Appl. Sci., 10, 4958, 2020. [CrossRef]

[30] Chen, C.C., Huang, P.T., Deng, Y.Y., Chen, H.C., Wang, Y.C. 'A Secure Electronic Medical Record Authorization System for Smart device application in cloud computing environments', Hum. Centr. Comput. Inf. Sci., 10, 21, 2020 [CrossRef]

[31] Roy, S., Das, A.K., Chatterjee, S., Kumar, N., Chattopadhyay, S., Rodrigues, J.J. 'Provably secure fine-grained data access control over multiple cloud servers in mobile cloud computing-based healthcare applications', IEEE Trans. Ind. Inform., 15, 457–468, 2018. [CrossRef]

[32] Kuldeep Singh Kaswan, Jagjit Singh Dhatterwal, Nitin Kumar Gaur', Blockchain of IoT Based Earthquake Alarming System in Smart Cities', book entitled "Integration and Implementation of the Internet of Things Through Cloud Computing", Published in IGI Global, ISBN 13: 9781799869818, ISBN10: 1799869814, EISBN14: 9781799869832, 2021.

[33] Suresh Kumar, V., Amin, R., Vijaykumar, V.R., Sekar, S.R. 'Robust secure communication protocol for smart healthcare system with FPGA implementation', Future Gener. Comput. Syst., 100, 938–951, 2019. [CrossRef]

[34] Kuldeep Singh Kaswan, Jagjit Singh Dhatterwal, Nitin Kumar Gaur, 'Smart Grid Using IoT', book entitled "Integration and Implementation of the Internet of Things Through Cloud Computing", Published in IGI Global, ISBN 13: 9781799869818, ISBN10: 1799869814, EISBN14: 9781799869832, 2021.

[35] Sood, K., Seth, N., & Grima, S. (2022). Portfolio Performance of Public Sector General Insurance Companies in India: A Comparative Analysis. In Simon Grima, Ercan Özen, Inna Romānova (Ed.), In *Managing Risk and Decision Making in Times of Economic Distress, Part B*. (pp. 215–229). Emerald Publishing Limited,UK.

[36] Grima, S., & Marano, P. (2021). Designing a Model for Testing the Effectiveness of a Regulation: The Case of DORA for Insurance Undertakings. *Risks*, *9*(11), 206.

11

Blockchain Application with Specific Reference to Smart Contracts in the Insurance Sector

Vandana Sharma[1], Prerna Ajmani[2], and Celestine Iwendi[3]

[1]CHRIST (Deemed to be University), Delhi-NCR, India
[2]Vivekananda Institute of Professional Studies-TC, GGSIP University, India
[3]School of Creative Technologies, University of Bolton, UK

Email: Vandana.juyal@gmail.com; prerna.ajmani@vips.edu; celestine.iwendi@ieee.org

Abstract

The term blockchain was coined in 2008 by Satoshi Nakamoto. Initially, it was used for carrying out decentralised transactions to solve the problem of fake transactions. In the past few years, this was explored extensively for cryptocurrency only, but, over some time, its potential has been explored in many areas. The major reason for the growing interest in this particular technology is that it provides a secure, reliable, and trusted platform to perform digital activities. This is executed without the involvement of any third party. Once the data is entered into the nodes, it is impossible to tamper it. Though blockchain is costly, it provides better solutions to many research problems in real time. In recent times, researchers have explored blockchain in deep and used it in many applications such as building smart contracts, supply chain management, digital identity providers, voting systems, banking, and finance applications, P2P learning, and insurance sectors. Through this chapter, the readers will get a systematic and detailed study of blockchain in the insurance sector and smart contracts and its current applications in the insurance sector. This chapter will also provide a fair idea of blockchain technology in the insurance sector and additionally its usage in specific applications. In the end, a relevant set of further reading references will be provided.

11.1 Introduction

In the modern digital world, technological advancements are rapid and evolving to provide prime solutions to the most complex real-time problems. In 2008, an anonymous researcher passionate about technology to solve research problems developed a revolutionary technology named blockchain [32, 33]. It came into existence to provide a feasible solution to the Bitcoin industry. This further facilitated the decentralised peer-to-peer currency transfer and mitigated the issue of the double spending problem [32, 34]. Within a few years, blockchain technology has ratcheted up and drawn the attention of industrialists, academicians, and researchers [35]. In 2018, blockchain was listed among the top five newest technology trends [36, 37]. Blockchain evolved over some time and is classified into three generations. In 2009, Blockchain 1.0 technology came with only one purpose, i.e., to promote digital currency named Bitcoin [32, 38]. As blockchain technology came into practice, its potential has been realised and in 2014, the subsequent generation – Blockchain 2.0 – evolved and laid emphasis on innovatively applying smart contracts in numerous applications, promoted by Ethereum [39]. In 2017, many Hyperledger businesses led to the evolution of Blockchain 3.0 resulting in the construction of a decentralised application system serving innumerable applications. In recent times, the blockchain horizon has broadened serving applications in academics, medical systems, farming, Internet of Things (IoT), and even in governance [40, 41]. Some other implementations of blockchain include electronic voting, autonomous vehicle system, trading, supply chain, smart grid, networking, and many more [42–45]. The authors in [46, 47] explained that emergency insurance is imperative and helps people counterpoise their financial loss. The biggest hurdle in insurance is to detect counterfeit claims. After careful thought to find an effective solution for the above problem, many big insurance companies realised the potential of blockchain technology to overcome the challenges [48, 85, 86]. Blockchain Insurance Industry Initiative (B3i) advertises the use of blockchain in the insurance industry to refine transactions, identify false claims, and make the entire system decentralised, automatic, and transparent [49].

11.2 Applications of Blockchain

11.2.1 Blockchain-based smart contracts

Blockchain has introduced the concept of smart contracts. The smart contract is an executable piece of code that runs automatically on meeting a certain precondition. All the parties involved are certain of the contract outcome

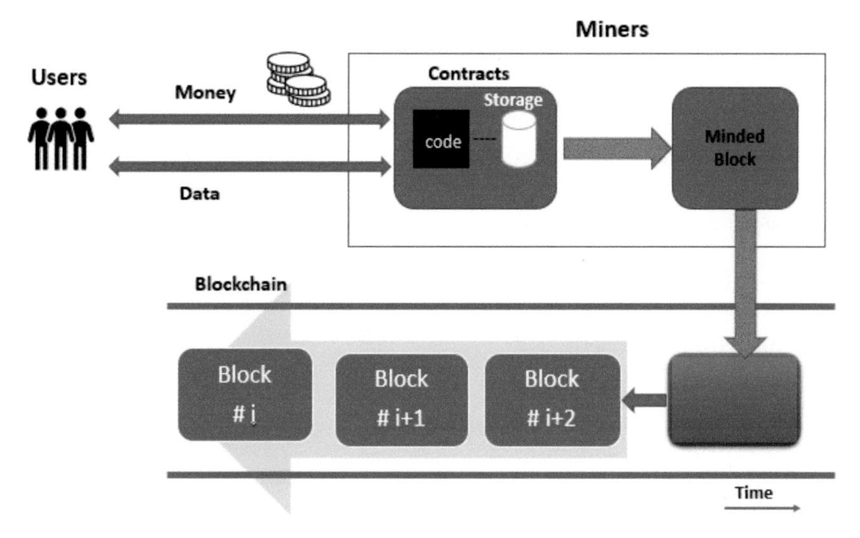

Figure 11.1 Smart contract system [2].

without any delays involved. Smart contracts are recorded on distributed set-ups and released to all the stakeholders. The authenticity of the transaction remains on the distributed servers and any change to the transaction is not possible [1–3]. Figure 11.1 is illustrating the concept of the smart contract system.

11.2.2.1 Phases for setting up smart contracts

- Identification: The first and most crucial phase of a smart contract is identification. In this step, all the parties are made aware of the scope, guidelines, associated rights, and other business legalities related to the contract.
- Set-up conditions: For a smart contract to be executed, some predefined condition is to be met. In this phase, the conditions that will trigger the smart contract must be identified. These triggers could be specific or general conditions. For example, if the insurance company sets up the condition of giving a discount to the insurer near his/her birthday if the insurance smart contract is initiated near this date, then the smart contract would automatically initiate a discount on buying the insurance policy.
- Coding: In this phase, the scripting of the smart contracts is done. A code based on predefined conditions is written and checked thoroughly before putting on the blockchain technology.

- Selection of blockchain technology and encryption: Blockchain technology to be used such as Ethereum, NXT, Bitcoin, etc., is selected in this step, and encryption is done accordingly.
- Execution: This step necessitates the smart contract transactions upon the validity, verification, and authenticity of each transaction. Predefined conditions are also checked thoroughly.
- Update the network: On the successful execution of the transaction, each node is updated on a different state. This is the last stage where the transaction is closed.

11.2.2.2 Benefits of smart contracts

- Trust and transparency: It ensures trust and transparency as there is no involvement of any third party. When putting a smart contract into practice, every piece of information is transparent, and once this contract is set up, it cannot be changed. Therefore, it ensures a high level of trust among the contract partners.
- Digital and fast: Smart contracts are digital and automatic and blockchain speeds up the process of creating a smart contract efficiently with accuracy. Once the contract's predetermined condition is met, it will be created automatically without any delays and human interference.
- Tamper free and secure: The contracts are secure. This is because once the smart contract enters the blockchain, it cannot be changed further. Under all circumstances, the false mutation will lead to another set of blocks that will reflect the change. This non-mutating feature of blockchain provides the highest level of security to financial transactions. Smart contracts provide means to execute secure contracts between non-trusting parties [4].
- Cost effective: Smart contracts eliminated the need for a conciliator, thus in turn saving the time and cost incurred in paying the fees of any mediator.

11.2.2.3 Demerits of smart contracts

In real-time applications, smart contracts have their demerits. The technological framework often separates and disassociates from the legal framework [5].

- Immutable: Once created, smart contracts cannot be changed; thus, there is hardly any scope for editing the contract if required. In practical scenarios, the conditions keep on changing from time to time, but a smart contract is inflexible in incorporating any changes. The unalterable condition of smart contracts after creation leads to a large number of practical issues. One of the problems faced by smart contracts is non-adaptiveness. The modifications in the contract are difficult to undergo. This is completely unlike the real paper contracts that are executed. Also, the security aspect cannot be ignored [5–7].

- Contractual privacy: Although the basic idea of the smart contract is to keep the ledgers in the public domain, there is a need to develop a protocol that can aid the transaction and verify it without using its content [7]. There is a need to keep all the blocks in the blockchain network anonymous. Also, the contact ledgers are kept open to be visible to all the other nodes in the network. It is a challenging issue to keep the private ledgers of transactions in public domains. In short, only the metadata of the contract shall be visible in the public domain [8].
- Legal adjudications and enforceability: Legal elements are imperative and still smart contracts are not in sync with contract laws. The key concept of any contract is an agreement between parties that legally binds them to the said conditions [9]. For example, in the finance sector, government put some regulations in the form of a license while creating the contract, but this is missing in smart contracts. Since smart contracts are generated by executing the code, it is high time that the legal parameters needed in contracts must be translated into code and get self-executed at the time of the creation of smart contracts. Presently, blockchain is being used to generate smart contracts, but, still, the validity and enforceability of the contract are in hassle. All smart contracts are condition driven and none of the nodes are capable of executing and adding a new block to the blockchain trail [10].

11.2.2.4 Platforms for smart contracts

Blockchain offers various platforms in which smart contracts can be created and implemented. Some platforms are public and open, whereas others are private. All the platforms have different features for creating smart contracts. In this section, we will focus only on some public blockchain platforms for building smart contracts.

a. NXT: It has many templates to build smart contracts but does not allow any changes or additional customisation in smart contracts due to its simple scripting language.

Figure 11.2 Asset management [11].

b. Bitcoin: It is again a public blockchain platform that was initially just used for transferring cryptocurrency. It also uses the bytecode language for scripting transactions. Bitcoin scripting language is rare to permit a change in the transactions [10]. It is useful for performing simple transactions and any complex execution is not supported by the Bitcoin scripting language. For example, enriched programming features like executing loops are not possible in the Bitcoin scripting language.

c. Ethereum: It is a very popular and public blockchain platform that can support advanced and customised smart contracts with the help of Turing-complete programming language [2]. The code is developed using stack-based bytecode language and compiled and run on Ethereum virtual machine. Languages like Solidity, LLL, and Serpent can be used to code Ethereum smart contracts. Due to its ability to provide an efficient coding platform, it is one of the popular platforms to develop smart contracts.

11.2.2 Blockchain in asset tracking and identity management

In recent times, blockchain usability has been explored beyond cryptocurrency applications. Asset tracking and identity management are two imperative and emerging applications of blockchain. Blockchain technology can significantly help to remove conflicts by keeping a fair track of assets and can manage the identity of users required for authentication and verification. This technology can speed up the process of asset tracking and identity management with high accuracy and minimal flaws.

a. **Asset tracking using blockchain**: It is the process of managing the tangible and intangible assets of a person. This also involves recording and keeping track of all assets from obtainment to giving away. Figure 11.2 illustrates a difference between tangible and intangible assets.

There are several reasons for considering asset management very important:

- Assets directly or indirectly involve a lot of investment, which must be tracked and protected. Many countries across the globe are struggling to maintain assets because of flaws in their tracking and they are being misused.
- Second, most of the services are dependent on assets to do effective delivery.

Advantages of using blockchain in asset management:

- **Portfolio management**: It is very crucial to organise portfolios for asset management. Blockchain offers an effective structure that will ease the process of managing portfolios.
- **Saves time:** Blockchain can help in accelerating the checking of the credentials of new clients with much ease and transparency, thus enabling faster access to asset management for the clients.
- **Immediate transaction clearance:** This technology can accelerate the immediate transaction consequentially reducing risk, increasing trust, and increasing money flow.
- **Smart contracts**: These play a very important role in asset management and can be used to automate tasks such as initiating sales based on some predefined conditions. The smart contracts are cost-effective and do not need the involvement of any third party.
- **Protection and security:** Blockchain makes records immutable by encrypting them. It eliminates the scope of hacking or modifying the data. Thus, this technology provides high-end security to the data. Additionally, this technology speeds up the process of asset tracking.

Figure 11.3 illustrates blockchain in asset management.

b. Identity management (IdM): Identity management is important concerning taking into account the users involved in the blockchain transaction. ACCESS management evaluates the attributes of identity and executes the access based on the decision policy based on these

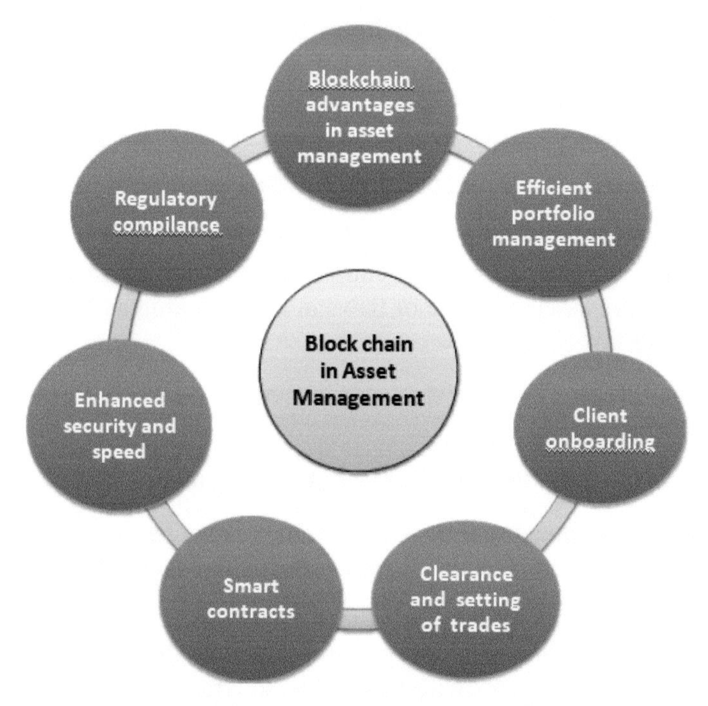

Figure 11.3 Blockchain in asset management [11].

attributes. Virtual identity is made by the user in an online space and various identifiers such as mail account, domain name, etc. are used to identify digital identities. This virtual identity information is stored on the web at some central repository, which is maintained by a third party. Thus, this model makes virtual identity information susceptible to modification or alteration. Blockchain is one of the solutions to stop this tampering of digital identities. In the information age, protection of virtual identities is crucial. Furthermore, let us first understand the deep concept of identity management and its challenges and further on how blockchain can be used to build strong fraudulent proof identity management systems.

a. Identity management system (IdMS): According to [12, 13], IdMS can be categorised into three sub-categories, namely isolated IdM, federated IdM, and centralised IdM [14].

   i. Isolated IdM: In this type of model, each client is given a distinctive identifier by the provider such that the user can access every internet service associated with identity in isolation. Figure 11.4

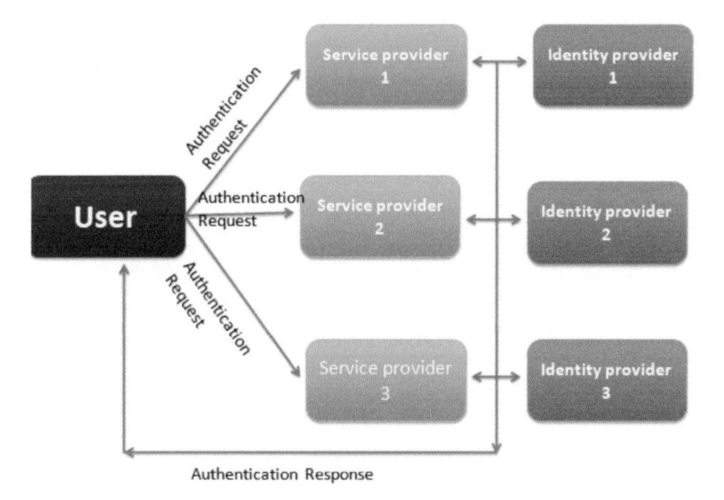

Figure 11.4 Isolated IdM [14].

demonstrates its architecture. Practically, with an increase in the number of online accesses, this type of model is not useful.

ii. Federated IdM: In this, different service providers group all the identity related services together based on certain rules. Figure 11.5 gives the architecture of a federated IdM model. Suriadi et al. [16] extended this model with user-centric approach. Camenisch et al. [15] proposed a project PRIME, which has implemented a framework for processing personal data [14].

iii. Centralised IdM: In this, there is one central identity provider. Each service provider uses the same credentials to get connected with the central identity provider. Moreover, the end user also must use all the online services using the same login information. Jøsang et al. [12] further sub-categorised this model as meta-identifier model, common identifier model, and single sign on (SSO) model. Figure 11.6 represents the architecture of centralised IdM.

b. Identity management challenges: There are several problems associated with IdM such as security, trustworthiness, privacy, affordability, integrity, anonymity, and interoperability.

   i. Security: In cyberspace, it is quite easy to copy or steal digital identities for malicious purposes. It is difficult to stave off the

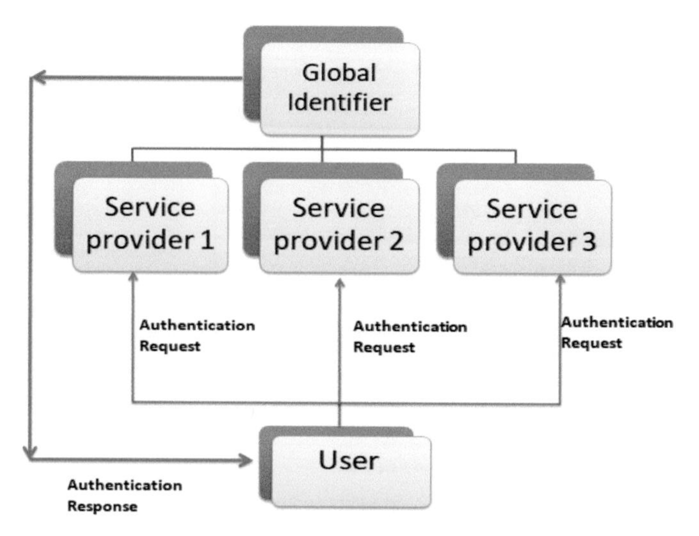

Figure 11.5 Federated IdM [14].

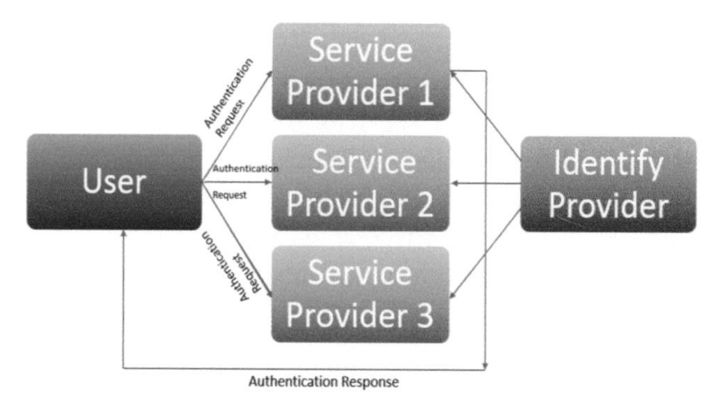

Figure 11.6 Centralised IdM [14].

hacking of this information. A single security breach or point of failure can give an easy way to hack user identity. AlpAlpárr et al. [17] have extensively researched the identity crisis due to insufficiency of security.

ii. Trustworthiness: Jøsang et al. [12] have deeply investigated on trust issues and requirement. It is a very crucial factor that must be taken care of while building the IdM system. Traditionally, a third party is responsible for ensuring trust by authenticating and ensuring the integrity of any transaction. However, each IdM model has different trust requirements.

iii. Privacy: It is one of the crucial requirements for any IdM system. The user trusts the IdM system and shares personal information with the service provider. Any breach of privacy can lead to unauthorised access to the personal data of the user. IdM also must give more authorisation to the user to access personal information and control its privacy as long as there is no compromise in security. Rossudowski et al. [18] suggested the usage of a single smart card for various applications [14]. Amo et al. [19] designed and applied a plugin in a Moodle system, a learning management system, which can resolve confidentiality and identity management issues [14].

iv. Affordability: It is a basic condition while building an IdM model. Affordability signifies that the model must be cost-effective and can be maintained in the future as well. The model must be effective and simple so that it can be reused further in extended or other applications.

v. Integrity: To safeguard the important and sensitive attributes of identity integrity must be ensured. It also helps in finding any fraudulent acts done on identity attributes. To preserve the integrity, Diffie and Hellman proposed to apply public key cryptography to identity attributes. Bertino et al. [20] proclaim that using only the cryptography method on identity attributes does not ensure integrity because any malicious act on security can consequently result in a stolen identity.

vi. Anonymity: Anonymity must be maintained at all levels while building the IdM model. However, the user must be able to access his/her personal information. Ahn and Lam [21] proposed the concept of pseudo-anonymity for securing the identity attributes of the user. Further, the user must have the right to decide which identity attributes he/she would like to share. Leenes [22] and Pérez Méndez et al. [23] discussed different identity management alternatives in their research paper, "Future of Identity in the Information Society," which primarily focus on solutions to manage and secure identities.

vii. Interoperability: With the advancement in technology, it is an essential requirement that digital attributes given to identify a person's identity must work well even if the system or platform of the IdM model changes. Torres et al. [24] argued that though digital identity interoperability is difficult technically and legally, yet this is to be achieved. So, to make digital identities interoperable, they should be created in a standard format and must be in sync with the current protocols.

c. Blockchain technology in building IdM: Though blockchain is still naïve, yet researchers have found that it is effective to use this technology while constructing IdM. Rathee and Singh [14] have done extensive research and found that blockchain technology can satisfactorily address IdM challenges. The authors have summarised their studies in a tabular form presented below in Table 11.1.

   Blockchain ensures security and removes the need for any third party. It can be effectively used to authenticate digital identities. The privacy challenge of IdM models can be easily handled using the blockchain because it uses hash for every transaction, which is difficult to calculate or imitate. Further, its decentralised nature ensures genuine users and eliminates the chances of any fake transaction. Zyskind et al. [25] have created a blockchain-based decentralised personal data management platform named Enigma, which ensures privacy. Blockchain ensures anonymity as it is automated and does not need any third party in any process of the transaction. The authors Heilman et al. [26] in their research paper have mentioned that in blockchain, blind signatures and smart contracts can be used to ensure anonymity. Blockchain eradicates the need of establishing trust among participating users. Since this technology is decentralised and distributed, there is almost negligible scope for any fake transaction approval; thus, this technology itself is trustworthy. Several IdM issues/challenges are taken care of easily by the blockchain; yet, interoperability needs to be developed and addressed by the blockchain.

   i. Various initiatives that are using blockchain for creating an IdM system: Government organisations are taking a lot of effort and using authorised blockchain to store virtual identities. Sullivan and Burger [27] have mentioned that the Estonian government has started using blockchain since 2014 to preserve national identities. Further, the "ID2020 alliance" [28] uses multi-stakeholder partnerships to preserve digital ID [14]. Turkanovic et al. [29] have developed a blockchain-based model "EduCTX" for credit management of students. Table 11.2 proposed by

Table 11.1 Blockchain analysis based on IdM challenges [14].

Publication with year	Security preserved	User privacy	Anonymity concern	Cost effectiveness	Data integrity	Trust factor	Portability
(Ma et al. 2018)	Yes	Yes	No	Highly effective	Yes	No	No
(Sullivan and Burger 2017)	Yes	Yes	No	Highly effective	Yes	Yes	No
(Hussein et al. 2018)	Yes	Yes	Yes	Highly effective	Yes	Yes	No
(Qin et al. 2017)	Yes	No	No	Highly effective	Yes	Yes	Yes
(Zhang et al. 2018)	Yes	Yes	No	Highly effective	Yes	Yes	Yes
(Azouvi et al. 2017)	Yes	No	Yes	Highly effective	Yes	Yes	No
(Augot et al. 2017b)	Yes	No	No	Highly effective	Yes	Yes	No
(Sarangal et al. 2019)	Yes	No	No	Highly effective	Yes	Yes	No
(Sarier, 2018)	Yes	Yes	Yes	Less effective	Yes	Yes	No
(Shetty et al. 2019)	Yes	Yes	Yes	High efficiency at low cost	Yes	Yes	No
(Elsa et al. 2018)	Yes	No	Yes	Highly effective	Yes	Yes	Yes
(Loukil et al. 2018)	Does not discuss security	Yes	Does not pertain to this topic	No information available	Yes	No information available	No information available
(Buccafurri et al. 2018)	Yes	Yes	No	Highly effective	Yes	Yes	No information available
(Duaddah et al. 2016)	Yes	Yes	Yes	No information available	Yes	Yes	No
(Liu et al. 2019)	Yes	No	Yes	No information available	Yes	Yes	No
(Abbasi et al. 2017)	Yes	Yes	Yes	No information available	Yes	Yes	No
(Cao et al. 2013)	Yes	No	No	Highly effective	Yes	Yes	No
(Al Bassam 2017)	Yes	No	No	Highly effective	Yes	Yes	No
(Liu 2016)	Yes	No	Yes	No information available	Yes	No	No

(Continued)

Table 11.1 *Continued*

Publication with year	Security preserved	User privacy	Anonymity concern	Cost effectiveness	Data integrity	Trust factor	Portability
(Friebe and Zitterbart 2018)	Yes	Yes	Yes	Depends on system	Yes	No	No
(Lee and Member 2018)	Yes	No	Yes	Highly effective	Yes	Yes	No
(Mikula and Jacobsen 2018)	Yes	Does not discuss user privacy	No	Highly effective	Does not relate to this	No information available	No information available
(Alansari et al. 2017)	Yes	Yes	No	Highly effective	Yes	No	No information available
(Sun et al. 2018)	Yes	Yes	Yes	Highly effective	Yes	Yes	No
(Du et al. 2018)	Yes	Yes	Does not pertain to this topic	Highly effective	Yes	No information available	No
(Kaaniche et al. 2018)	Yes	Yes	Yes	Highly effective	No	No	Yes
(Wang et al. 2017)	Yes	Yes	Does not pertain to this topic	No information available	Yes	No	No
(Augot 2017)	Yes	Yes	Yes	Fluctuates based on market forces	Yes	No	No
(Tian et al. 2019)	Yes	Yes	No	Highly effective	Yes	Yes	Yes
(Shen et al. 2020)	Yes	Yes	Yes	Effectiveness depends on computational cost	No	No	No

Table 11.2 Blockchain-based IdM initiatives [14].

Proposed work	Findings	Sector	Blockchain implementation	Network
Authentecq for ensuring the customers trust	It is a platform where we can create and verify unique identities	Proposed company	Permissioned implementation	Lapo blockchain network used
BlockAuth for decentralised authentication	An OpenID portal that will let the user take control of their own identity	Start-up proposed	Permissionless implementation	Bitcoin network used
BlockStack ("Blockstack technical white paper")	A decentralised computing network and app ecosystem that lets the user to control their identity	Company proposed	Permission less implementation	Stack blockchain based on Bitcoin network used
BlockVerity	Used to implement anti counterfeit solutions	Start-up proposed	Permissioned implementation	Bitcoin network used
Bloom Civic (Leimgruber)	A blockchain protocol for identity verification and credit scoring. A platform to protect user identities using multi-factor authentication and biometrics	Open-source start-up proposed	Public permissioned implementation	Ethereum network used
Cryptid	Eliminates counterfeits in identification by adding factors like encryption and identification	Hackathon project proposed	Permissioned implementation	Ethereum network used
Cambridge blockchain	Identity management by adding the facility of KYC and privacy requirement of the users	Start-up proposed	Permissionless implementation	Ethereum network used
CredyCo	Document verification Sans to ensure indisputable and credible statements	Company proposed	Permissioned implementation	Bitcoin network used

(Continued)

Table 11.2 *Continued*

Proposed work	Findings	Sector	Blockchain implementation	Network
ConsenSys	Building platform for government for identity management. A global decentralised identity network that promotes the use of self-sovereign identity	Open-sourced company proposed	Public permissioned implementation	Ethereum Hyperledger Indy network used
Existence ID	A platform for global identity storage providers the users to have control over their private information	Open-source platform proposed	Permissioned implementation	Bitcoin network used
10 digitals (UO Foundation)	Provides secure last and user-friendly decentralised identity management over the web	Start-up proposed	Private implementation	10-coin Blockchain network used
Jolocom	Provides self-sovereign digital identity by providing the users control over their identity	Open-source project proposed	Public implementation	Ethereum network used
My Health My Data	An EU funded project building dynamic consent interface and personal data account using smart contract	Government project proposed	Private permissioned	Hyperledger Indy network used
Netki	A platform for remote digital identity verification intended to facilitate compliance with KYC and anti-money laundering regulations	Start-up proposed	Private permissioned implementation	Hyperledger Indy

Namecoin	Improves decentralisation of internet services and identities. A decentralised naming and storing system for user identity having the user control of their identity	Open source proposed	Permissionless implementation	Bitcoin network used
PeerMountain	Provides ownership and control over user digital identities using cryptography. Additionally, it promotes secure commerce	Open source proposed	Public implementation	Blockchain network used
SelfKey	A self-sovereign identity platform providing full ownership of the user's digital identity	Nonprofit foundation proposed	Public permissioned implementation	Ethereum network used
Shocard	A platform to establish the identities between user and enterprise securely	Start-up proposed	Public permissioned implementation	Blockchain agnostic network used
Sovrin (Tobin and Reed)	A platform for self-sovereign identity management	Nonprofit organisation proposed	Public permissioned implementation	Hyperledger Indy network used
UniquiD	Provides identity as service for IoT	Open source proposed	Public implementation	Litecoin network used
uPont (Hecket et al.)	Provides platform for self-sovereign identity by adding identity and messaging protocols	Company proposed	Public/private implementation	Ethereum network used

T. Rathee and P. Singh [14] gives details about various blockchain-based IdM initiatives.

General Data Protection Regulation (GDPR), which was implemented in 2016, found that blockchain cannot ensure the privacy of the digital user. Hence, users themselves ensure their privacy, and authorised blockchains are used in the IdM system rather than the public blockchain.

ii. Consensus algorithms of blockchain used in creating IdM: Blockchain has several consensus mechanisms such as Proof of Work (POW), Proof of Stake (POS), practical byzantine fault tolerance (PBFT), and Proof of Concept (POC) [14]. However, after an extensive study done by T. Rathee and P. Singh [14], it has been found that Proof of Work (POW) is widely used while creating an IdM system. Further, this has been depicted in Table 11.3 proposed by T. Rathee and P. Singh [14].

iii. Future concerns to be taken care of while creating IdM using blockchain: Interoperability is yet to be solved while creating IdM using blockchain. Also, with the growing social media demand, digital identities can be used for various applications; therefore, scalability and cost incurred should also be taken into account while constructing an automated blockchain-based IdM system. Another major concern is that if the supposed user is a criminal, the automated system using blockchain could raise concerns. Adding to the above challenges, in reality, sometimes the identity of the user is not static and if identity reversal is needed, then unlike traditional systems, blockchain does not have any mechanism to alter the identity of the user entered into its system.

11.2.3 Specific potential implications of Blockchain: vehicle, marine, life, fire, disability, motor insurance

Mahmoud et al. [29] and Mohamed [30] stated that in unprecedented times, insurance acts as a solution to cover monetary losses in case any unwanted catastrophe occurs. The insurance sector requires to undergo a lot of processing while generating, maintaining, checking, and closing various policies and providing necessary funds to the policyholders. Thus, it takes quite a long time to process, settle, and securely pay the genuine claimant's claim.

Table 11.3 Popularity of blockchain-based consensus mechanism while creating IdM [14].

Publication with year	Proposed framework	Protocols
(Ma et al. 2018)	DRM chain is used	POW protocols used
(Sullivan and Burger 2017)	E-Residency is used	No information available
(Hussein et al. 2018)	Medical record sharing and access control system is used	MDS hash protocols used
(Qin et al. 2017)	CeCoin is used	POW protocols used
(Zhang et al. 2018)	FHIR chain is used	Not mentioned
(Azouvi et al. 2017)	Extension of frameworks is used	POW protocols used
(Augot et al. 2017b)	No proper information	Nit protocols used
(Sarangal et al. 2019)	Insurance platform is used	PRFT protocols used
(Sarier 2018)	Privacy preserving biometric identification is used	No information available
(Shetty et al. 2019)	Mobile HealthCare System is used	No information available
(Elisa et al. 2018)	Government system is used	DPOS protocols used
(Loukil et al. 2018)	End-to-end privacy preserving framework is used	POC protocols used
(Buccafurri et al. 2018)	Crowd shipping is used	NIL protocols used
(Quaddah et al. 2016)	FairAcess for IoT is used	POC protocols used
(Liu et al. 2019)	Identity authentication scheme is used	No information available
(Abbasi et al. 2017)	VeidBlock is used	POC protocols used
(Gao et al. 2018)	Blockid is used	No information available
(Rassm 2017)	PKI system is used	POW protocols used
(Liu 2016)	Smart contract is used	POW protocols used
(Friebe and Zitterbart 2018)	DecentID is used	POW protocols used
(Lee and Member 2018)	BIDAAS is used	POFT protocols used
(Mikula and Jacobsen 2018)	Hyperledger framework is used	POC protocols used
(Alansari et al. 2017)	Access control is used	No information available
(Sun et al. 2018)	DABS is used	PRFT protocols used
(Du et al. 2018)	No proper information	PRFT protocols used
(Kaaniche et al. 2018)	Consortium is used	POC protocols used
(Wang et al. 2017)	Ethereum blockchain is used	POC protocols used
(Augot 2017)	Bitcoin blockchain is used	POW protocols used
(Tian et al. 2019)	No proper information	Optimised PRFT protocols used
(Shen et al. 2020)	BASA using Corsrtises blockchain is used	No information available

Raikwar et al. [31] listed that the main challenge in the insurance sector is that a claimant has to wait for a longer period for receiving the claim. Moving ahead, blockchain has immense potential to resolve the above conflicts and can be applied in different types of insurance sectors. Many insurance companies with the help of blockchain are able to provide insurance facilities in the following sectors.

11.2.3.1 Vehicle insurance

It was introduced in the UK to safeguard the rights of the vehicles on the move [50]. In the vehicle insurance policy, the user is provided with financial support if any accident occurs while traveling [51]. Additionally, it also covers unwanted tragedies such as vehicular theft or damage due to violence or natural catastrophe. Increasing cases of accidents at an alarming rate necessitate the need to take vehicle insurance. It has been reported that in 2016, yearly road deaths were close to 1.35 million [52]. Apart from the deaths and injuries in road accidents, the economic losses due to medical and legal costs, travel delays, loss of life, and productivity are huge [53]. In the absence of vehicle insurance, compensation to injured persons in an accident is not possible, and drivers will have to pay hefty medical bills and repair bills by themselves [51]. Several nations impose the payment of insurance premiums irrespective of the vehicle used or not [54]. Considering the above scenario, vehicle insurance companies have created a discount program, usage-based insurance (UBI), for their customers [55–58]. UBI requires users to pay a premium for a vehicle insurance policy based on their usage [57, 58]. UBI needs to use the onboard devices [59] to find relevant parameters such as mileage, time, location, and trip duration needed to calculate the premium amount [56, 58, 60, 61]. UBI is also known as the Pay as You Drive and Pay How You Drive scheme [62, 63] which allows drivers to pay a lesser amount for insurance [64, 65]. Blockchain along with smart contracts allows us to process claims quickly. Blockchain requires predefined conditions/quantifiers to be satisfied for the claim to be approved and automatic payments to be made. Therefore, any rejection of a claim cannot be further challenged consequently, saving an additional amount spent on such false litigations on insurance companies [66, 67]. Additionally, blockchain in conjunction with IoT devices can capture, cache, and process real-time vehicle movement data [68], which helps collect premiums as well as makes reimbursement in usage-based insurance [54]. The author in [54] created a data analysis system for pay-as-you-go car insurance, which allows users to pay premiums when they use vehicles. According to [69, 70], blockchain with smart contracts and real-time data (collected by IoT sensors in customers' cars) can be used to automatically and effectively settle insurance claims in lesser time and cost than the traditional method [69, 70].

11.2.3.2 Marine insurance

Shipping goods through vessels and ships suffer various challenges and losses due to theft, replacement of original goods while on-the-go, natural disasters, and cross-border shootouts, consequently causing huge monetary losses to companies [71]. The above-mentioned risks entail the evolution of marine insurance to cover losses incurred. Presently, many companies, such as EY, Microsoft, ACORD, XL Catlin, Willis Towers Watson, and MS Amlin – Insurewave, are trying on a blockchain system to insure marine cargo and vessel loss. For marine insurance, blockchain must be prepared to execute millions of transactions automatically to support only a few vessels [72]. Blockchain has brought many companies in a single framework and its distributed ledger helps to effectively execute the transactions and store a permanent record for the audit. Marine insurance is bygone business insurance that is running for ages; so its processes and policies are well-defined and can be easily converted into programmable code for blockchain. Thus, marine insurance can improve transparency and can save money by eliminating the role of any third party. Further, blockchain will speed up the task of premium payments and transact money directly into the account of the policyholder. Additionally, any disputes could be settled in no time because blockchain maintains immutable records [73]. In cases of complete loss, product loss, and contamination of pharma goods or food, smart contracts can automatically trigger claims as soon as the proof has been submitted and verified.

11.2.3.3 Life insurance

It is the biggest insurance where the policyholder buys the life insurance policy and pays the premium to get a sum assured in the event of his/her life risk or death [74]. This policy provides the dependents with the insurer's required amount, which will be of immense help in case of the collapse of income. There are several benefits (such as death benefits, life risk cover, tax benefits, assured income after death, and loan benefits) of having life insurance. Life insurance policies are lawful agreements that specify the terms and conditions [79]. It may put a limit on events insured such as life insurance might not be taken in case of suicide, riots, or civil war. One of the biggest tasks is that beneficiary must submit the death certificate of the policyholder [75], which is time-consuming and may get destroyed due to natural calamity or other unwanted events. Additionally, sometimes, the beneficiary is unaware of the life insurance policies made by the policyholder, which can result in getting no claims due to missing documents [76]. It has been found that in the United States, 7.4 billion dollars of money is present because of unclaimed life insurance [77]. To mitigate the above problems, blockchain can extend its accurate, transparent, and security services to life insurance companies [78]. Smart contracts can eliminate paperwork and initiate claim money transfer into beneficiary account as soon as the condition is met and proof of work is done to validate the transaction. Further, blockchain can transparently store a copy of policies and make it accessible to the beneficiary [76]. The policy document stored in the blockchain is immutable and cannot be tampered with. Smart contracts can also automatically obtain the necessary documents such as a death certificate from the required system, and after validation, through a consensus mechanism, it can execute the required transaction.

11.2.3.4 Fire insurance

It is a high-risk activity and most of the infrastructure such as hospitals, hotels, tourist places, etc., insure themselves in case of fire. Fire insurance (FI) is property insurance that gives benefits to the policyholder in case of building damage or destruction due to fire. Fire insurance policy will give coverage in case property needs to be repaired, reconstructed, or replaced due to fire but it will not cover damage due to war or other such perils. In this type of insurance, blockchain plays a vital role by automating the transaction as soon as the predefined condition is met and verified by the smart contract (made between the policyholder and insurer). By using blockchain, one can be assured that the right amount of money is transferred based on the degree of damage and that audits are transparent. Blockchain technology is clubbed with IoT sensors installed in buildings, sensing real-time data and storing immutable reports of destruction in the form of records in the blockchain. Thus, a transparent system can be built so that a fair system can be established, which is also automatic.

11.2.3.5 Disability insurance

This insurance secures the policyholder's earnings loss emanating from disability. It is a societal scheme that intends to help persons with disability by giving them monetary and health benefits because they cannot earn a livelihood [80]. Disability insurance may insure sponsored medical leaves, short-term disability, and persistent disability [81]. In a survey, conducted in the United States of America in 2015, it has been found that there is a sharp surge in the ratio of the disability population from 11.9 to 12.6 when compared to the previous year, i.e., 2014 statistics [82]. In America, nearly 18% of citizens come under the disabled category and this number is surging due to deadly accidents occurring frequently. It has been found that in nearly 4 minutes, 3 out of 10 employees become disabled due to life-threatening accidents [83]. In 2012, the disability insurance scheme in America reimbursed beneficiaries approximately $136.9 billion. Disability insurance suffers from many challenges such as a large number of false rejections and negligible false acceptances [80, 84]. Innumerable issues about disability insurance can be resolved by blockchain. A database of disabled citizens indicating their extent of seriousness must be maintained. Further, smart contracts based on some predefined conditions can make an accurate amount of transactions to the beneficiary at relevant time intervals. The entire process using blockchain will be automatic and transparent, saving a lot of time, effort, and cost. Furthermore, blockchain rejects false claims with appropriate proofs so that they cannot be challenged.

11.3 Conclusion

This section provides a concise summary of the findings and a clear identification of the advances in the field of blockchain. This chapter provides a deep understanding of blockchain. It also highlights a detailed study of smart contracts, their usage, applications, and limitations. The current chapter also assesses the usage of blockchain for different applications in the insurance domain including specific potential implications of blockchain.

Acknowledgements

The authors wish to express their gratitude toward the management of CHRIST (Deemed to be University), Delhi-NCR, India and Vivekananda Institute of Professional Studies-TC for encouraging and giving research-oriented environment.

References

[1] V. Buterin, "A next-generation smart contract and decentralized application platform.," Available online at: https://github.com/ethereum/wiki/wiki/White-Paper.

[2] Maher Alharby and Aad van Moorsel, "Blockchain-based Smart Contracts: A Systematic Mapping Study", AIS, CSIT, IPPR, IPDCA - 2017 pp. 125–140, 2017. DOI: 10.5121/csit.2017.71011

[3] N.Szabo, "Formalizing and securing relationships on public networks.," Available online at: http://rstmonday.org/ojs/index.php/fm/article/view/548/4691.

[4] Marino, B., & Juels, A. (2016, July). Setting standards for altering and undoing smart contracts. In International Symposium on Rules and Rule Markup Languages for the Semantic Web (pp. 151–166). Springer, Cham.

[5] Kolvart, M., Poola, M., and Rull, A., 2016. Smart contracts. In The Future of Law and technologies (pp. 133–147). Springer, Cham.

[6] Huckle, S., Bhattacharya, R., White, M., and Beloff, N., 2016. Internet of things, blockchain, and shared economy applications. Procedia computer science, 98, pp.461–466.

[7] Silas M Nzuva," Smart Contracts Implementation, Applications, Benefits, and Limitations", Journal of Information Engineering and Applications, ISSN 2224-5782 (print) ISSN 2225-0506 (online), Vol.9, No.5, 2019. DOI: 10.7176/JIEA

[8] Buterin, V., 2014." A next-generation smart contract and decentralized application platform". White paper.

[9] Kim, H.M., and Laskowski, M., 2018. Toward an ontology-driven blockchain design for supply-chain provenance. Intelligent Systems in Accounting, Finance, and Management, 25(1), pp.18–27.

[10] Vukolić, M., 2017, April. Rethinking permissioned blockchains. In Proceedings of the ACM Workshop on Blockchain, Cryptocurrencies and Contracts (pp. 3–7). ACM.

[11] https://www.scalablockchain.com/blog/blockchain-benefits-asset-management/

[12] Jøsang, A., Fabre, J., Hay, B., Dalziel, J., Pope, S., 2005. Trust requirements in identity management, in: Proceedings of the 2005 Australasian Workshop on Grid Computing and E-Research-Volume 44. pp. 99–108.

[13] Ahn, G.J., Ko, M., 2007. User-centric privacy management for federated identity management. In: Proceedings of the 3rd International Conference on Collaborative Computing: Networking, Applications and Worksharing. https:// doi.org/10.1109/COLCOM.2007.4553829.

[14] T. Rathee and P. Singh, A systematic literature mapping on secure identity management using blockchain technology, Journal of King Saud University – Computer and Information Sciences, https://doi.org/10.1016/j.jksuci.2021.03.005

[15] Camenisch, J., Shelat, A., Sommer, D., Fischer-Hüubner, S., Hansen, M., Krasemann, H., Lacoste, G., Leenes, R., Tseng, J., 2005. Privacy and identity management for everyone, in. In: Proceedings of the ACM Conference on Computer and Communications Security, pp. 20–27. https://doi.org/10.1145/ 1102486.1102491.

[16] Suriadi, S., Foo, E., Jøsang, A., 2007. A User-centric Federated Single Sign-on System 101–108. https://doi.org/10.1109/NPC.2007.64.
[17] AlpAlpárr, G., Hoepman, J.-H., Siljee, J., 2013. The Identity Crisis Security, Privacy and Usability Issues in Identity Management. Journal of Information System. Security 9.
[18] Rossudowski, A.M., Venter, H.S., Eloff, J.H.P., Kourie, D.G., 2010. A security privacy aware architecture and protocol for a single smart card used for multiple services. Comput. Sec. 29, 393–409. https://doi.org/10.1016/j.cose.2009.12.001.
[19] Amo, D., Alier, M., Garc\'\ia-Peñalvo, F.J., Fonseca, D., Casañ, M.J., 2020. Protected users: A moodle plugin to improve confidentiality and privacy support through user aliases. Sustainability 12, 2548.
[20] Bertino, E., Martino, L., Paci, F., Squicciarini, A., 2009. Security for web services and service-oriented architectures. Springer Science & Business Media.
[21] Ahn, G., Lam, J., 2005. Managing Privacy Preferences for Federated Identity Management 28–36.
[22] Leenes, R.E., 2014. PRIME white paper v2.
[23] Pérez-Méndez, A., Torroglosa-Garc\'\ia, E.-M., López-Millán, G., Gómez-Skarmeta, A. F., Girao, J., Lischka, M., 2010. SWIFT–Advanced Services for Identity Management. Serbian Publication InfoReview joins UPENET, the Network of CEPIS Societies Journals and Magazines 13.
[24] Torres, J., Nogueira, M., Pujolle, G., 2012. Future Network 1–16.
[25] Zyskind, G., Nathan, O., Pentland, A., 2015. Enigma: Decentralized computation platform with guaranteed privacy. arXiv preprint arXiv:1506.03471.
[26] Heilman, E., Baldimtsi, F., Goldberg, S., 2016. Blindly signed contracts: Anonymous on-blockchain and off-blockchain bitcoin transactions, in: International Conference on Financial Cryptography and Data Security. pp. 43–60.
[27] Sullivan, C., Burger, E., 2017. E-residency and blockchain. Comput. Law Sec. Rev. Int. J. Technol. Law Pract. 1–12. https://doi.org/10.1016/j.clsr.2017.03.016.
[28] I/O Foundation [WWW Document], n.d. URL https://iodigital.io/ (accessed 3.14.21). id2020, n.d. We need to get digital ID right [WWW Document]. URL https://id2020. org/ (accessed 12.3.21).
[29] O. Mahmoud, H. Kopp, A. T. Abdelhamid, F. Kargl, — Applications of Smart-Contracts: Anonymous Decentralized Insurances with IoT Sensors, Proc. - IEEE 2018 Int. Congr. Cybermatics 2018 IEEE Conf.

Internet Things, Green Comput. Commun. Cyber, Phys. Soc. Comput. Smart Data, Blockchain, Comput. Inf. Technol. iThings/Gree, 2018.

[30] H. Mohamed, —Takāful (Islamic insurance) on the blockchain, Growth Islam. Financ. Bank., 2019.

[31] M. Raikwar, S. Mazumdar, S. Ruj, S. S. Gupta, A. Chattopadhyay, K. Y. Lam, —A Blockchain Framework for Insurance Processes, 9th IFIP Int. Conf. New Technol. Mobil. Secur. NTMS 2018 - Proc., 2018.

[32] S. Nakamoto, —Bitcoin: A Peer-to-Peer Electronic Cash System, Manubot 2008. [Google Scholar].

[33] C. S. Wright, —Bitcoin: A Peer-to-Peer Electronic Cash System, (August 21, 2008).

[34] T. Ahram, A. Sargolzaei, S. Sargolzaei, J, Daniels, B. Amaba, —Blockchain technology innovations. IEEE Technol. Eng. Manag. Soc. Conf. TEMSCON 2017.

[35] S. Feng, Z. Xiong, D. Niyato, P. Wang, S. S. Wang, Y. Zhang, —Cyber Risk Management with Risk Aware Cyber-Insurance in Blockchain Networks,2018 IEEE Glob. Commun. Conf. GLOBECOM 2018 - Proc., 2018.

[36] J. Kietzmann, C. Archer-Brown, —From hype to reality: Blockchain grows up, Business Horizons. 2019 Jan 1; 62(3):269–71.

[37] K. Panetta, —5 trends emerge in the Gartner Hype Cycle for emerging technologies, Gartner. accessed December 8, 2020 unpublished.

[38] E. Kapsammer, B. Pröll, W. Retschitzegger, W. Schwinger, M. Weißenbek, & J. Schönböck, —The Blockchain Muddle: A Bird's-Eye View on Blockchain Surveys‖, In Proc of the 20th Int Conf on Infor Integ and Web-based App & Ser (pp. 370-374).

[39] K. Wang, & A. Safavi, —Blockchain is empowering the future of insurance‖. Available at https://techcrunch.com/2016/10/29/blockchain-is-empowering -the-future-of-insurance unpublished.

[40] M. A. Ali, B. Balamurugan, R. K. Dhanaraj and V. Sharma, "IoT and Blockchain based Smart Agriculture Monitoring and Intelligence Security System," 2022 3rd International Conference on Computation, Automation and Knowledge Management (ICCAKM), Dubai, United Arab Emirates, 2022, pp. 1–7, doi: 10.1109/ICCAKM54721.2022.9990243.

[41] J. Mendling, I. Weber, W. V. Aalst, J. V. Brocke, C. Cabanillas, F. Daniel, S. Debois, C. D. Ciccio, M. Dumas, S. Dustdar, A. Gal, —Blockchains for business process management-challenges and opportunities, ACM Trans. on Mgt Inf. Sys (TMIS). 2018 Feb 26;9(1):1–6.

[42] A. Singh, R. K. Dhanaraj, M. A. Ali, B. Balusamy and V. Sharma, "Blockchain Technology in Biometric Database System," 2022 3rd International Conference on Computation, Automation and Knowledge

Management (ICCAKM), Dubai, United Arab Emirates, 2022, pp. 1–6, doi: 10.1109/ICCAKM54721.2022.9990133.
[43] K. Yeow, A. Gani, R. W. Ahmad, J. J. Rodrigues, K. Ko, —Decentralized consensus for edge-centric internet of things: A review, taxonomy, and research issues. IEEE Access. 2017 Dec 6; 6:1513–24.
[44] Singh, Anamika, et al. "Blockchain: Tool for Controlling Ransomware through Pre-Encryption and Post-Encryption Behavior." 2022 Fifth International Conference on Computational Intelligence and Communication Technologies (CCICT). IEEE, 2022.
[45] Ali, M. A., Balamurugan, B., & Sharma, V. (2022, April). IoT and blockchain based intelligence security system for human detection using an improved ACO and heap algorithm. In 2022 2nd International Conference on Advance Computing and Innovative Technologies in Engineering (ICACITE) (pp. 1792–1795). IEEE.
[46] M. Sharifinejad, A. Dorri, J. Rezazadeh, —BIS-A Blockchain-based Solution for the Insurance Industry in Smart Cities, arXiv, 2020.
[47] H. Kim, M. Mehar, —Blockchain in Commercial Insurance: Achieving and Learning Towards Insurance That Keeps Pace in a Digitally Transformed Business Landscape,‖ SSRN Electron J 2019.
[48] M. Mainelli, B. Manson —Chain reaction: How blockchain technology might transform wholesale insurance, How Blockchain Technology Might Transform Wholesale Insurance-Long Finance. 2016 Aug 1. Available at SSRN: https://ssrn.com/abstract=3676290.
[49] L. S. Howard, —Blockchain insurance industry initiative B3i grows to 15 members,‖ Insurance Journal. 2017.
[50] Road Traffic Act 1930 unpublished.
[51] R. Roriz, J. L. Pereira, —Avoiding Insurance Fraud: A Blockchain-based Solution for the Vehicle Sector, Procedia Comput. Sci., 2019.
[52] World Health Organization, https://www.who.int/gho/road_safety/mortality/traffic_ deaths_number/en/ unpublished.
[53] S.L. Poczter, L. M. Jankovic, —The Google Car: Driving Toward A Better Future?, J Bus Case Stud 2013.
[54] H. T. Vo, L. Mehedy, M. Mohania, E. Abebe, —Blockchain-based data management and analytics for micro-insurance applications, Int. Conf. Inf. Knowl. Manag. Proc., 2017.
[55] L. M. Palma, F. O. Gomes, M. Vigil, J. E. Martina, —A Transparent and Privacy-Aware Approach Using Smart Contracts for Car Insurance Reward Programs, InInternational Conference on Information Systems Security 2019 Dec 16 (pp. 3–20). Springer, Cham.
[56] S. Husnjak, D. Peraković, I. Forenbacher, M. Mumdziev, —Telematics system in usage based motor insurance, Procedia Eng., 2015.

[57] P. Handel, I. Skog, J. Wahlstrom, F. Bonawiede, R. Welch, J. Ohlsson, ―Insurance telematics: Opportunities and challenges with the smartphone solution, IEEE Intell Transp Syst Mag 2014.

[58] J. Wahlström, I. Skog, P. Händel, ―Driving behavior analysis for smartphone-based insurance telematics, Proc. 2nd Work. Phys. Anal., 2015.

[59] Lin W-Y, Lin FY-S, Wu T-H, Tai K-Y. ―An On-Board Equipment and Blockchain-Based Automobile Insurance and Maintenance Platform, In Int Conf on Broadband and Wireless Comp, Comm & Appl pp. 223–232, 2020.

[60] F. Li, H. Zhang, H. Che, X. Qiu, ―Dangerous driving behavior detection using smartphone sensors,IEEE Conf. Intell. Transp. Syst. Proceedings, ITSC, 2016.

[61] R. Harbage, ―Usage-based Auto Insurance (UBI),‖ 2011 https://www.casact.org/community/affiliates/sccac/1211/Harbage.pdf unpublished.

[62] A. Kumar, A. Prasad, R. Murthy, ―Application of blockchain in Usage Based Insurance, Int J Adv Res 2019.

[63] V. Aleksieva, H. Valchanov, A. Huliyan, ―Application of smart contracts based on ethereum blockchain for the purpose of insurance services,‖ Proc. Int. Conf. Biomed. Innov. Appl. BIA 2019, 2019.

[64] H. Qi, Z. Wan, Z. Guan, X. Cheng, ―Scalable Decentralized Privacy Preserving Usage-based Insurance for Vehicles, IEEE Internet Things J 2020.

[65] Z. Wan, Z. Guan, X. Cheng, ―Pride: A private and decentralized usage based insurance using blockchain, In 2018 IEEE Int Conf Internet of Things IEEE Green Comp & Comm (GreenCom) and IEEE Cyber, Phys Soc Comput (CPSCom) IEEE Smart Data 2018 Jul 30 (pp. 1349–1354). IEEE.

[66] M. Demir, O. Turetken, A. Ferworn, ―Blockchain based transparent vehicle insurance management, 6th Int. Conf. Softw. Defin. Syst. SDS 2019.

[67] M B. Abramowicz, ―Blockchain-Based Insurance, SSRN Electron J 2019

[68] P. K. Singh, R. Singh, G. Muchahary, M. Lahon, S. Nandi, ―A Blockchain-Based Approach for Usage Based Insurance and Incentive in ITS, IEEE Reg. 10 Annu. Int. Conf. Proceedings/TENCON, 2019.

[69] F. Lamberti, V. Gatteschi, C. Demartini, M. Pelissier, A. Gomez, V. Santamaria, ―Blockchains Can Work for Car Insurance: Using Smart Contracts and Sensors to Provide On-Demand Coverage,IEEE Consum Electron Mag 2018.. 2816247.

[70] C. Oham, R. Jurdak, S. S. Kanhere, A. Dorri, S. Jha, ―B-FICA: BlockChain based Framework for Auto-Insurance Claim and Adjudication,‖ Proc.

- IEEE 2018 Int. Congr. Cybermatics 2018 IEEE Conf. Internet Things, Green Comput. Commun. Cyber, Phys. Soc. Comput. Smart Data, Blockchain, Comput. Inf. Technol. iThings/Gree, 2018.

[71] Karan C, —What is Marine Insurance,‖ 2019 available online at https://www.marineinsight. com/know-more/what-is-marine-insurance/ accessed November 1, 2020 unpublished.

[72] EY. World's first blockchain platform for marine insurance now in commercial use, 2018, unpublished.

[73] C. Reed, —Blockchain and the World of Marine Insurance, 2017. available online at https://www.marinelink.com/news/blockchaininsurance429260 unpublished.

[74] B. Beers, —A Brief Overview of the Insurance Sector, 2020, unpublished.

[75] S. Lesley, —Life Insurance Industry under Investigation, CBS NEWS (Apr. 17, 2016) unpublished.

[76] A. Cohn, T. West, C. Parker, —Smart after all: Blockchain, smart contracts, parametric insurance, and smart energy grids, Georgetown Law Technology Review, 1(2), 273-304.

[77] D. A. Disparte —Blockchain Could Make the Insurance Industry Much More Transparent, unpublished.

[78] M. Lounds, —Blockchain and its Implications for the Insurance Industry, 2020 unpublished.

[79] Anokye Acheampong AMPONSAH and Professor Adebayo Felix ADEKOYA, Blockchain in Insurance: Exploratory Analysis of Prospects and Threats, (IJACSA) International Journal of Advanced Computer Science and Applications, Vol. 12, No. 1, 2021.

[80] H. Low, L. Pistaferri, —Disability insurance and the dynamics of the incentive insurance trade-off, Am Econ Rev 2015.

[81] B. Olmsted, —How to Buy Disability Insurance, 2019. unpublished.

[82] L. Kraus, —2015 Disability Statistics Annual Report. A Publication of the Rehabilitation Research and Training Center on Disability Statistics and Demographics, 2016, Institute on Disability, University of New Hampshire.

[83] Disability statistics and Facts, https://web.archive.org/web/20110727133227/http://nteu-chapter78.org/documents/member_benefits /Disability%20Statistics.pdf unpublished.

[84] J. Gera, A. R. Palakayala, V. K. K. Rejeti, T. Anusha, —Blockchain technology for fraudulent practices in insurance claim process,‖ Proc. 5th Int. Conf. Commun. Electron. Syst. ICCES 2020, 2020.

[85] Sood, K., Seth, N., & Grima, S. (2022). Portfolio Performance of Public Sector General Insurance Companies in India: A Comparative Analysis.

In Simon Grima, Ercan Özen, Inna Romānova (Ed.), In *Managing Risk and Decision Making in Times of Economic Distress, Part B*. (pp.215–229). Emerald Publishing Limited,UK.

[86] Grima, S., & Marano, P. (2021). Designing a Model for Testing the Effectiveness of a Regulation: The Case of DORA for Insurance Undertakings. *Risks*, *9*(11), 206.

12

A Review and Application of Blockchain Technology in Health Insurance

Bhavna Sharma[1], Girish Garg[1], and Fisnik Morina[2]

[1]School of Finance and Commerce, Galgotias University, India
[2]Faculty of Business, University "Haxhi Zeka" in Peja, Kosovo

Email: Bhavna656@yahoo.com; Girish.garg@galgotiasuniversity.edu.in; fisnik.morina@unhz.eu

Abstract

Repositioning the technology to transform the insurance sector will help in bringing radical growth. A blockchain is an emerging form of the latest technology that provides more innovation in health insurance. Health insurance is amongst the top payer scrabbling its way to ascertain records, complete transactions, and communicate with other parties. The application of blockchain leads to an exchange of patient records, risk reduction, and better customer relations. This analysis covers a variety of perspectives on the current state of blockchain applications. This business research analyses blockchain possibilities in insurance. Blockchain technology was tested to ease health insurance paperwork. The research examined how organizations that employ blockchain technology in marine, life, and general insurance are tackling the obstacles of increasing its usage to simplify and speed up claim settlement and improve insurance sector administration. The paper shows distinct blockchain health insurance alternatives and regulatory hurdles in different nations. This system helps resolve claims and monitor insurer status. Intermediaries cannot do illegal acts for insurers either. This study examines blockchain's operational relevance in insurance. Insurance businesses and software developers may employ blockchain applications to construct system bases based on industry needs.

12.1 Introduction

A distributed ledger of records is known as a blockchain [1]. The information ensures data security by storing it cryptographically. It is essentially stable and unchangeable, i.e., data on the blockchain can only be assessed once it has been recorded. It is not possible to make changes after the fact. Parties involved can see all the data to ensure transparency in smart contracts through blockchain [2]. A new approach to designing a safe distribution system has numerous benefits. By contributing to data verification via cryptography, participating nodes may work together to maintain the common shared ledger functioning smoothly. The blockchain is the collection of transactions that are recorded in chronological order attached to the previous one. Any part must be double-checked before being added to the chain. Contribution from participating nodes enables this process to happen as it comprises different partners to take better decisions with the right flow of information [3].

12.2 Why Blockchain?

The enhanced security of blockchain and its capacity to create trust between entities are two reasons why it may be able to tackle the interoperability challenges more effectively than current solutions. An interoperable and complete health record on the blockchain would be most likely derived from current EHRs in hospitals and physician offices. The majority of today's health records are kept in a single-provider system [4]. When a patient event happens, providers might choose whether information is to be published to a shared blockchain or continually uploaded to the blockchain. Data entering the blockchain via smart contract (decentralized programmer that automatically acts depending on blockchain activity) would be subjected to previously agreed-upon data standards, resulting in understandable and consistent data from all sources. Using blockchain automatic data verifications, trust concerns between entities might be handled [5]. There is no need for a middleman, and blockchain users are not required to communicate. Participants would be in charge of who had access to the data, which would be tamper-proof once within the blockchain. Sharing data is hampered by concerns about privacy and security. The Health Insurance Accountability Act, portability, and related rules are major source of concern for the healthcare business. Data storage and access might be more secure with blockchain. Patients can allow access to their information to physicians, insurers, and others with authorized users [6].

12.3 Employment of Blockchain System in Healthcare

Smart contracts on blockchain will automate and make contract compliance transparent throughout the continuum by applying a distributed and centralized ledger. It will minimize the frequency of erroneous insurance claims by recognizing the submission of repeated fraudulent claims.

12.3.1 Blockchain applications in healthcare

Healthcare is one of the industries where healthcare is thought to have a lot of potential. In 2016, the national coordinator of Health Technology stated the possible use of blockchain in healthcare, recognizing the relevance and value of technology. As a result of this difficulty, a few blockchain-based healthcare applications have been presented.

12.3.2 E-medical record

To revolutionize this field, the target sector would be health data management, which might benefit from the ability to integrate heterogeneous systems and improve the accuracy of electronic health records (EHRs), whereas electronic medical and health records are used synonymously. Electronic medical records (EMR) were coined initially in which medical and treatment histories of the patients are recorded in EMR in a single practice. Med Rec, an EHR, is a decentralized way to manage authorization, permissions, and data sharing between healthcare stakeholders [8]. It makes use of the Ethereum platform to provide patients with their medical records. With the help of a health chain platform that is managed by a consortium network made up of different entities, genuine information can be accessed by patients and healthcare professionals. Research Foundry is a novel blockchain application that manages consent and rights for sharing and access of health data, information, software codes, and other items relevant to health study, including research on the COVID-19 pandemic [9]. Its goals were to make cross-border collaboration easier while ensuring that partners follow all applicable data privacy laws. For instance, collected data in the online storehouse needs to be communicated with all parties, but data related to the health store may be kept private and inaccessible to blockchain nodes. An essential step is efficiently sharing the data collected in the online repository with all relevant parties. However, it is crucial to maintain confidentiality and limit access to blockchain nodes for any health store data. When two parties agree to share and exchange all or part of the repository, it must be done with consent [18,19].

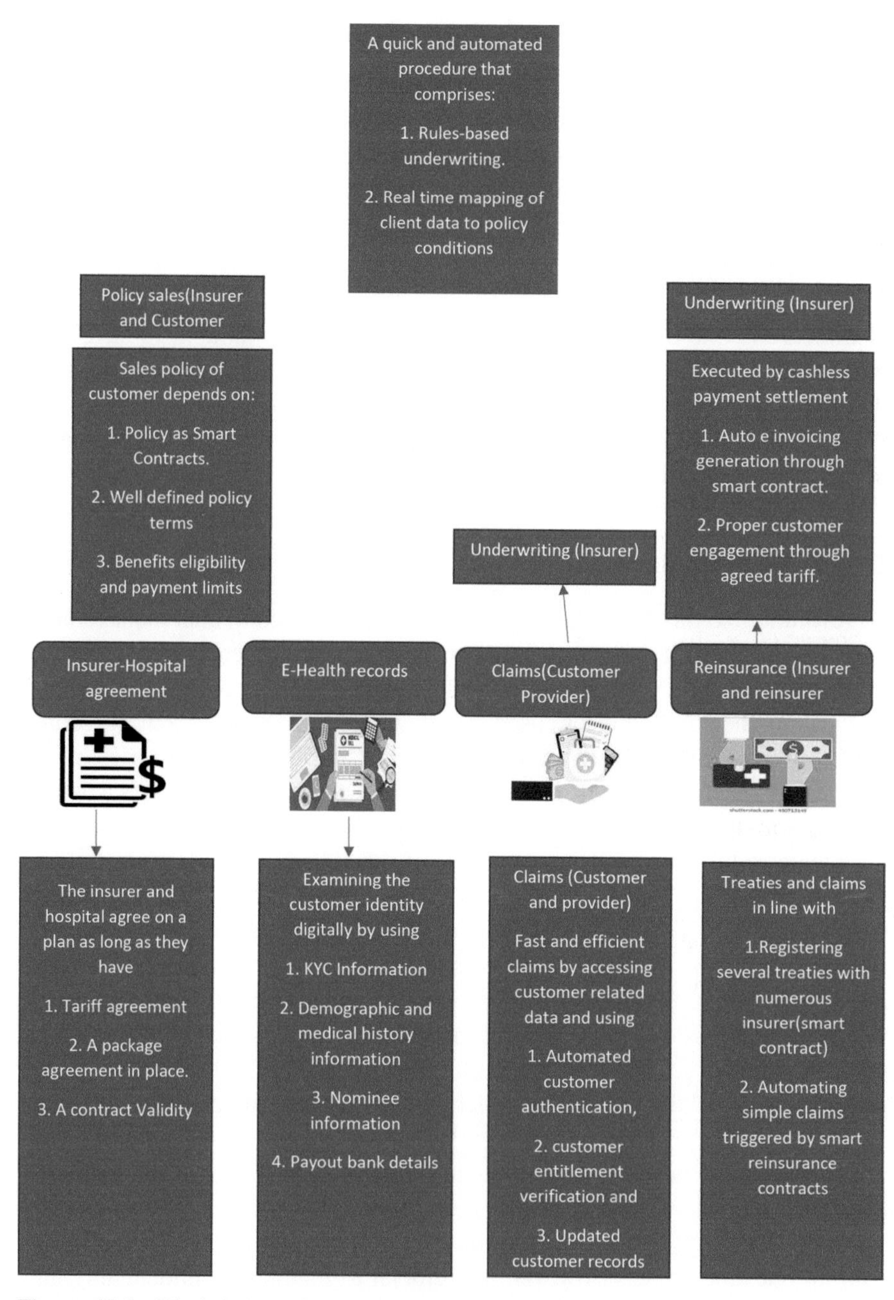

Figure 12.1 Blockchain and smart contract: process of health insurance value chain. Source: revamping-indias-health-insurance-sector-with-blockchain-and-smart-contracts.pdf [7].

12.3.3 Clinical trial research

Recruiting patients for clinical trials is one of the most difficult aspects. There are various challenges that must be addressed from sponsor patients, and main investigation viewpoints, resulting in majority of clinical trials falling to fulfil their recruitment target on demand. Conducting clinical trials with insufficient participants reduces the power of definitive conclusions or leads to early trial terminations. Because they have a comprehensive collection of participants' medical history, blockchain plays a critical role in future clinical studies; for instance, if a participant has an allergy to certain types of medicinal substance. The patient's comprehensive history of a transaction can be used in research trials and provide the additional details like patient's history of adverse response to particular pharmaceutical components. This enables the clinical researchers to make an educated trial choice. The comprehensive transaction history of the patient can be utilised in research trials to gather additional information, such as the patient's history of adverse reactions to specific pharmaceutical components. It allows clinical researchers to make an informed decision regarding the choice of trials. Blockchain technology is characterised by its nature as a decentralised, peer-to-peer distributed system that facilitates the storage and retrieval of data in a publicly accessible manner. Clinical trials are well-suited for audited transactions, ensuring data security and safeguarding patients' privacy [10].

12.3.4 Medical drug supply chain

In the healthcare business, a faulty supply chain can have a serious consequence. The blockchain technology aids in the tracking of medicine manufacture. In the first stage, the blockchain allows you to track the proper components through manufacturing process. It allows the total medication tracing the product starting from the producer to finish end user. This involves five departments: the manufacturer, distributors, transporter, pharmacy, and hospital. The journey of the medications may be watched and monitored as they transit through each department. Due to high demand, the healthcare system is finding it difficult to obtain medical equipment and supplies to combat COVID-19. The collapse of supply chain with well-known providers causes trust concerns. Compliance with standard, custom certifications, timely delivery of goods, and frauds are all concerns of new vendors. These trust concerns are exacerbated by vendors demand for upfront payment. Blockchain is a potential solution for ensuring suppliers legitimacy and tracking goods [11]. In April 2020, IBM unveiled IBM rapid supplier connect, a blockchain-enabled network that would connect government and healthcare organizations with non-traditional equipment, device, and supply vendor to tackle COVID-19 [12].

12.3.5 Medical insurance

Medical insurance protects you from the financial consequences of a serious accident or a medical emergency. Three entities handle medical claims, which include healthcare providers, insurance companies, and a clearing house. The clearing house acts as a middleman between healthcare providers and insurance companies when a claim is filed. This gives the clearing house control and allows it to behave maliciously on its own. With the use of blockchain technology, this may be prevented. The necessity for a clearing house to act as a middleman is gone, and blockchain provides a transparent system in which no one has ultimate control. Because the code once registered on the chain cannot be charged, smart contracts may be used in the insurance claiming process. Wireless body area networks, or WBANs, were offered as a blockchain solution for monitoring the patient's essential information [13].

12.4 Literature Review

Researchers examined the current healthcare system to determine what changes are required to improve interoperability. They examined the present system with regard to integration, operations, and interoperability, as well as changes that would be needed to be made to improve their qualities [14]. L. Zhou, L. Wang, and Y. Sun (2018) conducted a comprehensive review of details of technology aspects in health, stressing on its quality, safety, efficiency, as well as its contextual and implementation implications. Completion of details at the specific point of service and access restrictions related to individual's personal health data were among the challenges faced during the trial. Interoperability enables software apps and technology platforms to connect in a secure and smooth manner, share data, and use that data across health organizations and app vendors, resulting in a more efficient and effective healthcare system [15]. Nowadays, healthcare suffers from compartmentalized and fragmented data, delayed connections, and distinct workflow due to lack of interoperability. The blockchain allows access to longitudinal, comprehensive, and tamper-aware medical records that were housed in fragmented systems in a secure and anonymous way. The consolidation of health records from many sources has been a key stumbling block to increased medical use. In general, unconnected data sources can be used to improve medical data integration and aggregation. Researchers have pointed out that due to the distributed nature of blockchain, there are storages of data and management concerns in scientific studies, insurance, and personal health situations [16].

A universal shared ledger based on blockchain immutable and distributed properties might help with health data exchange. For instance, different researchers had conducted experimental research to explore patients' perspectives of various exchange systems, targeted on online health monitoring [17]. Moreover, while few conducted reviews of literature to lighten the outlook and limitation of blockchain implementation, others reviewed identity management solutions and developed evaluation frameworks for the evaluation of blockchain performance. Researchers have suggested various blockchain-based prototypes to give answer to the current system since access control and authentications were significant protective needs for maintaining the health data. When several parties such as clinicians, payers, and researchers seek to engage with patient data, the digitalization of electronic medical record may create significant threats to data security and privacy. Through a friendly approach, blockchain-enabled systems may be able to retain patient-sensitive data.

12.5 Methodology and Discussion

India's health insurance market received INR 51,400 crore in premiums, accounting for 28% of the total premium collected by insurance industry financial year 2020. Despite steady growth over the last decade, private health insurance is expected to cover just around 10% of India's population. The Indian health insurance ecosystem includes insurer, companies, intermediaries, value-added service providers, government regulators, and the government as a buyer of social insurance schemes. The fragmented flow of information, on the other hand, makes it difficult for many stakeholders to participate. The way information is shared can be difficult for many people to engage with and take part in. To improve this process, it is suggested that a trustworthy blockchain system be put in place to allow processes to be carried out across different nodes in a decentralized manner.

Blockchain and cryptographic algorithms may be the most efficient approach to store records and conduct rational commerce with all participants. Blockchain may gather transaction paperwork, terms, negotiated tariffs, and pertinent information, as well as connection information, and execute smart contracts in a transparent, immutable, and irrefutable way based on pre-defined business rules. Because the documents are tamper-proof, they will be made public, and concerns about data privacy will be addressed. Blockchain use might go a long way towards assuring confidence, enhancing consumer experience, and assisting in the expansion of health insurance penetration. Blockchain has remained popular as a global platform for medical data analysis and transfer. The use of blockchain-based health records (EHRs) in Estonia is an illustration just because it maintains data integrity, eliminates internal data risks, and enables multiple public and private sector entities to transfer data quickly. Data correctness, standards, norms, and dependability are critical from the start of a blockchain cycle and must be improved on a periodic basis to identify efficiency.

12.6 Findings and Conclusion

Given the fact that digital has benefited India's medical insurance industry and interaction among numerous actors in improving difficult procedures, the country still faces challenges. As a process, the entire industry, including insuring data, assert files, and medical records, changed from manual to digital data gathering and storage. The fragmented flow of information, on the other hand, makes it difficult for many stakeholders to participate. Through a participatory approach, blockchain can facilitate the frictionless sharing of information and build confidence among all parties. Blockchain-based transactions on a highly consensus track had the power to change the way the health insurance ecosystem stores records and executes through a participatory manner including key parties acting as nodes.

- The health and automobile insurance segments account for over two-third of India's insurance sector.
- Within the general insurance sector, the health insurance segment's share of gross written premium (GWP) has risen from roughly 11% in the financial year 2005 to over 28% in the financial year 2020.
- An aspiration Ayushman Bharat – Pradhan Mantri Jan Arogya Yojana (PMJAY) programme, which began in September 2018, covers around 40% of the country's population.
- It is vital that interactions of one of the many actors be swift and viewable in an irreversible and timely way, while also fulfilling all data vulnerabilities.

The sectors will implement distributed ledger devices and innovative protocols, thanks to the responsible stakeholders. The primary takeaways from the NDHM framework deployment may be used to examine and improve the usefulness of blockchain across the health insurance industry. An active knowledge repository that would benefit the business would be a blockchain-based, shared database of fraudulent companies generated from all insurers. Blockchain networks on distributed ledger will streamline and make transaction complying accessible across the timeline by using a distributed, centralized ledger. As a consequence, it will speed up more reliable reimbursements by eliminating the conflicts that might arise from vagueness in billing contractual terms in regular contracts.

12.7 Future Application of Blockchain

The blockchain will be utilized for birth and death registration, donation management, land records, credit worthiness of individuals, and other applications that require transparency. We establish plug-ins (APIs) to transmit mortality and birth certificate statistics after the AO approves the form. A money transfer hash is returned once the data is placed into the block chain. The QR code generated from this hash is written on the forms. We are looking to launch an app that will scan the QR code, retrieve data from the server, and inform the person about the validity of the certificate. For additional security (data integrity evident and robust), land title mutation provenance, higher efficiency, and cheaper service delivery costs, the customer wanted to install land title mutation on a blockchain.

References

[1] C. Agbo, Q. Mahmoud, and J. Eklund, “Blockchain technology in Healthcare: A systematic review,” *Healthcare*, vol. 7, no. 2, p. 56, 2019.

[2] M. S. Ali, M. Vecchio, G. D. Putra, S. S. Kanhere, and F. Antonelli, “A decentralized peer-to-peer remote health monitoring system,” *Sensors*, vol. 20, no. 6, p. 1656, 2020.

[3] M. S. Ali, M. Vecchio, G. D. Putra, S. S. Kanhere, and F. Antonelli, “A decentralized peer-to-peer remote health monitoring system,” *Sensors*, vol. 20, no. 6, p. 1656, 2020.

[4] “Home,” *BurstIQ*, 28-Dec-2022. [Online]. Available: https://www.burstiq.com/. [Accessed: 23-Nov-2022].

[5] Y. Chen, S. Ding, Z. Xu, H. Zheng, and S. Yang, “Blockchain-based medical records secure storage and Medical Service Framework,” *Journal of Medical Systems*, vol. 43, no. 1, 2018.

[6] D. V. Dimitrov, “Blockchain applications for healthcare data management,” *Healthcare Informatics Research*, vol. 25, no. 1, p. 51, 2019.

[7] N. Degnarain, "Five ways blockchain can unblock the Coronavirus Medical Supply Chain," *Forbes*, 23-Mar-2020. [Online]. Available: https://www.forbes.com/sites/nishandegnarain/2020/03/22/5-ways-blockchain-can-unblock-the-coronavirus-medical-supply-chain/#25d72dd11380. [Accessed: 23-Nov-2023].

[8] Ekblaw, A., Azaria, A., Halamka, J.D., Lippman, A.: A case study for blockchain in healthcare: "MedRec" prototype for electronic health records and medical research data. In: Proceedings of IEEE Open & Big Data Conference, vol. 13, p. 13 (2016).

[9] P. Esmaeilzadeh and T. Mirzaei, "The potential of blockchain technology for Health Information Exchange: Experimental Study from patients' perspectives," *Journal of Medical Internet Research*, vol. 21, no. 6, 2019.

[10] R. Guo, H. Shi, Q. Zhao, and D. Zheng, "Secure attribute-based signature scheme with multiple authorities for blockchain in Electronic Health Records Systems," *IEEE Access*, vol. 6, pp. 11676–11686, 2018.

[11] Y. S. Hau, J. M. Lee, J. Park, and M. C. Chang, "Attitudes toward blockchain technology in Managing Medical Information: Survey Study," *Journal of Medical Internet Research*, vol. 21, no. 12, 2019.

[12] A. Vazirani, O. O'Donoghue, D. Brindley, and E. Meinert, "Implementing blockchains for Efficient Health Care: Systematic Review," *Journal of Medical Internet Research*, vol. 21, no. 2, 2019.

[13] D. M. Maslove, J. Klein, K. Brohman, and P. Martin, "Using blockchain technology to manage clinical trials data: A proof-of-concept study," *JMIR Medical Informatics*, vol. 6, no. 4, 2018.

[14] R. Zhang, A. George, J. Kim, V. Johnson, and B. Ramesh, "Benefits of blockchain initiatives for value-based care: Proposed framework," *Journal of Medical Internet Research*, vol. 21, no. 9, 2019.

[15] L. Zhou, L. Wang, and Y. Sun, "MIStore: A blockchain-based medical insurance storage system," *Journal of Medical Systems*, vol. 42, no. 8, 2018.

[16] Sood, K., Seth, N., & Grima, S. (2022). Portfolio Performance of Public Sector General Insurance Companies in India: A Comparative Analysis. In Simon Grima, Ercan Özen, Inna Romānova (Ed.), In *Managing Risk and Decision Making in Times of Economic Distress, Part B*. (pp.215–229). Emerald Publishing Limited,UK.

[17] Grima, S., & Marano, P. (2021). Designing a Model for Testing the Effectiveness of a Regulation: The Case of DORA for Insurance Undertakings. *Risks*, *9*(11), 206.

13

Blockchain-based Applications in Insurance

Neeti Misra[1], T. Joji Rao[2], Sumeet Gupta[3], and Mark Laurence Zammit[4]

[1]Uttaranchal University, India
[2]O.P. Jindal Global University, India
[3]UPES, India
[4]Department of Insurance and Risk Management, Faculty of Economics, Management and Accountancy, University of Malta, Malta

Email: neeti.cm@gmail.com; jojirao2003@gmail.com; sumeetgupta@ddn.upes.ac.in; mark.l.zammit@um.edu.mt

Abstract

Blockchain has the power to change how insurance firms conduct business. Blockchain can help operators to save money and time, improve transparency, abide by rules, and develop better goods and markets. Insurers are inherently competitive in a highly competitive market where both individual and corporate customers expect the highest profitability and a superior online experience. In the insurance sector, blockchain technology represents an excellent transformational and growth opportunity.

Insurance can be purchased through a blockchain account, which provides greater automation and a tamper-proof audit trail through the use of smart contracts and decentralised applications on Ethereum. The low cost of smart contracts and associated fees, in particular, means that many items will be more competitive in limited markets in developing countries. Ultimately, the burgeoning blockchain economy will need its own insurance. Cyber insurance covers coverage (technology and software projects) with extensions and approvals for monetary losses (hot wallets and exchanges), species and crime (cold wallets and vaults), professional liability (developers), and guarantors can be used as a template for [1].

This study reveals the most fascinating aspects of blockchain technology that have paved new opportunities for the insurance industry using blockchain technology. This research addresses the use of smart contracts in blockchains, where information is secure and automated so that a specific type of payment contract can be entered into without human intervention. It also discusses the challenges and concerns that insurance companies may have with regard to the use of blockchain technology. Finally, barriers and obstacles to the adoption of blockchain technology in the insurance industry are discussed through a number of case studies.

13.1 Introduction

13.1.1 Blockchain technology

Blockchain technology offers significant efficiency gains, cost savings, transparency, faster payments, and fraud prevention, while enabling trusted and traceable real-time data exchange between various parties. Blockchain technology will also allow new insurance practices to develop better products and markets. Another type of distributed ledger technology (DLT) is blockchain. DLT consists of a ledger or database into which data is entered and maintained on a peer-to-peer (P2P) basis. Because of the P2P nature, there is no central trusted party or intermediary that manages the ledger, making it decentralised. Blockchain DLT technology got its name because the ledger is structured. Here, ledger entries are grouped into transaction blocks, verified, and sent to the network.

13.1.2 Blockchain and insurance

Blockchain is an emerging technology that has "quickly become a fixture in the financial services industry" because of its ability to do the following:

- Conciliation
- Better data reconciliation
- Effective firm model transformation

The insurance industry has a favorable reputation for being conservative. Insurance companies, in comparison to other financial sectors, particularly banks, are slow to respond to technological advances and changing consumer preferences.

Many factors have shaken the underlying long-term business model, but upheaval is only a matter of time. The industry is "uniquely positioned to benefit from blockchain technology" because it opens up significant digital opportunities to reduce costs, increase efficiency, improve the customer experience, improve data quality, and acquire and improve analytics. Through innovation, insurers can increase market share, gain a competitive advantage, and develop an integrated service approach. According to a recent survey, 46% of insurers expect to integrate blockchain within the next two years, and 84% believe blockchain and smart contracts will revolutionise the way they work with new partners. They say it is attractive. Blockchain technology can be used to improve existing insurance processes or develop entirely new insurance practices. Using blockchain to improve existing insurance processes means adding to existing business practices. As a primary use case, you can use blockchain to pay fees, salaries, bonuses, and benefits at a lower cost per transaction. Blockchain as a more complex bet could serve as a common database for multiple insurance companies. Introducing a new method of insurance, blockchain represents a more fundamental implementation of new technologies through smart contracts and real-world insurance through decentralised applications (dApps).

This results in cost savings, efficiencies, and other benefits. Blockchain insurance is the next logical step in the evolution and expansion of cyber insurance. It is not a blockchain application per se but rather an attempt to extend insurance coverage to the growing infrastructure around the distributed ledger. Blockchain insurance presents a market opportunity for the insurance industry. The following pages describe some use cases for blockchain-based insurance extensions and emerging practices [26, 27].

13.1.2.1 The first question arises is: what is digital insurance?

Digital insurance refers to several technologies that have transformed the way insurance service providers work. This is a reference to insurers adopting a technology-based operating model for selling and managing insurance policies. Most insurers have their own digital lines of business for this while following traditional processes. Digital insurance service providers have the following different approaches.

- A customer-centric approach to business transactions: A multi-channel presence allows potential customers to explore and understand your brand without waiting to speak to a representative.
- An insurance technology ecosystem where pricing, underwriting, and claims are handled through an open and connected software platform.
- A coverage plan for customers who need little coverage.

A digital insurance model offers speed, agility, easy availability, and an easy-to-use interface. This digitisation enables the creation and implementation of new and better services.

13.1.2.2 Blockchain enhancing current insurance workflows

Utilising blockchain to improve current insurance procedures entails leveraging the technology to supplement current corporate procedures. Blockchain technology, for example, enables parties to exchange data in real time in a traceable and secure manner. Every time a file is added to the blocks or later updated, a new transaction is layered and time-stamped by the network in this occurrence. As a result, the history of each file is completely clear from the beginning to the end, "allowing detection and prevention of unauthorised interference from within or outside the system." Private strings, or a combination of public and private strings, can be used to grant access to files as desired. Additionally, since blockchain is a distributed ledger, there is no single point of failure, and the database is more reliable and safer than other databases. Blockchain's ability to share data could significantly improve insurance operations. At the same time, it improves data quality and reduces costs.

13.2 Digital Ledger Product for Insurance Policies and Some Case Studies

With blockchain technology, insurance companies can create smart contracts to track claims, automate outdated administrative processes, and protect sensitive information [7].

What blockchain does is that it is generally a record or ledger of digital events – one that is distributed or shared between many different teams in an organisation. It can only be updated with the consent of the majority of system participants. And once information is entered, it can never be transformed or modified. Blockchain contains a definite and verifiable record of every single transaction ever made.

Blockchain technology can help reduce costs, improve risk assessment, and improve client onboarding in underwriting.

Once a transaction is made, insurers can use blockchain distributed ledger technology to update the information and match it with other records on the network. Reduce contract, billing, and relationship management costs. Optimise operations and increase customer satisfaction. Businesses can also take advantage of opportunities and returns through new business models and new insurance products. Blockchain's ability to inspire trust in trusted ecosystems through public ledgers and advanced cybersecurity protocols will have a positive impact on the future growth of the insurance industry. With the help of artificial intelligence and big data, the feasibility of using blockchain for insurance largely depends on three unique features [7].

13.3 Blockchain-based Smart Contract

Smart contracts are part of blockchain technology and can be used at any time by multiple processes running in different systems and databases. They are executed in blockchain technology as a program for automated confirmation and similar tasks where manual authentication can have a high risk of error and misuse. The implementation of such technology could bring dramatic changes to the insurance industry, because you can freely translate quotes and policies into computer code. To safeguard a participant's information, including their identity and personal information, to secure transactions, and to verify the authenticity of transactions, we employ sophisticated digital procedures that are comparable to cryptographic methods. He identified four critical properties of smart contracts, which Murphy and Cooper (2016) identified.

Digital form: Code, data with program execution.

Embedding: Legitimate concepts or practical outcomes derived from computer programming in software.

Once started and running, the effects that smart contracts encode cannot be over unless the end result depends on unseen conditions. Therefore, a contract between two or more parties that is safely kept digitally and carried out by secure code constitutes a smart contract-based blockchain. Users of the blockchain can transfer assets in a transparent manner without the use of a middleman, thanks to smart contracts. Smart contract scanning, like a physical contract, automatically populates the rules between two parties. Unlike physical contracts, smart contracts allow for the tracking of insurance payments and claims and the accountability of both parties. The corporation offers to pay for any future medical bills for anyone willing to pay the insurance company under the terms of the policy, which can be implemented as a coded decentralised smart contract. Blockchain smart contracts can create immutable data from policyholder records. Policyholders can approve or reject claims presented to the company.

Using IBM Blockchain, AIG and Standard Chartered transform multiple policies into "smart contracts" to provide a single, unified aspect of data and policy documents in a real hour. The solution provides coverage and premium payment summaries and provides network participants with automatic notification of payment events.

13.4 Blockchain in Asset Tracking and Identity Management

A transparent blockchain solution that allows multiple companies to create linked records together can make collections more efficient. General ledger capabilities help insurers settle claims, build trust in evidence sharing, and improve the overall customer experience.

Anti-money laundering (KYC/AML) Know Your Customer rules pose specific challenges for banks and insurance companies. Financial to institutions, including insurance companies, can use a blockchain-enabled shared database streamline and reduce KYC/AML compliance costs. Client input is done by only one institution. If the client wishes to contract with a new agency, for the purposes of due diligence, the agency may request access to already-existing documents in the chain. Institutions can only access papers that have been authorised, thanks to encryption, and changes to client files are always visible to whoever makes them. Transaction review and monitoring can also be automated completely. Goldman Sachs estimates that by "sharing" information about financial transactions and simplifying KYC onboarding via a distributed ledger, US banks could reduce their AML compliance costs from $3 billion to $5 billion. These banks' AML compliance costs in 2016 totaled over $10 billion, and including regulatory penalties, he exceeded $18 billion. Organisations can benefit from deduplication of integration efforts and improved data quality. Both measures aim to reduce the need for employees who make up 80% of their budget for KYC/AML compliance. Fines may also be reduced if the "catch rate" is improved. Finally, improved information security should lower the cost of hacking and nullify cyber incidents.

13.4.1 Fraud mitigation

Fraud is a major problem in the insurance industry. Honest policyholders end up paying high costs associated with false or exaggerated claims. In Ireland, it is estimated that insurance fraud costs insurers £200 million annually. According to the FBI, in countries like the United States, non-health insurance fraud costs more than $40 billion in annual premiums or between $40 and $700 per family per year. Blockchain could help the insurance industry fight fraud by improving coordination among insurers. Insurance companies can prevent double bookings and the processing of multiple claims for the same occurrence by using a centralised blockchain-backed database with several levels of access and management. To reduce counterfeiting, digital certificates give ownership of expensive goods. Reduce premium only (e.g., when an unlicensed broker sells insurance and collects premiums). Fraud prevention is an attractive use case for blockchain for the insurance industry.

13.4.2 Collateral

Reinsurers is a large Irish company, which has the second most reinsurers in Europe. Blockchain technology can help insurers and reinsurers exchange information and payments more easily. It is possible to take data from the blockchain for automated modeling, auditing, and compliance. By providing all stakeholders with automatic notifications, risks and claims can be shifted, as well as automated payment and reconciliation processing. Blockchain technology enhances data quality while decreasing costs, errors, and time. This method of implementation necessitates close collaboration between insurers and reinsurers, but the end result is significant cost savings.

13.5 Special Potential Implications of Blockchain: Vehicle, Marine, Life, Fire, Disability, and Motor Insurance

13.5.1 Autonomous Vehicles

Unbeknownst to many, blockchain technology has been central to the upshot of autonomous vehicles. Is it true? Autopilot mode enables a car to easily park itself or perform various other tasks. The car's digital system can be instructed to perform many activities using voice commands. This capability is made possible by blockchain technology.

13.5.1.1 Benefits of blockchain in autonomous industry

13.5.1.1.1 Ownership transfer

This is where blockchain technology can make the process incredibly easy and possibly eliminate the need to deal with agents. It can be easily processed. Smart contracts can also be used to enable buyers and sellers to seamlessly complete the sales process. The fact that middlemen can be eliminated is the biggest advantage of smart contracts when transactions can be made legally binding. This is not yet a reality due to inadequate laws around the world, but change is underway.

13.5.1.1.2 Automation of vehicle maintenance

Vehicles currently using blockchain technology include advanced on-board computers such as EDU (Engine Management). The EDU contains an error log that service professionals can use to quickly repair or diagnose the vehicle. Maintenance status is monitored in seconds; there is no need to go to car service. This also saves money as the vehicle does not end up in an undesirable situation of being towed.

13.5.1.1.3 Funding

This is another advantage of blockchain technology that is bound by the laws of various countries. However, if cryptocurrency payments are possible, this is a great advantage to consider. The entire auto finance industry is gearing up for the inevitable future of using digital coins to buy cars [8].

13.5.1.1.4 Security

Finally, we can see that this technology has played a very important role in the automotive industry, which is now hardly considered. Vehicles are becoming more and more automated. This means more digital technology is involved, automatically making cars more vulnerable to cyberattacks. As a result, the use of blockchain becomes mandatory as it uses very strong cryptography. Reverse engineering is not possible. Distributed ledger technology makes storing data easier. Immutability is practically achieved [11].

13.5.1.2 Blockchain technology has an impact on the automotive industry

When we talk about blockchain technology, we rarely think of the automotive industry. Blockchain is already in the market and has undergone major changes; so this could be a big mistake. As we will see in the next section, different parts of the automotive industry are actually adopting blockchain to improve their services.

13.5.2 Case study: Real-world manufacturing companies using blockchain technologies:

13.5.2.1 Volkswagen Group

The Volkswagen Group tracks its mineral supply chain using blockchain in its manufacturing. In fact, the company tries to source materials efficiently and keep them in good condition. They are even exploring cobalt for use in lithium-ion batteries. Testing these minerals is very resource intensive, but the blockchain can save time and money.

13.5.2.2 Toyota Group

Toyota is another company using blockchain for the automotive industry. In fact, they are studying the impact of blockchain on industry to identify where it can maximise productivity within organisations. It also manages the network using blockchain to get the information it needs for autonomous vehicles [9].

13.5.2.3 Samsung

Samsung is using blockchain very effectively in the manufacturing market. In particular, the company hopes to use blockchain to track shipments from factories to stores with hourly tracking. The company says it can reduce shipping costs by up to 20% once it adopts blockchain. This is a very important aspect for manufacturing companies like Samsung

13.5.2.4 Ford

Another car manufacturer employing blockchain to acidify minerals is Ford. They actually use IBM technology to validate mineral shipments for the sector. Their primary objective is to offer a public network that enables people to verify and examine all minerals in order to save time and money.

13.5.2.5 Nestlé

Blockchain technology is being used in Nestlé's production process. Here, the business is collaborating with OpenSC to enhance its own supply chain and process monitoring. However, this project is distinct from what was done at IBM Food Trust.

13.5.2.6 Unilever

Blockchain will be used in a year-long trial project by Unilever to trace transactions in the tea supply chain. Today, the company coordinates the supply of tea to farmers and the full production process with the help of big banks.

13.5.2.7 Merck & Co.

Another pharmaceutical business adopting this technology in production to address issues with medication storage, regulatory compliance, clinical trial efficiency, and cold chain effectiveness is Merck & Co. They currently participate in the group established by the Drug Supply Chain Security Act (DSCSA).

13.5.2.8 Foxconn

One of the largest technology corporations today, Foxconn, is in charge of producing the Apple iPhone. In reality, the usage of blockchain has made it possible for non-bank lenders to make loans directly to the supply chain. Without additional paperwork, this makes it easier for lenders and suppliers to interact.

13.5.2.9 Pfizer

Pfizer also employs blockchain in its manufacturing processes. These are real pharmaceutical companies that want to make sure they have the appropriate products in their portfolio.

13.5.3 Marine insurance

Marine insurance covers loss or damage to ships, cargo, terminals, and any transportation by which property is transferred, acquired, or held between the point of origin and destination.

How blockchain is reducing the fluidity of risk in marine insurance: Numerous potential advantages of blockchain technology exist for the insurance sector. The power of blockchain ledgers lies in their fundamental capability to connect all stakeholders and make the entire database completely clear to all. It provides a solution that goes beyond the specifics of marine insurance and is applicable across the insurance industry [12].

Blockchain allows these documents to be digitised into secure, smart B/Ls, speeding up shipping approval processes while reducing fraud risks. Administrative delays at ports and customs can be reduced by using smart B/L technology to create a more efficient global shipping system.

The time has come for transformation to manage dynamic risks and also to empower global trade.

Near-real-time data visualisation, related directly to smart contracts, also enhance decision-making, data security along with transparency, with parties such as moderators or auditors able to outlook the database [12].

"This more accurate information means less liquid capital on paper and frees brokers who spend most of their time on costly administration to work on value-added services," says Crawford. "But for a shipping company that spends hundreds of millions annually on insurance, a blockchain solution offers an opportunity for tighter premiums and better claims processes. They have a much clearer value. We go to the end client to transform the entire industrial process."

However, successfully blending blockchain technology to its full capability over the industry is extremely complex. Many jurisdictions as well as regulatory bodies are integral to marine insurance, making it difficult to implement solutions for large-scale ecosystem-wide changes.

Benefits of blockchain in marine insurance:

13.5.3.1 Claims handling

Blockchain allows every parties, including insurance companies, to read blockchain data such as lading bills, charter parties, and reports, and to reduce the time it takes to collect relevant documents from stakeholders. Furthermore, this technology can reduce human error when inspecting damage documents and improve the efficiency of damage decision-making.

13.5.3.2 Underwriting/risk assessment

By connecting brokers, insurers, and third parties to a distributed ledger that captures identity, risk, and exposure data and integrates it into insurance contracts, blockchain technology has the potential to streamline processes. It is concealed. The collected data can then be linked to policy agreements, allowing blockchain platforms to: receive and act on information that leads to price changes and business processes.

13.5.4 Life and health insurance

The impact of blockchain technology on life and health insurance:

Blockchain technology may be applied by health and life insurance companies to improve stakeholder communications, corporate processes, and record-keeping. How blockchain's special qualities may assist insurers with cost-savings, risk management, customer service improvement, business expansion, and eventually revenue growth is one of the important challenges – whether or not connected. The reasons for the adoption of blockchain technology in health and life insurance are as follows.

13.5.4.1 Moving toward interoperable and detailed health records

This technology tackles interoperability difficulties better than existing solutions because to increased security and the capacity to build trust between entities.

13.5.4.2 Using smart contracts to support administrative and strategic needs

Blockchains automatically gather all contract and transaction records, as well as other crucial information. They then combine this data and employ smart contracts and take appropriate action in light of the information.

13.5.4.3 More effective fraud detection

Smart contracts assist in determining whether a communication of fraudulent information to a life or health insurance provider through false claims, false claims, or other channels is genuinely valid.

13.5.4.4 Improved provider directory accuracy

Distributed consensus system on the blockchain will make it quicker and simpler for providers and insurers to change their quotes.

13.5.4.5 Make the application process easier by emphasising the needs of the applicant

It can bring convenience and security to any process.

13.5.4.6 Enabling dynamic insurer/customer relationships

A foundation for incorporating a variety of health-related behaviors into insurer/customer dynamics could be provided by electronic health data that are securely kept in smart contracts. Nature is there.

13.6 Motor vehicle insurance

Blockchain technology is impacting motor vehicle insurance:
Technology has grown by leaps and bounds over the last decade and has permeated almost every aspect of our lives. The multi-trillion dollar global auto insurance industry is no stranger to technology either. Thanks to the breakthrough blockchain technology, motor insurance is no longer as we remember it. Motor insurance companies are now able to cryptographically secure all insurance related information on a secure and reliable record keeping platform using blockchain technology [11].

13.6.1 How can blockchain technology be used at insurance organisations?

13.6.1.1 Fraud prevention

Families may pay an additional premium of $400–$700 per period as a result of insurance fraud. According to an FBI report, experts calculated that fraud against medical insurance is worth around $40 billion yearly in the US.

Using blockchain insurance claims, users can get rid of these common types of insurance fraud.

How can blockchain be used to solve the problem of insurance fraud?

Once insurance claims are moved to a blockchain-based ledger shared between insurer teams, the immutability of the blockchain prevents changes and updates, reducing fraud. Blockchain also makes it much easier to coordinate and connect insurers. By viewing the shared blockchain ledger, insurers can instantly get if a particular payment has been paid. It uses the same historical billing information; so you can quickly identify suspicious behavior [10].

Combined with encryption for data security, blockchain will now allow insurance claims information to be exchanged between insurance companies without exposing sensitive, personally identifiable information [10].

13.6.1.2 Customer engagement

Insurance companies face many challenges when gathering information. This is because customers are reluctant to share sensitive responses with insurance teams and agents as they are concerned about protecting sensitive data [10].

Blockchain ensures better protection and control of personal data. This specialised technology allows customers to share personal and sensitive information securely and confidentially [10].

It is easy to request data from Know Your Customer (KYC) data services that allow customers to easily share their identity information based on smart contracts. It is also easy to reuse for the most secure authentication of other companies [10].

13.6.2 Some insurtech organisations that use blockchain technologies

13.6.2.1 Ryskex

Insurtech company Ryskex helps users assess and manage risk more easily through its blockchain-based platform [10].

13.6.2.2 B3i

Founded in 2018, B3i is to support the insurance market by providing end customers and users with superior solutions through faster access to insurance and reduced administrative costs.

So we can finally say that blockchain could be a game changer for today's insurance industry. Blockchain has several useful use cases and can transform the way physical assets are managed, tracked, and protected using technology in digital form.

It is time to adopt blockchain as it offers various benefits such as increased cost efficiency, reduced risk, and more [10].

13.7 Secured Insurance Framework: Utilising Blockchain Technology and Smart Contracts for Insurance

Immutable data can be provided in terms of data storage, security, transparency, authenticity, and security during the execution of a process or job using blockchain and this contemporary contract technology. Blockchain implementation makes the entire insurance process from verification to claims settlement more secure and transparent. The architecture for using smart contracts for insurance contracts and storing them on the blockchain is described in this study. A transaction occurs when all pre-determined conditions are met in the case of a claim. Otherwise, it will be turned down. The terms of service cannot be altered. In other words, both sides have room for improvement. The private Ethereum network serves as the foundation for this blockchain and smart contract framework. To create smart contracts, the Solidity programming language is used. To validate transactions, this framework employs the Proof of Authority (PoA) consensus algorithm. In the event of an incorrect transaction request, the consensus algorithm will respond to the request and cancel it. By employing blockchain technology and smart contracts, this framework can address all trust and security difficulties that rely on traditional guarantees. Participants buy insurance and then use the proceeds in case of an emergency. Excellent money makers that produce employment and pay taxes like any other business are insurance against it. Consequently, it cannot be presumed that the insurer's concern is the settlement of the claim or the fraudulent claim. A quick and secure claim process is therefore crucial.

13.7.1 How blockchain works in this case?

Data is stored in blocks using the independent and secure verification mechanism of blockchain technology. Unofficial changes are almost impossible since transaction requests for each process or change within a block must pass through a consensus procedure. In order to make the entire transaction process safe and immutable, it additionally encrypts data during transactions and embeds time-stamp blocks. The use of transparent and extremely secure smart contracts streamlines the entire process. Consider a scenario where a test case matches a trigger for a transaction request, passes through a consensus mechanism, and then is eventually executed. The entire insurance system is made more dependable, effective, and safe for all parties – the insured and the insurer – by putting this full insurance procedure into practice. The primary objective is to begin the entire implementation and keep the job.

13.7.1.1 Overview of the entire system (flow)

Customers can register and issue insurance policies, and apply and reimburse with the help of their respective agents. With the help of suitable agents, information is placed on the Ethereum private network. Agents are responsible for sending all client requests over the network. When a transaction occurs, the validator is responsible for validating the transaction. Note that this is an insurance company framework. Validators pre-select and validate transactions that follow the proof-of-authorisation algorithm's guidelines.

13.7.1.2 For verification

The majority system is designed and processed algorithmically, but there is a minimal front end that also includes the authentication process of the system's stakeholders. Google's Firebase system handles authentication. Tenant and authorisation both use the same login name. A hard-coded filter is activated after login to filter users and administrators. Agency members are manually registered due to the need to maintain confidentiality and discreetness within the Agency. Registration is therefore only possible for relevant users and customers [13].

13.7.1.3 Agent panel

Agent panel, be it desktop, console, or web application, is an integral part of your system; so it is essential. It manages critical functions such as policy issuance, policy initialisation, and various client issues. The portal is shared by both agents and validators. Because the agency in some cases allows user to choose one for both roles. You can manage all operations through the agent panel interface, which includes the client list view, agency policy list, task and transaction history, and others [13, 15].

13.7.2 Character participation in this model

The protagonist, who is the client, is directly involved in the structure. Contracts, insurance claims, claims submissions, refund requests, and other functions are available. All documents and customer requests are processed by an intermediary or agent and placed on the network's blockchain. They are also present in this system as part of the self-certification team and are in charge of storing contracts in the general ledger and validating contract policies and transactions [13, 14].

13.8 Blockchain-based Data Exchange Platform

Blockchain helps insurance businesses succeed by addressing two critical needs of the insurance industry: operations and opportunities. At the same time, it can pave the way for new business opportunities, such as expanding into specialised areas such as parametric insurance and event insurance. Blockchain technology can also support new business models in which companies partner with other insurers to share political risk.

Blockchain enables a new insurance practice that uses smart contracts and dApps to execute real-life insurance transactions using blockchain accounts. While such implementations are more radical and, at times, speculative than blockchain insurance, they provide exciting innovation opportunities that lead to new products and markets. Blockchain, in particular, has the potential to lower the identity threshold required for insurance. Index or parametric insurance, as well as peer-to-peer (P2P) insurance, are examples of areas where blockchain insurance has been used.

13.8.1 Insurance that is parametric or index-based

Index-based insurance bases coverage on the performance of each underlying index. Temperature, precipitation, and wind speed, among other factors, may influence the index. Net losses are not covered by this type of insurance.

A certain lump payment is paid if an objective predefined parameter happens throughout time. Thirty days during a drought year in a particular area to make up for lost crops or herds. Index-based insurance is less expensive to manage than conventional insurance products, but blockchain can greatly automate each stage of the contract to claim processes for index-based insurance, allowing you to further cut expenses. By using oracles to feed outside data into smart contracts, index coverage on the blockchain makes it possible to automate the initiation and settlement of transactions between blockchain accounts in a matter of minutes.

Because of its simplicity, index insurance is particularly effective in developing countries. In much of Africa and Asia, the value of insurance claims tends to be very low, but administration costs remain as high as in developed countries. Index-based insurance can address the imbalance, and running on technology allows for even more innovation. By reducing overhead, automating billing, and speeding up payments, blockchain technology makes these products more cost-effective and more convenient for consumers, making them more accessible to consumers. You can make it attractive.

13.8.2 P2P insurance

Peer-to-peer (P2P) insurance is a disruptive form of claims management that allows policyholders to pool capital and self-manage their insurance. To prevent or mitigate the potential for fraud, a key aspect of P2P, this type of insurance creates a group of people who share mutual trust and common interests. Fundamentally, P2P insurance is very different from traditional insurance where insurance companies leave premiums unpaid. Peer-to-peer (P2P) insurance is a type of mutual property and casualty insurance that is funded by a group of people with specific links (family, friends, business partners, etc.). This mutual fund boasts low loss rates and cheap fees for members because it was carefully chosen. Additionally, money that is still in the Fund at the end of the coverage period may be given back to subscribers and policyholders. Although it may also call for the signatures of several members of the claim verification team, the smart contract code may designate a certain external reviewer as the signatory. Members can rely on a voting system since the blockchain retains an immutable record of everyone's decisions.

13.8.3 Claims handling

Collecting and processing damaged data can be difficult and expensive. Because data is manually entered and exchanged between different parties using different systems, transmission errors may occur. Blockchain technology has the potential to greatly automate the payment process and shorten processing times. A smart contract notification sensor, for example, could be installed in a car that runs for a day. The smart contract automatically notifies emergency services, verifies insurance, and initiates an insurance claim in the event of a collision. This type of automation results in improved service for policyholders and lower costs for insurers. Manual processes are being reduced or phased out.

13.9 Blockchain Disruption in Various Insurance Domains: Automation in Claim Submission and Processing, Fraud Detection, and Prevention

The insurance industry could be significantly disrupted by blockchain technology in the following ways:

- Smart contracts and triggered events
- Better back-end efficiency
- Conciliation
- More accurate price and risk assessment
- New insurance products

A significant advantage that blockchain can provide is the ability to achieve insufficient cost savings. It is obvious that blockchain use can have an impact on billing, governance, underwriting, and product development, and majority of blockchain applications currently are geared toward reducing costs. One of the first areas that insurers are investigating is the use of blockchain to automate claim payments. By validating insurance coverage between companies and reinsurers, it is the potential of blockchain to automate functional claims. This automates payments between claimants while also lowering administrative costs for insurers. Future changes in pricing and other areas of product development are expected as it becomes easier to integrate digital proof into underwriting. The combination of the Internet of Things (IoT) and artificial intelligence (AI) will lead to the automation of insurance processes, and in the near future, our industry will be completely different. However, these are still new technologies that must be thoroughly researched before they can be fully utilised by the insurance industry [20].

Cognizant, an IT multinational, collaborated with 14 Indian insurance companies in 2018 to develop a blockchain-enabled shared database. The platform is intended to reduce the risk of data breaches, fraud, and money laundering while also providing a better customer experience through increased process efficiency, improved records, and quick turnaround times. Blockchain's decentralised and immutable nature makes it more resistant to document tampering and accounting fraud.

EY and Guardtime launched Insurwave in 2018 in collaboration with leading insurers such as Maersk, Acord, Microsoft, MS Amlin, Willis Towers Watson, and XL Catlin. To support hull insurance, the platform makes use of blockchain technology. New vessels are registered on-chain, premiums are calculated using algorithms, and insurance policies are automatically distributed to carriers. The path of the train is tracked in real time, from location to weather conditions. This information is recorded on file and used for future underwriting as the vessel passes through the risk area. Claims are evaluated and approved more quickly, but decisions are more precise. The data quality has greatly improved, allowing various stakeholders to access immutable records of vessel life in real time for increased reliability and transparency.

13.10 Blockchain-based Health Data Collection (Electronic Health and Prevention)

One of our most effective weapons in the fight against the COVID-19 pandemic is data; yet, health data, like consumer data, is now kept in the safes of numerous organisations. Although this data may be tracked, purchased, sold, and shared by several organisations and businesses, humans remain the most potent resource we currently have for solving significant issues. We require technology to manage our health staff and medical supply chains, track and monitor infections, and plan for the future of our economy.

Data is one of our most powerful tools in the war against the COVID-19 pandemic, but like consumer data, health data is now housed in the archives of various organisations. Although numerous organisations and businesses can track, purchase, sell, and exchange this data, the most effective resource that we still have to address significant issues is people. Technology is necessary for us to identify and monitor infections, manage our medical supply chains and personnel, and plan for the future of our economy. But how can we balance privacy rights with data collection? Big Data laws and regulations, such as the General Data Protection Regulation (GDPR) of the European Union and the Consumer Privacy Act of California, are at best hazy. Yippee! Regulators and industry professionals have paid particular attention to the demand side of the data equation. Corporate apps from companies like Facebook, Google, and Uber use or sell personal data without the user's knowledge. Who developed, owns, and originally acquired the data? From where does it originate?

The answer is indeed "yes." This data is only a portion of your digital identity. In other words, "virtual pals" are made through Internet data traces. The majority of businesses and organisations advertise themselves in this way. Carlos Moreira, CEO of WISeKey, stated that although that identity is now yours, the data collected from that engagement in the world belongs to someone else.

With the help of blockchain technology, we can recognise the genuine worth of every element of our lives and use information to proactively better our health. Data management may enhance many aspects of your life, but one sticks out as a good place to start. It is health information.

Why are health data important?

According to economist Robert J. Shapiro, "Imagine General Motors not paying for steel, rubber, or glass - that's its contribution." "The major Internet businesses operate in this manner. It is a blessing." Executives who need as much data for their companies as possible while also respecting privacy and individual rights face a real challenge.

It is possible that we will not be able to use our data to plan for our future health or lives.

There are additional resources for treatment schedules, prescriptions to take, medical equipment for Medicare, insurance, and supplemental insurance, and advice on how to manage your funds. All of this data on us is kept in various databases that are kept by various healthcare providers, pharmacies, insurance providers, as well as local, state, and federal organisations. We do not have access to any of these holdings, with the exception of the American Medical Collection Agency (AMCA) execution. We might not be aware of this happening.

We do not charge for using this data, but we do bear all risk and liability for data erasure in the event of loss or misuse. When AMCA was breached in 2019, the personal data of about 5 million people whose lab tests were performed by AMCA clients Quest Diagnostics, LabCorp, BioReference Lab, and others were at risk. These clients are not inundated with individualised fraud alerts and online con artists. None of these businesses, unlike Alectra, Amazon, and Tesco, are leveraging data to enhance health or cut expenses. We consider this to be data misuse. We are unable to commercialise or regulate the information in these datasets for the benefit of our families or future generations. Take Henrietta Lacks, for example. Unintentionally, Henrietta Lacks' cancer cells were employed to treat cancer. Without access to the Internet, people are frequently unable to access formal identities, home addresses, or bank accounts. As a result, they are unable to engage in the global economy. They have records. These are those who lack knowledge.

We are unable to create or support the proposed health policy from elected officials, to successfully advocate for the improvements that our families need, or to engage in collective bargaining. All of this can be achieved by utilising our data as a component of the system, not only to advocate for business-friendly regulation but also to negotiate coverage and rates with governments.

With the use of wearable technology and the IoT, we can progressively track BP, step count, daily step elevation, and insulin levels. By controlling our health information and other personal data, we can address the following five issues.

Considerations including advocacy, commercialisation, access, security, and privacy are crucial. Utilising current technology to manage our data in compliance with our own terms of use is essential.

How can patient control over their health information speed up the application of data to medicine?

Innovative companies like Canada's University Health Network (UHN) have used blockchain technology, which functions as a distributed shared ledger, to develop win–win solutions. This form of ledger differs from an interface to a conventional database or health data shop in that it is both decentralised, allowing peer-to-peer transactions to be verified, and immutable, preventing transactions from being changed or reversed without the majority of users' approval.

Applications like contact tracing can use this kind of capabilities. Think about the following example: With the UHN solution, all Canadian patients can share personal information over time, including location, with healthcare providers across the country. Additionally, workers may demonstrate their fitness and return to work, thanks to Vital Chain's blockchain-based transformation of clinically confirmed results into health and safety data.

Applying abilities internationally offers a single, comprehensive view of the global influence and consequences that can be relied upon. The startup Hacera is aiming to accomplish that. He launched his MiPasa with assistance from the Linux Foundation, IBM, Microsoft, Oracle, and others. A global initiative called MiPasa aims to combine, gather, and distribute data from various verified sources (Centers for Disease Control or World Health). There is an agent panel interface where you can manage all of the features, including client list view, agency policy list, task and transaction history, etc. Send data via your network's integration of data providers and the IBM Blockchain Platform and IBM Cloud. For programmers to use while creating apps for this platform, Hacera has created a handbook. Collaboration between organisations is greatly influenced by this kind of value creation in order to track down contacts, lessen infections, save lives, and return more individuals to employment. The need for a COVID-19 vaccination is urgent. The blockchain company Shivom is embarking on a global project to gather and exchange data on virus servers in response to the Improved Medicines Initiative's call to action.

Shivom scientists established the Multi-Omics Data Center Foundation, an international partnership of universities, hospitals, and businesses. Many of them have experience with blockchain and artificial intelligence to battle COVID-19. The consortium's data center was constructed using a portion of Shivom's blockchain-based platform for precision medicine. Shivom built the platform Axel Schumacher, which leverages using blockchain to dynamically manage patient permission and transfer genomic data and data analysis to third parties without requiring access to raw genomic data safely and secretly. Dr. Schumacher claims that researchers can use the data to run algorithms to "Give a brief summary of the data set's statistics. No patient-specific information may be gathered without that person's agreement and with a specific goal in mind."

To make autonomous future a reality, we need to solve a real problem.

I do not exist virtually for you. We need a self-contained, non-transferable digital identity that may be used offline, online, or anywhere in the world. It cannot be granted or canceled by an administrator. It lacks the technical tools to wield this kind of sovereignty. There is now a solid technical base organisation who want to use secure messaging multi-party encryption, homomorphic encryption, smart contracts, and zero-knowledge proofs. Organisations that want to build on top of public key infrastructure decouple identification and authentication from transactions.

Consider your digital identity to be stored in a blockchain-based digital wallet. Every day, the wallet gathers and safeguards all of your biometric, financial, and geographic information. You have complete control over how you use it. This identification will require your medical records. Your body generates health information. There is your heartbeat and body temperature, not big corporations or governments. They are providing services when doctors measure you and perform various types of tests. The result is the capital that comes from your body. You need to control it.

Our goal is to fundamentally change the way ownership of data assets is defined and assigned, and the way identities are created, managed and protected in the digital world. Changing those rules ultimately changes everything.

13.11 Examples of How the Insurance Business is Utilising Blockchain

Blockchain in the insurance market: Insurance companies are entities that process a large amount of data as part of their activity. This is often sensitive and confidential data. As a result, they are usually the target of all sorts of cyberattacks aimed at stealing them. Undoubtedly, a factor increasing the risk of loss or leakage is the storage of all or a significant part of the data in the same place (e.g., on a single server) [21].

Blockchain technology in the insurance industry – use case:

Fraud detection – Finds and helps with fraudulent claims that are common in the insurance industry. Blockchain ensures that every transaction made is permanent and time-stamped. This makes it impossible for any party, including insurers, to modify the data to prevent a breach [21].

Automation: The insurance sector can save a lot of money with blockchain technology by reducing administrative costs through automation. Blockchain technology can automate death claims. Smart contracts allow life insurance companies to automatically receive information and verify the death of an insured when a hospital uses blockchain-based systems. Thus, recipients can grieve without further worry [21].

Smart contracts are self-executing contracts that take the form of apps, which work more or less in such a way that when a condition A occurs, B also occurs. Software algorithms in the smart insurance code can eliminate administrative hurdles, develop insurance payment scenarios, and automatically fulfill contract terms, while they leave no room for manipulation and at the same time increase trust [21].

Almost 80% of insurance executives either already use blockchain technology in their business divisions or intend to do so. Blockchain (distributed ledger technology) provides the ideal level of trust, transparency, and immutability.

Five real-world uses for blockchain in the insurance industry are as follows:

- detection of fraud;
- utilising blockchain and IoT to structure data;
- several forms of reinsurance;
- instantaneous insurance;
- micro-insurance.

13.11.1 Fraud detection

Fraudulent claims, including health insurance, cost the United States more than $40 billion each year. Despite digitisation, fraud cannot be detected by standard methods. Blockchain is very useful for fraud detection and prevention.

Blockchain ensures that all transactions performed are durable and time-stamped. This means that no one, including insurance companies, can change the data to prevent a breach. This data also helps insurers define fraudulent transaction patterns that can be used in anti-fraud algorithms.

Blockchain fraud detection use case: Etherisc independently verifies claims using multiple data sources using the Etherisc smart contract. For insurance claims, for example, we compare satellite imagery, weather forecasts, and drone imagery to images of the perpetrator.

13.11.2 IoT and blockchain together to structure data

As IoT connects several devices and the amount of data produced by each device will accelerate dramatically. For example, there were 26.66 billion UK IoT hardware active in 2019, with approximately 127 His IoT devices connected to the internet per second.

The above data is of great value to insurers in developing accurate, use-based actuarial insurance framework. In the auto industry, data collected on driving time, distance, acceleration, brake pattern along with other behavioral statistics can help identify great risk drivers.

However, the problem is how to control the amount of data with several devices communicating per second. And the answer for it is blockchain!

This helps users (insurance companies) to manage large as well as complex networks on a P2P basis. Instead of building high cost data centers, blockchain provides a decentralised place for storing and processing data.

13.11.3 Reinsurance and multiple risk involvement

Reinsurance is insurance provided to insurers. It safeguards insurers against large volumes of claims.

Due to information silos and drawn-out procedures, the existing reinsurance system is incredibly inefficient. Blockchain can help reinsurers in two different ways. One – secure and, second, by automating the procedure, the process can be sped up by using records for precise damage analysis.

Exchange of data and information: According to PwC, the reinsurance industry may save up to $10 billion by improving operational effectiveness.

For instance, a smart contract administration platform for Property Cat XOL contracts was introduced by the blockchain research collaboration B3i in 2017. This kind of reinsurance protects against catastrophic occurrences.

13.11.4 Immediate insurance

Immediate insurance is a flexible insurance framework that allows holders of policies to activate or deactivate policies with a single click. The more you are exposed to policy documents, the harder it becomes to manage records.

For example, on-demand insurance needs multiple underwriting/policy documents, buyer documents, calculations, risks, claims, and more compared to traditional insurance.

However, blockchain technology makes it easier to keep the books (records) of a business. On-demand players can use blockchain to efficiently note coverage from initiation to payment. An interesting use case for blockchain insurance, i.e., Ryskex, a German InsurTech founded in 2018, providing B2B insurers with a blockchain-based insurance platform for faster and transparent risk transfer more transparent.

13.11.5 Microinsurance

As an alternative to traditional insurance, microinsurance provides protection from certain risks at a much lower premium than traditional insurance. Microinsurance only benefits large retailers. However, despite the immediate benefits, microinsurance has been neglected due to low margins and high distribution costs. Blockchain will provide a parametric insurance program. As a result, insurers need lesser local agents as well as "card salespeople" can take place of local bankruptcy adjusters. For example, Surity utilises blockchain to provide insurance in micro-level, especially for those in Asia who do not have access to banks and other financial institutions. Blockchain's ability to inspire trust in an environment through a shared ledger and advanced internet security protocols will positively impact growth in future of the insurance industry [22]. Blockchain has the potential to penetrate insurance companies through three key functions, along with artificial intelligence and big data. Customer fraud is a big problem for businesses. Claims settlement in the insurance industry is costly due to loopholes, willful omissions, and fraud. By implementing blockchain technology, the entire insurance industry can work together to protect against fraud. If the policyholder provides incorrect/fraudulent information, or if the insurer refuses to cover the pre-agreed conditions, smart contract will be axiomatically terminated and the premium paid will be automatically terminated and be refunded to the individual. The whole process creates mutual trust between the two parties for two reasons. All data is provided transparently and minor breaches of contract lead to victim compensation. A previous report by Kim and Kang claimed as much blockchain presents a large market opportunity for the insurance industry in goods and services. According to the report, potential use cases for blockchain technology are included.

13.12 Conclusion

Blockchain-based insurance could change the industry. The ability to decompose the energy of data and move toward decentralisation of information and energy saves time and money when checking data quality. Smart contracts and oracles not only automate many of today's labor-intensive processes but also enable the development of new products designed to better meet customer needs. If insurance institutions want to utilise blockchain technology, please note the following: 1) collaboration and partnership are absolutely necessary; 2) collaboration and partnership are essential, and blockchain technology is the foundation for both; 3) to realise these possibilities, we need to focus on open architectures and microservices. Transition to new and improved business models. 4) Insurance companies are increasingly expected to send and receive information quickly in ways that are not possible with traditional ledger tools. Benefits include improved KYC/AML, fraud prevention, security, and new parametric and P2P products, more automated claim processing. This chapter, however, offers an illustration of what has been done in: Insurance and blockchain concepts under investigation. Case studies show the speed at which the theory is put into practice, and some major insurance companies are already playing a role in this area.

References

[1] Y. Zhang, "Increasing Cyber Defense in the Music Education Sector Using Blockchain Zero-Knowledge Proof Identification," vol. 2022, 2022.

[2] "Key Advantages of Blockchain Technology." https://imiblockchain.com/advantages-of-blockchain/

[3] S. Xu and X. Yin, "Recommendation System for Privacy-Preserving," vol. 2022, 2022.

[4] P. Tasca, "Insurance Under the Blockchain Paradigm: Volume I," 2019, pp. 273–285. doi: 10.1007/978-3-319-98911-2_9.

[5] P. S. A. Sreeramana, "Industry Analysis – The First Step in Business Management Scholarly y Analy ysis – The First St tep in Busin ness Manage," *Int. J. Case Stud. Business, IT Educ.*, vol. 2, no. 1, pp. 1–13, 2017.

[6] Mark Lounds, "Blockchain and its Implications for the Insurance Industry". https://www.munichre.com/us-life/en/perspectives/underwriting/blockchain-implications-insurance-industry.html

[7] Sam Daley, "How blockchain in insurance works, and a look at the companies leading the way." https://builtin.com/blockchain/blockchain-insurance-companies

[8] M. Georgiou, "Top 7 Use Cases of Blockchain in the Insurance Industry." https://imaginovation.net/blog/blockchain-insurance-industry-examples/#:~:text= Importance of Blockchain in Insurance 1,Plus%2C they can streamline the insurance... More

[9] Hasib Anwar, "blockchain in manufacturing Reviews Hasib Anwar on November 29, 2020 Blockchain In Manufacturing: A Guide To Industrial Empowerment." https://101blockchains.com/blockchain-in-manufacturing/

[10] "b33d2849d9a5053cde06bfca61053725692519f6 @ www.blockedge.io." [Online]. Available: https://www.blockedge.io/blog/blockchain-in-automotive-supply-chain/

[11] Hidyat, "No Title طرق تدريس اللغة العربية," Экономика Региона, p. 32, 2015.

[12] P. S. Aithal and S. Aithal, "Management of Core ICCT Technologies Used for Digital Service Innovation," Int. J. Manag. Technol. Soc. Sci., vol. 4, No. 2, pp. 110–136, 2019, doi: 10.47992/ijmts.2581.6012.0077.

[13] A. Hassan, M. I. Ali, R. Ahammed, M. M. Khan, N. Alsufyani, and A. Alsufyani, "Secure Insurance Framework Using Blockchain and Smart Contract," Sci. Program., Vol. 2021, 2021, doi: 10.1155/2021/6787406.

[14] J. Moormann, F. Pisani, and F. Holotiuk, "Impact of blockchain technology on business models in the payments industry", WI 2017 Proc., pp. 912–926, 2017.

[15] T. Shen, F. Wang, K. Chen, K. Wang, and B. Li, "Efficient Leveled (Multi) Identity-Based Fully Homomorphic Encryption Schemes," IEEE Access, vol. 7, pp. 79299–79310, 2019, doi: 10.1109/ACCESS.2019.2922685.

[16] J. E. B. and J. J. Little, "Biometric gait recognition," *Adv. Stud. Biometrics*, pp. 19–42, 2005,[Online]. Available: http://www.springerlink.com/index/UVXVNA70B62Y1V0B.pdf

[17] Mark Lounds, "Blockchain and its Implications for the Insurance Industry", [Online]. Available:https://www.munichre.com/us-life/en/perspectives/underwriting/blockchain-implications-insurance-industry.html#:~:text=Blockchain has the ability to,administrative costs for insurance companies.

[18] Paul DeCoste, "Peer-to-Peer Insurance – Blockchain Implications". https://blockchain.asu.edu/p2p-insurance/ (accessed 21 July 2022).

[19] S. Aithal and M. L. M., "Information Communication and Computing Technology (ICCT) as a Strategic Tool for Industries," Int. J. Appl. Ing. Manag. Lett., Vol. 3, No. 2, pp. 65–80, 2019, [Online]. Available: https://srinivaspublication.com/journal/index.php/ijaeml/article/view/245

[20] C. Liu, "Internet product design is the whole design around 'product strategy'," Lect. Notes Computing. Sci. (including Subser. Lect. Notes Artif. Intell. Lect. Notes Bioinformatics), Vol. 10288 LNCS, pp. 114–121, 2017, doi: 10.1007/978-3-319-58634-2_9.

[21] Blockchain in Insurance - Opportunity or Threat?, J.-T. Lorenz, B. Münstermann, M. Higginson, P. B. Olesen, N. Bohlken, and V. Ricciardi, McKinsey Co., no. July, pp. 1–9, 2016. [Online]. Available: http://www.mckinsey.com/industries/financial-services/our-insights/blockchain-in-insurance-opportunity-or-threat

[22] Satoshi Nakamoto, "blockchain.pdf."

[23] B. R and P. S. Aithal, "Blockchain-Based Service: A Case Study of IBM Blockchain Services & Hyperledger Fabric," Int. J. A case study. Business, IT, Educ., vol. 4, No. 1, pp. 94–102, 2020, doi: 10.47992/ijcsbe.2581.6942.0064.

[24] D. See Lovelock, J. and Furlonger, "Three things CIOs need to know about forecasting the business value of blockchain. Published by Gartner," 2017. https://www.gartner.com/en/documents/3776763

[25] Bernard Marr, "What is the Difference Between Blockchain And Bitcoin?" What is the Difference Between Blockchain And Bitcoin? %7C Bernard Marr

[26] Sood, K., Seth, N., and Grima, S. (2022). Portfolio Performance of Public Sector General Insurance Companies in India: A Comparative Analysis.In Simon Grima, Ercan Özen, Inna Romānova (Ed.), In *Managing Risk and Decision Making in Times of Economic Distress, Part B*. (pp.215–229). Emerald Publishing Limited,UK.

[27] Grima, S., and Marano, P. (2021). Designing a Model for Testing the Effectiveness of a Regulation: The Case of DORA for Insurance Undertakings. *Risks*, *9*(11), 206.

SECTION III

14

Blockchain Technology and Structural Change in Financial Services

Sonal Trivedi

School of Management, Birla Global University, Orissa, India

Email: trivedi.sonal86@gmail.com

Abstract

For various forms of cryptocurrencies, blockchain technology was initially deployed as the public transaction ledger. Apart from cryptocurrencies, blockchain has been focusing on a variety of other applications due to its notable benefits, which include anti-tampering, security, transparency, better services, decentralisation, instant settlement, and improved performance. As a result, the features that blockchain technology possesses are able to boost the Fintech industry by changing the way many previous services are taken in the financial industry. The current study looked at a Fintech use case and how it was altered by the use of blockchain in this study. The current study is developed based on secondary data such as previous research papers, newspaper articles, and interviews of experts available on YouTube. This study demonstrated how the R3 platform/Corda makes Know Your Customer (KYC) services more robust, simple, and dependable. Blockchain is gaining traction as a future technology that is affecting the Fintech industry. In this regard, this chapter first provides the potential and significant changes blockchain will bring to Fintech sector. Thus, this chapter is relevant for financial sector experts, financial businesses, policy-makers, academicians, and students.

14.1 Introduction

As of late, blockchain is acquiring consideration as there is by all accounts part of innovation potential open doors that can change our future in sure viewpoints. It rose as another regular illustration of the monetary market [49]. In 2016, the World Economic Forum (WEF) says that the blockchain has been named in the "Eventual fate of 10 new advancements," the report distributed by UN Future. That is the reason why numerous nations are putting a tonne in the blockchain sector, as subsidising expanded possibilities of quick development additionally increments [12].

The utilisation of blockchain helps Fintech associations to address many dependable issues that are looked into by the associations [10]. Out of this multitude of issues, one is KYC (Know Your Customer) strategy, which is solved by the utilisation of blockchain, and, furthermore, it is displayed in our utilisation case and its investigation is likewise done.

In this chapter, the KYC in finance services (banks) is finished with the assistance of blockchain, which gives distributed information assortment, complete verification cycle, and correspondence, and straightforwardness is inspected. This chapter adds to understanding the progressions in the monetary area utilising this application. The current study looked at a Fintech use case and how it was altered by the use of blockchain in this study.

The organisation of chapter is as follows. It covers the literature review explaining the benefits and drawbacks of blockchain in the financial sector and potential applications of blockchain in the financial sector. Then the chapter discusses the structural changes brought by blockchain technology in the financial sector. The chapter also presents a use case of "KYC in finance administrations (banks) utilising blockchain." Finally, the chapter presents findings and analysis followed by a conclusion.

The next section presents the literature review.

14.2 Literature Review

This section presents the literature review explaining the benefits and drawbacks of blockchain in financial sector on the basis of previous studies in the last decade. The literature review presents a brief introduction about blockchain and DLT followed by a brief introduction to Bitcoin and other cryptocurrencies.

Blockchain or distributed ledger technology (DLT) is a decentralised system that uses algorithms to eliminate the need for a central authority [36]. It offers many opportunities for cost savings, new markets for existing or innovative services, and start-ups adopting technology [40]. Blockchain eliminates trusted third parties through consensus within ledger participants [51]. The consensus consists of a time-consuming proof of work (PoW), with the majority of participants voting that the transaction is valid and processed [1].

The blockchain has different categories for each type of application. When a group of entities owns and controls a ledger, it is called an authorised private blockchain network [22]. If the ledger is owned by everyone and everyone can participate in the consensus, it is known as an unauthorised public ledger. This is a method of creating and manipulating with Bitcoin.

Bitcoin saw the first spread of this technology. This technology is mainly implemented in various cryptocurrencies such as Bitcoin, Ethereum, Litecoin, Primecoin, Namecoin, and Z Cash [37]. However, the underlying technology has great potential in other sectors such as decentralised identity management, credit systems, and banks.

After understanding the concept of blockchain, DLT, and Bitcoin, let us discuss the benefits and drawbacks of using this technology in the financial sector.

14.2.1 Benefits of blockchain in the financial sector

14.2.1.1 Instant settlements and reduced risk

Transactions are processed in few seconds rather than taking days in current systems. Need for much of the middle office and back-office staff can be eliminated by using blockchain because of the instant settlement of transactions [32].

14.2.1.2 Better financial products and services

Smart contract-based crowd funding provides transparency and reduces expenses with many innovative and unique ways of fundraising [33]. Money circles, where individuals can save and borrow money within the community, can be improved by using blockchain to improve transparency and trust among participants [1]. Outside communities are also accessible to raise funds from which was initially not promoted due to trust issues.

14.2.1.3 Improved performance of the contracts

Smart contracts can shorten the time needed to finish transactions. Complex financial asset transactions can be benefitted due to automatic settlement by the incorruptible set of business rules using smart contracts [25]. The only downside is that businesses will have to strictly follow certain regulatory compliances, across jurisdiction if needed.

14.2.1.4 Transparency

Entire monetary flow and transactions are recorded on the chain and can be audited by any party. Hence, transparency can be achieved across the system

[3]. Although not every data point can be exposed and so many solutions have been implemented like zero-knowledge proof, which only allows the required information to be seen by a party by the permission of the user, hiding other crucial information [45]. Enigma secret contracts encrypt the information to be shared in smart contracts between parties, to hide it from other members of the chain. Confidential transactions provide encrypted transactions between members. All these methods can be used as required in the system.

14.2.1.5 Reduced fraud by using self-sovereign identity

Blockchain prevents DoS attacks, hacking, and other types of frauds if used along with self-sovereign identity, which is a blockchain-enabled digital ID [42]. This ID helps in reducing identity fraud. The owner of the ID also has more control over it than the traditional IDs or Aadhar numbers. It will reduce duplication of information and massively reduce conflicts and confusion occurring in financial services.

14.2.2 Drawbacks of blockchain

Blockchain has major limitations like scalability, energy consumption, and concentration of mining pools. Money laundering and other criminal activities pose a threat if some form of real identity is not used to authenticate the participants [32]. There have been massive thefts of cryptocurrencies from the centralised entities used by many people to hold Bitcoins [19]. There is a widespread manipulation in the lightly regulated cryptocurrency exchanges [39]. Among promoters of initial coin offerings, there have been frauds to obtain real money in exchange for worthless cryptocurrencies [44]. Almost half of the Bitcoin transactions are by the tether "Stable Coin," which is said to be backed by US dollars [47]. It has never been audited and is believed to be involved in any suspicious activities.

After discussion of benefits and drawbacks of blockchain in financial sector, let us look into the potential applications of blockchain in the financial sector.

14.2.3 Potential applications of blockchain in the financial sector

14.2.3.1 Digital fiat currency or cryptocurrency

Bitcoin and Ethereum are two very popular and widely accepted cryptocurrencies. Many online vendors and online payment applications accept digital payments in Bitcoin and investors and investment firms are investing in cryptocurrencies [27]. Bitcoin has the potential to become the new world currency or even an intermediary for the new world currency although its possibility is not foreseeable in the near future practically [9]. But blockchain is set to disrupt the currencies of the world fundamentally if not completely replace them.

14.2.3.2 ICO (initial coin offering)

Initial coin offering or initial currency offering is a type of funding by using cryptocurrencies [41]. It can be public funding like crowdfunding or privately held funds for start-ups. A set quantity of cryptocurrency or "tokens" are sold to the investors or speculators, which can be exchanged for valuable cryptocurrencies like Bitcoin and Ethereum, or these tokens are realised in value when a set goal is met for the start-up and the project launches successfully [2]. Many innovative projects and ideas are seeking ICOs for their initial funding. ICOs have a potential for truly innovative ideas to be implemented and, if done correctly, can disrupt many financial services.

14.2.3.3 International market

Blockchain can fundamentally disrupt international money transfers. Banks and financial firms charge a fee of set percentage value to transfer the funds across borders [24]. With the help of blockchain, these fees can be eliminated and the money can directly be transferred to anyone across the border [16]. International banks require payments in gold while working with different fiat currencies. So instead of transferring gold now and then, they rely on third-party banks or firms to issue credits or contracts to the banks as payments [20]. This payment can be done using smart contracts on the blockchain and the need for the third firm can be eliminated by creating a trustless and robust system of contracts [15]. Even eliminating the gold transfers as the contracts would be reflected across the member banks and retains its true value throughout [52]. Many big banks are forming a consortium to combat disruptive technology, reduced customer turnover, and reduced fees for providing essential services. With the help of blockchain, these banks are forming private blockchain with the help of Hyperledger (a blockchain-based framework) and with the support of IBM run Linux Foundation, to work with each other and protect their firms from technological disruption.

14.2.3.4 Blockchain in banking services

Know Your Customer (KYC) and Know Your Business (KYB) require many documents each time a customer or a business goes to a different financial firm or bank [17]. Also, these documents are used to profile the data so that personalised services can be provided and future data points can be aggregated to [46]. Using blockchain to streamline the process can eliminate much of the paperwork for submitting documents every time a new account is created and also the process is instantly eliminating the wait for the approval of the documents [31]. Credit scoring is another parameter that can be streamlined by using common blockchain by the banks and financial firms [18]. All the firms participate in the same chain and so the credit scoring customers and businesses are transparent [23]. Insurance claims are also ripe for disruption by blockchain. Insurance claims management is a lengthy and complicated process involving many stakeholders and intermediaries participating to decide on the issuance of claims. Using blockchain, many of the intermediaries and stakeholders are eliminated, making the process fast and straightforward.

After looking into potential applications of blockchain in the financial sector, the next section discusses the structural changes brought by blockchain technology in the financial sector.

14.2.4 The structural changes brought by blockchain technology in the financial sector

It is important to understand the structural changes brought by blockchain technology in the financial sector. Additionally, it will help regulators to ensure effective regulation of financial sector after the inclusion of blockchain technology. This technology has changed the way financial sector operates and has also changed the structure of financial market.

Understanding the structure of blockchain technology system:

The blockchain generation gadget can sign up all transactions from extraordinary events without the use of the relied-on offerings providers [11]. Blockchain verification maintains the consistency and immutability of the transactions using consensus algorithms and cryptographic methods. [3]. It is regularly believed that monetary establishments, banking houses, institutions of commerce and land registry government must preserve an eye fixed in this generation [48]. It is even stated that those bodies (perhaps) will get replaced with the aid of using this developing disruptive generation. Blockchain has been defined as equivalence with the conventional system, known as ledger. This is a gadget of recording information (transactions, contracts, and agreements), whatever that is wished to be recorded one after the other and shown as having happened (BBC Business News). Originally conceived as the idea of cryptocurrencies, components of blockchain generation have far-accomplishing abilities in a lot of different areas [3]. To apprehend this ability, it is vital to differentiate centre blockchain components: dispensed-ledger generation and clever contracts [53, 54].

Blockchain works on the idea of proportion ledger that maintains a document of conserving for any sort of bodily or digital assets; a unit of currency, gold units, diamond, or any item [35]. Every player within the gadget has a duplicate of all transactions in a blockchain and replaces routinely each time a brand-new transaction occurs [4]. In blockchain, the database is represented by the shared ledger. As with relational databases, there is no standard format. There are two unbreakable principles for the blockchain shared ledger. It is tamper-proof because it is immutable in the first place. Next, it is organised. Time determines the order.

The shared ledger may have restrictions on individuals with accuracy. If individuals of the system are confined and preselected, that is permissioned sort of ledger; however, if absolutely every person has entry to the system, it is known as un-permissioned ledger [50]. Further, it can be a private, public, or hybrid dispensed community. A time-ordered record of immutable transactions is a property that gives blockchain its tremendous power which is produced by these two rules. The shared ledger produces a framework for delivering trust, transparency, and provenance when combined with consensus and cryptography [6].

The distributed ledger generation is, in a few manners, a superior model of the presently used centralised and decentralised ledger distribution gadget with the aid of banks.

The monetary transactions had a lot of successes that rely upon those approaches, especially within the fast payments services. The centralised distributions are obviously expensive, and all information in addition to processing is centralised to some extent with the aid of using integration with every player's personal systems [7]. On the other hand, many decentralised databases can take a seat down round the rims of a community whilst messages circulate among them [21]. Distributed gadget, in contrast, connects tens of thousands of computer systems in a community via the internet, with self-validation and reliability check [8]. DLT enables individual and companies around world to carry transactions on peer-to-peer and direct basis without any requirement to go via traditional banks.

In 1994, Nick Szabo, a researcher in the field of lawful issues and cryptography, concentrated on the decentralised conveyed record framework that could be utilised for the production of self-executing contracts, called advanced agreements, blockchain agreements, or brilliant agreements [38]. Digital contracts known as "smart contracts" are stored on blockchain and are executed automatically when certain criteria are met. The majority of the ordinary agreements have no straight association with the computerised code that executes them. In brilliant agreement design, agreements could be changed in a computerised PC decipherable code, put away, and reproduced in the organisation framework and constrained by the organisation of PCs of the blockchain [5]. In the event that clients go into a shrewd agreement, it will hold judgement that works on the information in all aspects of the disseminated record. Smart contract operated on simple "if/when...then" phrases that are typed into code and placed on a blockchain. When predefined circumstances have been verified and met, a network of computers will carry out the actions. These actions can be entail paying out money to the right people, registering a car, sending out notices, or writing a ticket [26]. When the transaction is finished, the blockchain is then updated. As a result, the transaction cannot be modified, and only parties to whom permission has been granted can view the outcome.

The smart agreements are being utilised in different fields, essentially for tyrant consistence, administration of the executives, and item recognisability, and, furthermore, to beat counterfeit items and misrepresentation in numerous areas: financial services, health, aviation, IT and communications, agriculture, oil and gas, and so on [43].

The circulated records framework, blockchain, has many advantages. The first advantage is the reduced dissatisfaction among customers and service providers. Blockchain helps in delivering services beyond trust as it provides instant traceability, increased transparency, automation, reduced cost and improved efficiency. Second, all exchanges by any members on the organisation are noticeable to any remaining members, which builds certainty, review, and trust. Third, changes to such close organisation are truly troublesome, and in the extremely rare case that such a change happened, it would be recognisable to any remaining clients [13].

After looking into structural changes brought by blockchain technology in the financial sector, the next section presents a use case to explain KYC in finance administrations (banks) utilising blockchain.

14.3 Use Case

KYC in finance administrations (banks) utilising blockchain:

Corda, an open source blockchain stage, empowers associations to acknowledge direct exchanges and with improved protection attributable to savvy contracts, lessening exchange, capacity costs, and pipelining different activities of business and associations [28]. A company relying on blockchain technology named R3LLC is developing an application with Corda which can be utilized by various sectors such as digital assets, trade finance, insurance and other financial services [14]. Its base camp is arranged in New York

City, USA, established by David E. Rutter in 2014. Synechron is a data innovation and counselling organisation arranged in New York [30]. It centres around the monetary administrations industry, which incorporates banking, capital business sectors, and insurance agency.

A KYC arrangement-based Corda blockchain was utilised for its most memorable worldwide preliminary in 2018 [29]. In June 2018, the preliminary was conducted, which ran for four days and involved the fruitful execution of north of 300 KYC exchanges in 19 nations all the while. While testing, banks have been able to demand consent for getting the client's KYC information. Subsequently which was automatically updated for all banks that have access permissions of user [34]. Around 39 banks partook in this preliminary, to be specific ABN AMRO, Alfa Bank, Bank ABC, ALD Automotive, CTBC Holding, BCI, Deutsche Bank, BNP Paribas, Bank of Cyprus, DNB, Hana Bank, ING, KB Kookmin Bank, Banca Mediolanum, Natixis, National Bank of Egypt, Qiwi, China Merchants Bank, RCI Bank and Services (the monetary administrations supplier for Groupe Renault), SBI Bank LLC, Commercial International Bank – Egypt, U.S. Bank, Shinhan Bank, Societe Generale, Raiffeisen Bank International, Woori Bank, and NH Nonghyup Bank. This large number of members in the preliminary sent and executed something like 45 hubs in Microsoft Azure climate and exhibited the circulated cooperative working over the Corda organisation.

The banks that took an interest in the test dealt with client's/clients' KYC information subtleties across the CorDapp planned stage, which is organised by Synechron.

The next section presents the finding and analysis, highlighting various advantages of blockchain to banks and their clients.

14.4 Finding and Analysis

The examination is done according to the point of view of key highlights given by blockchain. The elements all the while offer different advantages to both the clients and the banks.

1. **Distributed ledger:**
   This blockchain-based KYC utility framework takes into consideration the smooth distinguishing proof and check stage for the monetary and banking areas. That is on the grounds that buyer information is really gathered and kept within a dispersed design. To get to this information, KYC suppliers are expected to impart their client information to the organisations, associations, or banks requiring something very similar.

This blockchain KYC framework likewise offers much better information security by making it a requirement that information access is just made in the wake of conceding an affirmation or authorisation from the important element. This reduces the possibilities of unapproved access giving clients a more prominent command over their information.

2. **Secure authentication process:**
   A cryptographic confirmation helps monetary associations to check the personality of any substance quickly. This is fundamental for information assurance and forestalling cheats. The rising interest for banks and other monetary associations for creating applications that permit clients to manage online exchanges additionally presents novel difficulties. Security imperfections, bugs, and burglary of shrewd gadgets have made the general public demand for a safer decentralised arrangement that will briefly address the security worries, everything being equal.

   The security presented by R3's blockchain framework makes the possibilities of extortion probability considerably less. While a talented programmer might look through some private data assuming the PC is taken, because of the unchanging nature property of the blockchain, they cannot, in any case, change the information.

3. **Transparency:**
   R3's KYC in light of Corda's blockchain stage permits dynamic continuous observing of everything going from account openings to everyday exchanges. When joined with smart contracts which have predefined criteria to determine fraudulent activity, the corda platform of R3 has capability to alert banks of any suspicious activity.

   The blockchain's changelessness angle helps in gaining sound trust between concerned substances associated with the KYC cycle. The capacity to believe the information put away on a KYC blockchain programming arrangement eliminates the requirement for cross-checking, confirming, or validating components by auxiliary outsider organisations. Additionally, a dispersed record framework makes the revealing and correspondence processes more effective, which saves a tonne of time and assets. Since gatherings can, without much stretch, access trusted and solid information, dubious cycles and extortion can be identified in a more limited time.

14.5 Conclusion

Blockchain is receiving recognition as a future innovation. It is getting notable changes in the Fintech area. Connected with this, we first give an outline of the capability of blockchain and afterward the way in which it acquires a critical change the Fintech area. Thus, we talked about a few central questions that should be thought about while fostering any financial administrations with the utilisation of blockchain. Blockchain designers believe that these utilisation cases will become regular in the Fintech sector in the forthcoming years. We have shown how the Corda/R3 stage makes the KYC (Know Your Customer) administration become more simpler, solid, and hearty. The opposite side of blockchain is the pessimistic idea stimulating among individuals and their legislatures connected with cryptocurrency. Its future in long haul is still needed to be worked out as of now, and the above recorded use case is exceptionally productive with the assistance of blockchain and it is one of the most hopeful innovations in the area of Fintech.

References

[1] Ali, O., Ally, M., & Dwivedi, Y. (2020). The state of play of blockchain technology in the financial services sector: A systematic literature review. *International Journal of Information Management*, *54*, 102199.

[2] Benedetti, H., & Kostovetsky, L. (2021). Digital tulips? Returns to investors in initial coin offerings. *Journal of Corporate Finance*, *66*, 101786.

[3] Chang, V., Baudier, P., Zhang, H., Xu, Q., Zhang, J., & Arami, M. (2020). How Blockchain can impact financial services–The overview, challenges and recommendations from expert interviewees. *Technological forecasting and social change*, *158*, 120166.

[4] Chen, T., & Wang, D. (2020). Combined application of blockchain technology in fractional calculus model of supply chain financial system. *Chaos, Solitons & Fractals*, *131*, 109461.

[5] Chen, Y. (2018). Blockchain tokens and the potential democratization of entrepreneurship and innovation. *Business horizons*, *61*(4), 567–575.

[6] Chen, Y., & Bellavitis, C. (2020). Blockchain disruption and decentralized finance: The rise of decentralized business models. *Journal of Business Venturing Insights*, *13*, e00151.

[7] Chen, Y., Ma, X., Tang, C., & Au, M. H. (2020, September). PGC: decentralized confidential payment system with auditability. In *European Symposium on Research in Computer Security* (pp. 591–610). Springer, Cham.

[8] Chiesa, A., Green, M., Liu, J., Miao, P., Miers, I., & Mishra, P. (2017, April). Decentralized anonymous micropayments. In *Annual International Conference on the Theory and Applications of Cryptographic Techniques* (pp. 609–642). Springer, Cham.

[9] Diniz, E. H., Cernev, A. K., Rodrigues, D. A., & Daneluzzi, F. (2021). Solidarity cryptocurrencies as digital community platforms. *Information Technology for Development, 27*(3), 524–538.

[10] Dorfleitner, G., & Braun, D. (2019). Fintech, digitalization and blockchain: possible applications for green finance. In *The rise of green finance in Europe* (pp. 207–237). Palgrave Macmillan, Cham.

[11] Fanning, K., & Centers, D. P. (2016). Blockchain and its coming impact on financial services. *Journal of Corporate Accounting & Finance, 27*(5), 53–57.

[12] Fosso Wamba, S., Kala Kamdjoug, J. R., Epie Bawack, R., & Keogh, J. G. (2020). Bitcoin, Blockchain and Fintech: a systematic review and case studies in the supply chain. *Production Planning & Control, 31*(2–3), 115–142.

[13] Garg, P., Gupta, B., Chauhan, A. K., Sivarajah, U., Gupta, S., & Modgil, S. (2021). Measuring the perceived benefits of implementing blockchain technology in the banking sector. *Technological Forecasting and Social Change, 163*, 120407.

[14] Girasa, R. (2018). *Regulation of cryptocurrencies and blockchain technologies: national and international perspectives*. Springer.

[15] Grover, P., Kar, A. K., & Vigneswara Ilavarasan, P. (2018, October). Blockchain for businesses: A systematic literature review. In *Conference on e-Business, e-Services and e-Society* (pp. 325–336). Springer, Cham.

[16] Harris, W. L., & Wonglimpiyarat, J. (2019). Blockchain platform and future bank competition. *Foresight*.

[17] Hassani, H., Huang, X., & Silva, E. (2018). Banking with blockchain-ed big data. *Journal of Management Analytics, 5*(4), 256–275.

[18] Hassija, V., Bansal, G., Chamola, V., Kumar, N., & Guizani, M. (2020). Secure lending: Blockchain and prospect theory-based decentralized credit scoring model. *IEEE Transactions on Network Science and Engineering, 7*(4), 2566–2575.

[19] Hawlitschek, F., Notheisen, B., & Teubner, T. (2018). The limits of trust-free systems: A literature review on blockchain technology and trust in the sharing economy. *Electronic commerce research and applications, 29*, 50–63.

[20] Hooper, A., & Holtbrügge, D. (2020). Blockchain technology in international business: changing the agenda for global governance. *Review of International Business and Strategy*.

[21] Huberman, G., Leshno, J. D., & Moallemi, C. (2021). Monopoly without a monopolist: An economic analysis of the bitcoin payment system. *The Review of Economic Studies, 88*(6), 3011–3040.

[22] Hughes, A., Park, A., Kietzmann, J., & Archer-Brown, C. (2019). Beyond Bitcoin: What blockchain and distributed ledger technologies mean for firms. *Business Horizons*, *62*(3), 273–281.

[23] Jain, N., Agrawal, T., Goyal, P., & Hassija, V. (2019, October). A blockchain-based distributed network for secure credit scoring. In *2019 5th International Conference on Signal Processing, Computing and Control (ISPCC)* (pp. 306–312). IEEE.

[24] Joo, M. H., Nishikawa, Y., & Dandapani, K. (2019). Cryptocurrency, a successful application of blockchain technology. *Managerial Finance*.

[25] Karagiannis, I., Mavrogiannis, K., Soldatos, J., Drakoulis, D., Troiano, E., & Polyviou, A. (2019). Blockchain Based Sharing of Security Information for Critical Infrastructures of the Finance Sector. In *Computer Security* (pp. 226–241). Springer, Cham.

[26] Kewell, B., Adams, R., & Parry, G. (2017). Blockchain for good?. *Strategic Change*, *26*(5), 429–437.

[27] Kirkby, R. (2018). Cryptocurrencies and digital fiat currencies. *Australian Economic Review*, *51*(4), 527–539.

[28] Kolehmainen, T., Laatikainen, G., Kultanen, J., Kazan, E., & Abrahamsson, P. (2020, November). Using blockchain in digitalizing enterprise legacy systems: an experience report. In *International Conference on Software Business* (pp. 70–85). Springer, Cham.

[29] Malhotra, D., Saini, P., & Singh, A. K. (2021). How Blockchain Can Automate KYC: Systematic Review. *Wireless Personal Communications*, 1–35.

[30] Martens, D., Tuyll van Serooskerken, A. V., & Steenhagen, M. (2017). Exploring the potential of blockchain for KYC. *Journal of Digital Banking*, *2*(2), 123–131.

[31] Martino, P. (2019). Blockchain technology: Challenges and opportunities for banks. *International Journal of Financial Innovation in Banking*, *2*(4), 314–333.

[32] Niranjanamurthy, M., Nithya, B. N., & Jagannatha, S. J. C. C. (2019). Analysis of Blockchain technology: pros, cons and SWOT. *Cluster Computing*, *22*(6), 14743–14757.

[33] Osmani, M., El-Haddadeh, R., Hindi, N., Janssen, M., & Weerakkody, V. (2020). Blockchain for next generation services in banking and finance: cost, benefit, risk and opportunity analysis. *Journal of Enterprise Information Management*.

[34] Parra Moyano, J., & Ross, O. (2017). KYC optimization using distributed ledger technology. *Business & Information Systems Engineering*, *59*(6), 411–423.

[35] Poongodi, M., Sharma, A., Vijayakumar, V., Bhardwaj, V., Sharma, A. P., Iqbal, R., & Kumar, R. (2020). Prediction of the price of Ethereum blockchain cryptocurrency in an industrial finance system. *Computers & Electrical Engineering*, *81*, 106527.

[36] Priyadarshini, I. (2019). Introduction to blockchain technology. *Cyber security in parallel and distributed computing: concepts, techniques, applications and case studies*, 91–107.

[37] Rahman, A., & Dawood, A. K. (2019). Bitcoin and future of cryptocurrency. *Ushus Journal of Business Management*, *18*(1), 61–66.

[38] Raikwar, M., Gligoroski, D., & Kralevska, K. (2019). SoK of used cryptography in blockchain. *IEEE Access*, *7*, 148550–148575.

[39] Shanaev, S., Sharma, S., Ghimire, B., & Shuraeva, A. (2020). Taming the blockchain beast? Regulatory implications for the cryptocurrency Market. *Research in International Business and Finance*, *51*, 101080.

[40] Sharma, A., Tiwari, S., Arora, N., & Sharma, S. C. (2021). Introduction to Blockchain. In *Blockchain Applications in IoT Ecosystem* (pp. 1–14). Springer, Cham.

[41] Sharma, Z., & Zhu, Y. (2020). Platform building in initial coin offering market: Empirical evidence. *Pacific-Basin Finance Journal*, *61*, 101318.

[42] Singh, R., Tanwar, S., & Sharma, T. P. (2020). Utilization of blockchain for mitigating the distributed denial of service attacks. *Security and Privacy*, *3*(3), e96.

[43] Staifi, N., & Belguidoum, M. (2021). Adapted smart home services based on smart contracts and service level agreements. *Concurrency and Computation: Practice and Experience*, *33*(23), e6208.

[44] Tiwari, M., Gepp, A., & Kumar, K. (2020). The future of raising finance-a new opportunity to commit fraud: a review of initial coin offering (ICOs) scams. *Crime, Law and Social Change*, *73*(4), 417–441.

[45] Vincent, O., & Evans, O. (2019). Can cryptocurrency, mobile phones, and internet herald sustainable financial sector development in emerging markets?. *Journal of Transnational Management*, *24*(3), 259–279.

[46] Wang, H., Ma, S., Dai, H. N., Imran, M., & Wang, T. (2020). Blockchain-based data privacy management with nudge theory in open banking. *Future Generation Computer Systems*, *110*, 812–823.

[47] Wei, W. C. (2018). The impact of Tether grants on Bitcoin. *Economics Letters*, *171*, 19–22.

[48] Werth, O., Schwarzbach, C., Rodríguez Cardona, D., Breitner, M. H., & Graf von der Schulenburg, J. M. (2020). Influencing factors for the digital transformation in the financial services sector. *Zeitschrift für die gesamte Versicherungswissenschaft*, *109*(2), 155–179.

[49] Xu, M., Chen, X., & Kou, G. (2019). A systematic review of blockchain. *Financial Innovation*, *5*(1), 1–14.

[50] Yoo, S. (2017). Blockchain based financial case analysis and its implications. *Asia Pacific Journal of Innovation and Entrepreneurship*.

[51] Zachariadis, M., Hileman, G., & Scott, S. V. (2019). Governance and control in distributed ledgers: Understanding the challenges facing blockchain technology in financial services. *Information and Organization*, *29*(2), 105–117.

[52] Zhang, L., Xie, Y., Zheng, Y., Xue, W., Zheng, X., & Xu, X. (2020). The challenges and countermeasures of blockchain in finance and economics. *Systems Research and Behavioral Science*, *37*(4), 691–698.

[53] Sood, K., Seth, N., & Grima, S. (2022). Portfolio Performance of Public Sector General Insurance Companies in India: A Comparative Analysis. In Simon Grima, Ercan Özen, Inna Romānova (Ed.), In *Managing Risk and Decision Making in Times of Economic Distress, Part B*. (pp.215–229). Emerald Publishing Limited,UK.

[54] Grima, S., & Marano, P. (2021). Designing a Model for Testing the Effectiveness of a Regulation: The Case of DORA for Insurance Undertakings. *Risks*, *9*(11), 206.

15

Influence of Blockchain Technology in the Insurance Sector

Seema Rani[1], Nishtha Rawat[2], Vandana Sharma[3], and Carsten Maple[4]

[1]Amity Institute of Information Technology, Amity University, India
[2]IIT Roorkee, India
[3]CHRIST (Deemed to be University), Delhi-NCR Campus, India
[4]Secure Cyber Systems Research Group (SCSRG), WMG, University of Warwick, UK

Email: seema7519@gmail.com; nish.shiv@gmail.com; vandana.juyal@gmail.com; cm@warwick.ac.uk

Abstract

This chapter highlights the potential of blockchain technology in insurance sector. The technology simplifies and smoothens out the process from different insurers and insured viewpoint who interrelate for any communication or transaction in insurance sector. The insurance sector contributes as vital constituent towards economic growth. It increases the trust in individuals by augmenting the sense of financial stability and security to an individual, organisation, or business. The insurance companies' provider covers long or short-term risk activities encountered during the life span of an individual, organisation, or business. In situation of no loss or damage, insurance services offer considerable and unquantifiable peace of mind, which, in turn, leads to satisfaction or healthy life. In the digital era, e-insurance has taken up the segment to the level of incredible success. To deal with smooth delivery of insurance services along with security and confidence, adoption of emerging technologies like blockchain in the insurance sector will be an added advantage. The appearance of capital market and existence of computer networks, marketing, and distribution of products in online mode has increased the global market circumference. The insurance companies who have deployed the solution based on blockchain technology are benefitting their direct or indirect consumer or customer.

15.1 Introduction

The blockchain technology brings revolution in view of secure solution to many problems and opening new windows for research and innovation. The benefits of blockchain in segments like finance, Internet of Things, railway, and energy segment is immense and further researchers are continuing to make the best possible use of this technology for providing the solution to many problems. One such segment that has the potential of getting immense benefit from blockchain technology is the insurance sector. This paper introduces the basics of blockchain and its working in insurance sector along with benefits and challenges in implementation. All aspects related to insurance sector has been taken into consideration [42, 43].

Blockchain was coined by Satoshi Nakamoto in 2008 and his invention of cryptocurrency system is named Bitcoin. This was the beginning of the electronic payment scheme and it revealed the potential of blockchain technology. Subsequently, the blockchain technology has shown enormous return in different fields like business, insurance, and medicinal segment [1]. Blocks are connected in the blockchain network through a link where each block comprises valid transactions. These transactions exposed the progress of an asset either tangible (a product) or intangible (intellectual). Also, a larger block is required if the number of transactions inside a block increases. The transaction can be initiated by any node present in the network and then can be forwarded to all other existing nodes in that network. Validation of transaction is done through algorithms. Once the validation is accomplished, the respective transaction will be included to the existing blockchain [2].

Blockchain technology is capable of resolving challenges for the following reasons. (i) It is distributed and not owned by any single entity. This ensures decentralisation, robustness, and access control. (ii) It ensures that data is cryptically secure. (iii) Information in blockchain cannot be changed or altered. This ensures immutability of transactions and data integrity. (iv) Blockchain is transparent and thus ensures nonrepudiation [3].

This chapter is planned and structured in seven sections. Section 15.2 introduces very briefly the commonly used terminology in the blockchain technology. Section 15.3 converses how blockchain technology will play a role for abridging the operations involved in insurance sector. As a continuation, Section 15.4 highlights how blockchain is functional in various segments for insurance domain. Section 15.5 introduces the benefits of blockchain technology in agriculture segment from farmers' welfare viewpoint. Section

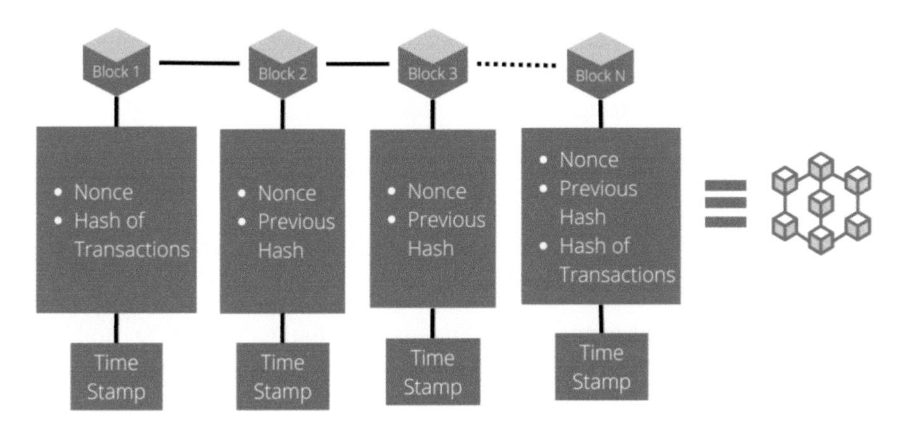

Figure 15.1 Blockchain elements. Source: https://arxiv.org [1].

15.6 discusses few use cases about some segments without blockchain and with blockchain, how and what advantages has been gained. Section 15.7 highlights challenges faced in the implementation of blockchain technology, which is followed by Section 15.8 with use cases and Section 15.9 with challenges in implementation followed by conclusion along with future directions in Section 15.10.

15.2 Essential Fundamentals of Blockchain

This section will help in getting acquainted with the essential elementary components of the blockchain domain, symbolised in Figure 15.1.

- Block: The collection of transaction is contained in a block. It is a unit of data or record. A blockchain is formed by combining many other blocks arranged in a specific order. Each block has three basic elements along with data [4].
- Hash: To make blockchain secure, hash is used, which identifies block and all its content uniquely.
- Nonce: It is an abbreviation for "number only used once," which is a number added to a block. It is a 32-bit whole number generated randomly at the time of creating a block.
- Time stamp: It indicates the time when the actual block was created.

At the time of creating the first block in a chain, a nonce was generated by the cryptographic hash. The data in the block is recognised as signed and endlessly tied to the nonce and hash unless it is mined [5].

Blockchain: It is a chain of blocks in which each block is linked to the previous block's data as depicted in Figure 15.1. The essential requirements for blockchain for its working are as follows [6]:

- Peer-to-peer (P2P) network: A network of computing nodes that collectively work on transaction in a network.
- Cryptography: For validation and authentication of transaction, cryptography encryption is used.
- Consensus algorithm: These ensure that the participating nodes in a network agree on some rules for adding a block. Thus, it ensures authenticity.
- Reinforcement in the form of reward and punishment: This ensures that all nodes follow the rules for the best outcome.

Miner**:** It is a blockchain user or nodes who along with other blocks do validation and solve the complex problem of creating new blocks in the chain through a process called mining; subsequently, it is financially rewarded [7].

Ledger: It is a record of transactions that can be tracked and analysed over a period of time. It documents the transfer of ownership and authenticating ownership. In other words, ledger is defined as a book of record keeping of all the financial transactions of the organisation. There are two types of ledgers: (i) centralised ledger and (ii) distributed ledger. A centralised ledger is one in which there is a central authority for controlling and recording transactions. The drawback of the centralised ledger is that dependency on the centralised authority increases and is prone to failure and is insecure [8]. A distributed ledger is a chain of blocks comprising explicit number of valid transactions with their time stamped. In distributed ledger, there is no central authority [9].

15.3 Types of Blockchain

Blockchain technology is divided as permissioned and permissionless based on its uses for different cases.

Permissioned blockchain: As the name suggests, permission is required for participating in the network. Some conditions are set up for any node to participate and make changes; thus, privacy is improved, and confidentiality is maintained as any changes made to the database are transparent. The private blockchain falls into the category of permissioned blockchain, in which members are private and belong to a single organisation [10].

Permissionless blockchain: As the name suggests, no permission is required for participating in the network. One benefit of permissionless blockchain is that more people can develop an application using smart contracts. The public blockchain falls into the category of permissionless blockchain [11].

15.4 Smart Contract

Smart contracts are one of the important constituents of blockchain for deploying in any domain. A smart contract is a type of agreement that is maintained as a computer code. Smart contracts can trigger certain events when predefined conditions are met. When trustworthy data is available, smart contracts can self-execute and self-verify reliable transactions, thus making the procedure fast and without the involvement of third parties [12].

A smart contract can be created by a node to initiate a transaction to the blockchain. The transactions using smart contracts are easy to track and are irreversible. The components that are part of the smart contract in blockchain are line of code and storage files. Storage files are used by smart contract for performing operations such as read and write. The line of code contains the actual operation mechanism for executing the transaction and it cannot be altered once it is created and is immutable [13].

Smart contracts are capable of handling the unexpected scenarios by automatically executing the preset term of agreement. Let's consider a fundamental example for smart contract i.e., travel insurance smart contract. After the posts cancel a covered train, it can automatically go for payment for those who have purchased insurance without claiming the railway department to verify the loss. From the smart contract, both the insurance company and the customer can save themselves from the hassle of further claiming process. It can help an insurance company to save money as they are not employing any person for this work; so the benefit of cost-saving directly shifts to the customers by lowering rates [14].

Section 15.5 delivers a brief discussion of the blockchain technology positioning in the insurance sector.

15.5 Blockchain in the Insurance Sector

The primary processes involved in the insurance sector are registration of client, assignment of policies, paying premium or claims, submission of claims, etc., and for settlement, it includes collecting the claim data, calculating the premiums or claims, confirming premiums or claims, reconciliations, dispute settlement, etc. [15]. The main pillars, i.e., immutability, traceability, transparency, and accuracy, are responsible for the insurance industry's success. The possibility of changing a client's contract policy without his knowledge or consent turns out to be a serious inconvenience towards situations where they might need to report the damages to the insurance company. The results and order of each transaction is maintained in blockchain. This ensures that false accusation by client is prohibited and similarly insurance companies are responsible for all the benefits and services they provide. Due to the situation, a system that grants integrity, confidentiality, transparency, and security for their clients is needed. Blockchain is a platform that can provide all these characteristics. Blockchain technology can make the insurance value chain more efficient from product management to customer service, which guarantees security to policyholders. This technology may change the organisation's role and helps in managing customer relationships in a better way [16].

Blockchain provides a function of trust, and it eliminates the role of the third party, i.e., brokers between an organisation and a customer. With blockchain, the consumer himself handles the data and controls it within the business system. This brings several benefits [17]:

- Insurance market can be autonomous, and services can become fast.
- It helps in maintaining data integrity and management in the insurance sector.
- Automatic verification of the policyholder is possible.
- The settlement and reconciliation time can be reduced.

Implementing insurance procedures in the blockchain platform requires smart contracts. These set up the rules for transactions. The following entities are involved: (1) primary entities – client and agent; (2) components of blockchain – ledger, database, endorsers, and validators [18].

Client: They are covered through insurance, they request for the insurance policies, and later for claiming, they submit the claim requests, and once the claim is verified, they receive the funds.

Agent: They act on behalf of the clients by forwarding their requests for completion. They are in contact with multiple clients and deal with their requests simultaneously [19].

Figure 15.2 represents the insurance process flow in the blockchain framework.

As the request is forwarded by the agents, in the insurance blockchain network, a block will be created once the linked nodes in the validator set

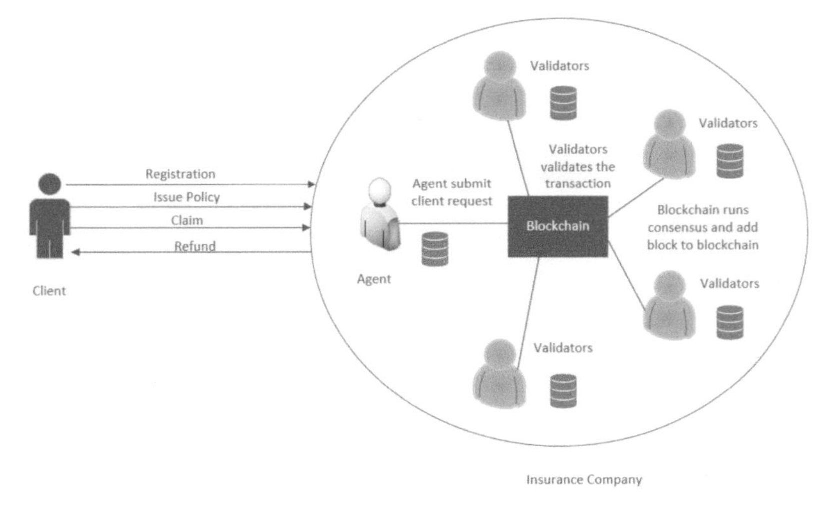

Figure 15.2 Insurance process flow with blockchain framework. Source: [12].

execute consensus over a set of transaction outcome. Each smart contract comprises an endorsement (or verification) logic, which stipulates under which conditions a transaction should be realised. The endorsement logic is executed by a set of endorsers who access the blockchain to determine whether contract conditions are fulfilled. The distributed blockchain ledger logs execution outcomes of all transactions and the database preserves the insurance contracts and transaction outcomes of all the clients [20].

15.6 Blockchain Functional in the Insurance Sector

The insurance provided by various companies and blockchain involvement into these peripheries is adding a value to the existing system of operations. Some of the segments are listed below.

- Vehicle insurance: This insurance policy covers vehicles that are for roads, e.g., trucks, cars, bus, motorcycle, and other vehicles. This insurance provides financial protection against bodily injury, physical or property damage, which can be caused due to accidental crash on the road and hostile traffic as well. It may include vehicular events. Under certain scenarios, vehicle insurance provides financial benefit for vehicle theft or damage due to natural disasters. Vehicle insurance can help injured people in case of an accident to get compensation and will help motor drivers to cope with charges for the destruction of their vehicle. For competence and appealing the customers by increasing their satisfaction level, some vehicle insurance companies have introduced customised payable premium programs like usage-based insurance (UBI). It is dependent on the way a vehicle is used on the road, i.e., some reward point system for good conduct and driving habits. A blockchain system implementation with smart contracts and customer's vehicle with appropriate sensors will assist in manually updating the policy coverage based on environmental conditions for claims reimbursement and decision [21].

- Marine insurance: This insurance provides protection against sea-based transport-related loss. Voyage transportation deals with massive threats such as piracy, cross-border shootouts, cargo loss, and products and goods contamination as in pharmaceuticals and food as well as natural occurrences that can harm the cargo, ships, terminals, and vessels and can cause enormous monetary loss to individuals and companies. Since marine insurance is one of the oldest commercial insurances, most of the processes, practices, customs, and policy terms are structured and can be easily translated into programmable lines of code. Marine insurance procedures would be significantly boosted with blockchain after the essential announcement and confirmation have been ended. Subsequently, the proof has been communicated to ledger in blockchain framework. This ensures that the process of premium payments is quick and directly transferred to the insurer account without third-party intervention. It enhances system transparency and solution for conflicts at the earliest [22].

- Life insurance: Under the life insurance scheme, companies provide a premium of an assured amount to the nominee of the insurer on the demise of the policyholder. This helps the family financially if the deceased person was the only bread earner of the family and ensures future stability. With blockchain, the accuracy of the information, transparency, and security can be ensured, which will benefit the life insurance industry to reduce fraud and for a faster claim process for genuine cases. Blockchain-related smart contracts will have all details related to consent or agreement documents. The policy can be logged in the distributed ledger of blockchain. All individuals related to policy such as insurer, beneficiary, and other entities will be able to access insured amount details, documents related to insurer birth, or death certificate from trusted sources [23].

- Home insurance: This insurance covers damage to homes, buildings, or domestic properties against events like fire, hail, wind, or natural disasters like hurricane. During the natural calamities, a huge number of people and societies' homes got destroyed or demolished. Then, they are destined to live sub-standard lives and to survive with lot of hardship. In the high-risk zones or areas of damage, governments mediate and take responsibility when the risks are considered uninsurable. In case of any event like natural disaster, the events can be confirmed by authorised state establishment about its intensity, influence level, and extent. With the employment of blockchain technology, a smart contract can be created to trigger payment transfer to the people or communities affected the most and those who require the help immediately, thus automating the claiming process [24].

- Disability insurance: This insurance offers income and health assistances to individuals who are unable to earn due to some form of disability. This safeguards the insurer from the fear of not getting regular income for livelihood due to their disability. The insurance scheme has provision for sick leave, temporary disability, and longstanding disability with paid benefits. With blockchain, the use of database of civilians containing information of people with disabilities along with the extent or severity by one who is suffering can be leveraged. The information in these databases embedded in a smart contract will be used to regulate the time and claim settlement amount with more transparency and automatically depending on predefined parameters [25].
- Health insurance: This insurance covers the monetary aspect of expenses from medical bills and surgeries. The medical expenses can be overburdening for an individual. The method of medical billing involves lot of complexity. For the claim processing, many steps need to be followed, which include confirmation, data entry, billing, sharing all the bills with the insurance company, and then reimbursing the bills after some time. With blockchain, trust and transparency can be ensured between the different stakeholders involved, such as patients, nurses, doctors, and insurers, and immutability of blockchain will provide all data from the ledger in a consistent form to all permissible members. Adoption of hybrid blockchain system will assist the patients to be familiar with the hospital services and treatments that come under claim reimbursement category [22].

Now, Section 15.7 reveals how blockchain technology implementation in agriculture sector will be advantageous. The technology may contribute for farmers in a way that they can get the government policy or scheme benefit in case of natural calamity at the right time. Ultimately, this will boost the farmers' spirit to continue with farming even in adverse conditions. Subsequently, it will add in increasing the economic growth of our country.

15.7 Blockchain in Agriculture for Farmers' Welfare

In India, the common trend is that people give their fertile farmlands on lease to farmers or labours for some specific period to utilise it for crop production. However, according to the policies and schemes offered by the government, it applies only to the land holders; thus, these people who are using the land do not get the benefit of using these schemes provided by the government.

The skewed ownership of land requires strengthening laws in respect of land reforms with attention to laws related to rental and leasing policy. Based on the present state of Indian economy, individuals are going towards tenancy and this number will increase further. Further, land scarcity is increasing due to urbanisation in rural area. This scenario introduces the demand of stringent laws to avoid the conflict. Consequently, rental, sharing, and leasing are becoming livelihood options in the state where agriculture is the main profession. On the other hand, the option of tenancy and leasing contributes in increasing the individual income and plays a role in strengthening the economy [26].

Agriculture with its associated sectors is India's largest livelihood provider, generally, in rural areas, as it is playing a vital role in augmenting a significant contribution to the gross domestic product (GDP). However, maintainable agriculture with respect to food security and inducing employability in rural areas along with ecologically bearable technologies also contributing in GDP. The factors such as soil conservation, natural resource management, and guarding biodiversity are crucial for development in rural areas from a broader perspective [27].

All India Rural Financial Inclusion Survey (2016–2017) points to the increasing inclination of families involved in business of agricultural to lease-in land instead of farming themselves. In India, 12% of families in agriculture business are in leasing the farming lands. The Government policies provide financial support to farmers in case of natural disasters/calamities, insect, pest diseases, and adverse weather conditions. Different attributes are taken care of by the government under insurance schemes for crops, tractor, and water pump [28]. The benefit of an insurance scheme is applicable according to their region. The farmers must be having the crop loan account based on notified crop and notified area. Risk details will be shared by the implementing agency [29]. They will be contacting the

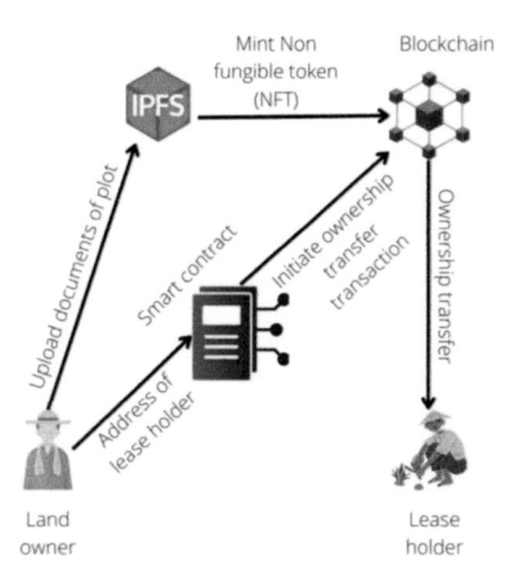

Figure 15.3 Transferring ownership of farmland by landowner to leaser. Source: Author's compilation.

authorities of state government or the empanelled bank or registered insurance companies. Land leasing act need to very stringent and important for land reforms through which the requirements of landlord as well as lease holder can be met and taken care of. With the help of smart contract, ownership can be identified easily, and it will be transparent for the government agencies as well. Figure 15.3 conveys ownership transfer process from landowner to an individual who does farming via smart contract. This is secure and transparent and helps in reducing the delay in transaction. At times, the lease holder does not receive loan, insurance, and disaster relief as per the government policy in time and they do not have sufficient money to deal with a situation and come out of it gracefully. Subsequently, they could not invest further in agriculture. For resolving the conflict and dispute among the landlord and the lease holder, "Special Lan Tribunal" is formed in Civil Court. By that time, loss of lease holder has already been escalated to a massive level. This drives them towards suicidal attempts and ultimately the loss in economy growth as there will be no further investment by the farmer. This loss is irreversible [30].

Gap in policy, deficiency of suitable credit flow, and poor insurance coverage for lease holder farmers hamper their growth. As per study, from total tenant farmers, 80% of them come to the thought of attempting the suicide. Hardly, they get loan from banks and subsidies as per the government policy, which is announced from time to time. Even though the tenant farmer holds a Kisan Credit Card, they hardly get credit for farming. As the identification of ownership for lease holder is impractical for government agency, they do not receive farmer loan waiver scheme benefit on crop loan insurance scheme. In many cases, the landowner is not involved in farming or

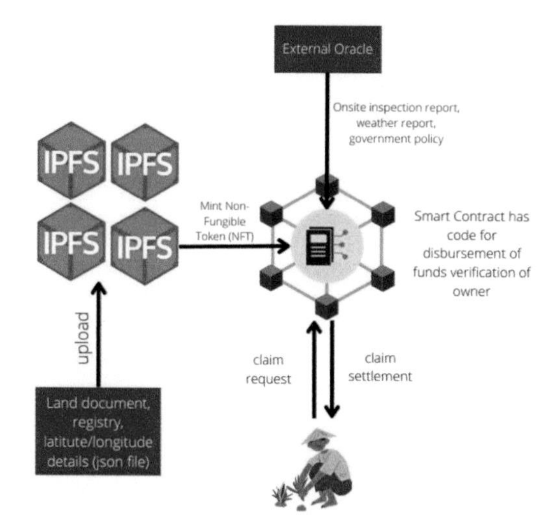

Figure 15.4 Insurance claim settlement with blockchain. Source: Author's compilation.

cultivation, but they receive the crop loan insurance benefit instead of land lease holder who is actually involved in farming [31]. Due to non-availability of no documentary evidence, the tenant farmer becomes ineligible for crop insurance under PM Fasal Bima Yojana (PMFBY) [28]. However, state level panel data of NABARD discloses that a 10% increase in agricultural growth leads to a 2.1% rise in GDP, but indirectly leading to slowing down the economic growth. Scarcity of suitable credit flow and deprived insurance coverage for the tenant farmers prevent the growth of country.

The blockchain technology improves the process of claim reimbursement to the right person – either owner or lease holder, whosoever holds the insurance policy for farmers' welfare. Further, related risks are covered through agricultural insurance schemes. With blockchain, smart contracts can help farmers insure their crops and claim damages. At the same time, for farmer's welfare, the pump set and the tractor can also be insured. To avoid room for fraud due to unpredictable weather anomalies and correct estimate of losses, applicability of smart contracts in blockchain framework will be beneficial [32]. With blockchain tracking and identification of farmer who has taken the land on lease becomes easy. This will create trust, transparency, and confidence between farmers and government. Blockchain technology allows farmers to store their data in one place easily and be accessed by those who need it, as depicted in Figure 15.4. This simplifies the entire process, saves valuable time and energy, and increases the efficiency of the farmer [31].

15.8 Use Cases in Insurance Domain

This section presents some case scenarios where blockchain technology has been applied under various arms of insurance segment. It reveals how different users are going to be benefitted after deploying the upcoming technology.

- **Claim processing**

Existing scenario:
The insurance claim process proceeds as follows:

- For the claim, the form is filled by the policyholder or alternatively a phone call is made to report the claim.
- Verification of proof is done by the insurer for the claim made by the policyholder, and to measure the damage, an additional adjuster might be sent out.
- After verification, the insurer pays out the claim. In case of any dispute, the process of claim can take longer.
- If there is gap or weakness in the claim process, then fraud can be committed. The policyholder could take advantage of this weakness (information asymmetry and data silos).
- Manual work for the insurer can be costly and time consuming like collecting the claim data, calculating claims, confirming claims, dispute settlement, etc.
- For the policyholder, the claiming process can become more inconvenient when in need or during times of suffering or distress [33].

With blockchain:
Smart contract can be used for automatic payment of claims when some conditions are met. The claim trigger events can be easily verified through trusted publicly available data like reporting of extreme weather or flight delays or natural disasters [34].

With blockchain, multiple claims on the same insured event can be prohibited.

Main beneficiaries and benefits:

- The main beneficiaries for claim processing are start-ups, officials, and customers.
- Positive quality of experience for customers.

- Elimination of re-claiming on the same event, thus preventing insurance fraud.

Main challenges:

- Issue related to standardisation.
- Existing technology may already be providing good claim processing benefit. Thus, blockchain may be difficult to introduce.
- Trustworthy third party (i.e., oracles) for complex or more events are few; thus, triggering a claim requires more data.

- **Tokenisation of insurance risk**

Existing scenario:
The capital raised gets reduced when intermediaries (e.g., investment banks) collect fees for providing security to insurance risks. This raised capital becomes a costly affair for insurers [2].

With blockchain:
Payments to investors can be generated when conditions are encountered with the use of smart contract.

Information flow to investors is increased by packaging insurance linked securities (ILS) into a "token," which represent the ILS digitally, and investors are more aware of the informed price in market [34].

Main beneficiaries and benefits:

- The main beneficiaries for tokenisation of insurance risk are start-ups, officials, and investors.
- No fees are provided to intermediaries (e.g., investment banks); thus, we get a suitable approach for raising capital as well as for risk transfer.

Main challenges:

- Issue related to standardisation.

- **Reinsurance and swaps**

Existing scenario:

- A time period of 2–3 months is required for the settlement of a reinsurance account. This process includes the time involved in collecting the claim data, calculating the premiums or claims, confirming premiums or claims, reconciliations, dispute settlement, etc.

- As data are stored in a centralised repository, the process of reinsurance gets slower and becomes prone to errors as this same data is required by several parties to work on.
- Cleansing the data may consume a lot of time, which may lead to missed opportunity for closing a deal (e.g., bulk annuity). As data is not ready, it becomes difficult to make a deal.
- Third parties are involved in some transactions to manage the collateral assets.
- Longevity risk transfer becomes difficult as full information of past claims is not available to new transacting parties [35].

With blockchain:

- The smart contract is added with terms reinsurance treaties/swaps; thus, payments to investors can be triggered when conditions are met.
- Blockchain is used for recording experience data, which is tamper-resistant and immediately auditable [34].
- With blockchain, everyone has access to a single version of information as all information is recorded there.
- To make faster deals, data that is cleansed can be stored in blockchain so that everyone could see the recent changes made.
- New transacting parties or future transacting parties can have full information regarding past claims as these are stored in blockchain and all transactions can be viewed transparently [36].

Main beneficiaries and benefits:

- The main beneficiaries for reinsurance and swaps are start-ups and officials.
- Complexities related to contracts are simplified through coded rules.
- More transparent and flowing market for deals related to reinsurance is available, which makes the market of insurance linked securities (ILS) much easier.

- Transparent data, reduced time in processing, and simplified procedure can be attained.

Main challenges:

- Issues related to standardisation; however, an initiative has been taken by B3i (Blockchain Insurance Industry Initiative) for establishing a standard.

- **Decentralised digital identity**

Existing scenario:

The data collected from the customers for Know Your Customer (KYC) procedure contains personal data and other relevant information. The protection of this data is costly and requires a lot of effort. Also, customers have minimum control over this data regarding its sharing [37].

With blockchain:

- Blockchain can help data flow from customers to insurers so that insurers do not have to store customers' private information locally, which can lead to malicious attack or data breach.
- Smart contracts can be used for KYC information and be further automated [34].

Main beneficiaries and benefits:

- The main beneficiaries for decentralised digital identity are start-ups, officials, and customers.
- The customer has full control of their data and sharing.
- As the customer's private data is no longer being stored locally by the insurer, this reduces the data breach and any malicious attack that can compromise the privacy of customers.
- Any changes in KYC can be done easily and timely.
- Sharing of data to multiple insurers by customers becomes easy. And advanced encryption techniques become useful.

Main challenges:

- Mindset and working of entities involved need to be updated. Customers can share their personal data by monetising and getting the benefit. This technology is emerging to develop the infrastructure private enterprise needs; however, the benefits are more for enterprises but less incentives leading in delaying the implementation.

- **Decentralised autonomous insurer (DAI)**

Existing scenario:
The existing models are not optimised according to the benefits of policy-holders. The reasons are high operating costs that exist in the insurance business as well as the capital that needs to be returned to the shareholders [38].

With blockchain:

- DAI is the sum of all the areas like artificial intelligence (AI), big data, and blockchain.
- Smart contracts are used for payments of claims and premiums.
- With blockchain, the KYC procedure is automated though digital identity. Also, it records all transactions in a transparent manner.
- Big data is used for adding a price to the insurance products.
- With blockchain, risks can be shared by communities, individuals, and corporations through pooling [39].

Main beneficiaries and benefits:
- The main beneficiaries for DAI are end users or customers.
- The high profit priced in insurance contracts is removed.
- The main benefit received is the cost benefit as operating cost gets reduced with DAI.
- The peer-to-peer model help businesses which are not established well to purchase the insurance cover for their development.

Main challenges:

- The maturity of the technology is under progress.

- **Decentralised data lake**

Existing scenario:
Digital data has increased due to the immense data collected from the use of sensors and other data sources or devices. This centralised repository of data is called data lakes, which are often handled by numerous entities [40].

With blockchain:

Blockchain-based platform can be used to store these data collected from various devices. With this, a data marketplace can be formed, which can further promote data collection and sharing from customers by providing them some incentives and monetary benefit. And this data can be directly purchased by start-ups removing the role of third parties like incumbent insurers [41].

Main beneficiaries and benefits:

- The main beneficiaries for decentralised data lakes are start-ups and customers.
- As blockchain is used for storing data, the surety of data source is confirmed.
- The data can be directly shared by customers to the start-ups with help of data marketplace; thus, high pricing of customers' data collected by third party insurers will be removed. And monetary benefit will be provided to customers for sharing their data.

Main challenges:

- Creating such a data marketplace requires a shift in the present operating environment.

15.9 Challenges in Implementation

Each technology has pros and cons or some limitation in the beginning phase of adapting it. This also has some challenges confronted while implementation in real-life scenario due to varying dynamics with time. The following addresses the challenges of blockchain technology:

- Cross-border data sharing may face problems due to government regulations and different policies may exist depending on different countries [4].
- A large amount of data can lead to mining delays in blockchain. The potential or capability of blockchain in storing and processing massive data needs to be checked. As the transaction will increase in volume for any blockchain, there would be delay of mining blocks. New transactions on blockchain may take longer time to get verified [5].
- In the real-world scenario, there are mixed communication systems; the working of blockchain in such network requires single global access

policy so that blockchain can work properly even with networks owned by different service providers.

- A transaction in blockchain requires a block to be created for which computing power needs to be supplied from a network of inter-connected nodes. Thus, blockchain consumes computing power to process transactions, and for operating such systems, there is always a hidden cost as well as an implementation cost. Also, the cost may further increase as the capacity and the size of transactions increase [6].
- The practical cost of set-up can include the hardware cost, the software cost, the implementation cost, and further support when performing different transactions in insurance segment. Thus, these costs need to be assessed and studied in detail before transiting.
- The blockchain technology does not work efficiently with multi-dimensional data. Thus, we have a disadvantage when working for data with high temporal resolution such as complex texts and images when using blockchain [7].
- One of the significant difficulties with technology is that it has a very harmful effect on the environment. Massive carbon footprints leading to high environmental cost. A site named "bitcoin energy consumption index" has alarming statistics, which shows that there are terrible repercussions of blockchain on environment in a long term [8].

The industry is growing, which increases the number of customers. To entertain a large group of customers, there is requirement for proper infrastructure for the optimum utilisation of this technology. As the data need to be decentralised, each node must be processing the required data for every transaction.

The other similar works influenced by Blockchain technology is implemented in sectors like biometric databases [44], smart agriculture [45], controlling ransomeware [46] and human detection [47] are currently prevailing. In near time many such secure implementation will be incorporated in all the sectorssuccessfully.

15.10 Conclusion

This chapter encompasses the status of applicability of blockchain in insurance industry. This is a guide for current risks and challenges in adopting blockchain technology for the insurance sector. This can bridge the gap between the insurance industry practitioners and blockchain technology by providing a simple way to understand the functional aspects.

The transparency of data sharing and its security becomes a major concern when these different applications play a role in the domain of insurance. Blockchain technology implementation from different perspectives of applications will deliver the solution to obtain the transparency and security in the system. For commercial segment and academia, this study offers different use cases with blockchain and viewpoint for adopting the blockchain technology in the agriculture sector for farmers' wellbeing.

The insurance sector has enormous scope, and it can be benefitted and evolved by adopting blockchain technology. It helps in the growth of insurance services by enhancing the working process and reducing faults in keeping records of money transactions. The study shows that four main pillars, i.e., immutability, traceability, transparency, and accuracy of blockchain, are responsible features for the success of the insurance industry. The present study's findings have its motive for deploying blockchain technology to be adopted by the insurance sector, decision-making process for employees of an insurance company, and other financial institutions.

References

[1] Sharifinejad, M., Dorri, A., & Rezazadeh, J. (2020). BIS-A blockchain-based solution for the insurance industry in smart cities. arXiv preprint arXiv:2001.05273. DOI: 10.48550/arXiv.2001.05273.

[2] Brophy, R. (2019). Blockchain and insurance: a review for operations and regulation. Journal of financial regulation and compliance, 28(2), 215–234. DOI:10.1108/JFRC-09-2018-0127.

[3] Halima, E. H., & Yassine, T. (2022). Insurtech & Blockchain: Implementation of Technology in Insurance Operations and its Environmental Impact. In IOP Conference Series: Earth and Environmental Science (Vol. 975, No. 1, p. 012010). IOP Publishing.

[4] Eckert, C., Neunsinger, C., & Osterrieder, K. (2022). Managing customer satisfaction: digital applications for insurance companies. The Geneva Papers on Risk and Insurance-Issues and Practice, 1–34. DOI: 10.1057/s41288-021-00257-z.

[5] OECD. (2017). Technology and innovation in the insurance sector.

[6] Kherbouche, M., Pisoni, G., & Molnár, B. (2022). Model to Program and Blockchain Approaches for Business Processes and Workflows in Finance. Applied System Innovation, 5(1), 10. DOI: 10.3390/asi1010000.

[7] Kar, A. K., & Navin, L. (2021). Diffusion of blockchain in insurance industry: An analysis through the review of academic and trade literature. Telematics and Informatics, 58, 101532. DOI: 10.1016/j.tele.2020.101532.

[8] Loukil, F., Boukadi, K., Hussain, R., & Abed, M. (2021). CioSy: A collaborative blockchain-based insurance system. Electronics, 10(11), 1343. DOI: 10.3390/electronics10111343.

[9] Cousaert, S., Vadgama, N., & Xu, J. (2021). Token-based insurance solutions on blockchain. arXiv preprint arXiv:2109.07902. DOI: 10.48550/arXiv.2109.07902.

[10] Chakravaram, V., Ratnakaram, S., Agasha, E., & Vihari, N. S. (2021). The Role of Blockchain Technology in Financial Engineering. In ICCCE 2020 (pp. 755–765). Springer, Singapore. DOI: 10.1007/978-981-15-7961-5_72.

[11] Kalsgonda, V., & Kulkarni, R. (2022). Role of Blockchain Smart Contract in Insurance Industry. Available at SSRN 4023268. DOI: 10.2139/ssrn.4023268.

[12] Raikwar, M., Mazumdar, S., Ruj, S., Gupta, S. S., Chattopadhyay, A., & Lam, K. Y. (2018, February). A blockchain framework for insurance processes. In 2018 9th IFIP International Conference on New Technologies, Mobility and Security (NTMS) (pp. 1–4). IEEE. DOI: 10.1109/NTMS.2018.8328731.

[13] Grima, S., Spiteri, J., & Romānova, I. (2020). A STEEP framework analysis of the key factors impacting the use of blockchain technology in the insurance industry. The Geneva Papers on Risk and Insurance-Issues and Practice, 45(3), 398–425. DOI: 10.1057/s41288-020-00162-x.

[14] Sun, J., Yao, X., Wang, S., & Wu, Y. (2020). Non-repudiation storage and access control scheme of insurance data based on blockchain in IPFS. IEEE Access, 8, 155145–155155. DOI: 10.1109/ACCESS.2020.3018816.

[15] Lepoint, T., Ciocarlie, G., & Eldefrawy, K. (2018, April). Blockcis—a blockchain-based cyber insurance system. In 2018 IEEE International Conference on Cloud Engineering (IC2E) (pp. 378–384). IEEE. DOI: 10.1109/IC2E.2018.00072.

[16] Amponsah, A. A., Adebayo, F. A., & Weyori, B. A. (2021). Blockchain in Insurance: Exploratory Analysis of Prospects and Threats. International Journal of Advanced Computer Science and Applications, 12(1). DOI: 10.14569/IJACSA.2021.0120153.

[17] Xu, J., Wu, Y., Luo, X., & Yang, D. (2020, June). Improving the efficiency of blockchain applications with smart contract based cyber-insurance. In ICC 2020-2020 IEEE International Conference on Communications (ICC) (pp. 1-7). IEEE. DOI: 10.1109/ICC40277.2020.9149301.

[18] Dhieb, N., Ghazzai, H., Besbes, H., & Massoud, Y. (2020). A secure ai-driven architecture for automated insurance systems: Fraud detection and risk measurement. IEEE Access, 8, 58546–58558. DOI: 10.1109/ACCESS.2020.2983300.

[19] Campanile, L., Iacono, M., Levis, A. H., Marulli, F., & Mastroianni, M. (2020). Privacy regulations, smart roads, blockchain, and liability insurance: putting technologies to work. IEEE Security & Privacy, 19(1), 34–43. DOI: 10.1109/MSEC.2020.3012059.

[20] Gera, J., Palakayala, A. R., Rejeti, V. K. K., & Anusha, T. (2020, June). Blockchain technology for fraudulent practices in insurance claim process. In 2020 5th International Conference on Communication and Electronics Systems (ICCES) (pp. 1068–1075). IEEE. DOI: 10.1109/ICCES48766.2020.9138012.

[21] Sadiku, M. N., Eze, K. G., & Musa, S. M. Blockchain in Agriculture.

[22] Aleksieva, V., Valchanov, H., & Huliyan, A. (2020, September). Implementation of smart contracts based on hyperledger fabric blockchain for the purpose of insurance services. In 2020 International Conference on Biomedical Innovations and Applications (BIA) (pp. 113–116). IEEE. DOI: 10.1109/BIA50171.2020.9244500.

[23] Meskini, F. Z., & Aboulaich, R. (2019, October). Multi-agent based simulation of a smart insurance using Blockchain technology. In 2019 Third International Conference on Intelligent Computing in Data Sciences (ICDS) (pp. 1–6). IEEE. DOI: 10.1109/ICDS47004.2019.8942270.

[24] Popovic, D., Avis, C., Byrne, M., Cheung, C., Donovan, M., Flynn, Y., ... & Shah, J. (2020). Understanding blockchain for insurance use cases. British Actuarial Journal, 25. DOI: 10.1017/S1357321720000148.

[25] Chishti, M. S., Sufyan, F., & Banerjee, A. (2021). Decentralized On-Chain Data Access via Smart Contracts in Ethereum Blockchain. IEEE Transactions on Network and Service Management, 19(1), 174–187. DOI: 10.1109/TNSM.2021.3120912.

[26] Swaminathan, M. S. (2016). National policy for farmers: ten years later. Review of Agrarian Studies, 6(2369-2020-2039).

[27] FICCI and PwC. (2020). Revamping India's Health Insurance Sector with Blockchain and Smart Contracts.

[28] Pradhan Mantri Fasal Bima Yojana. Available at: https://pmfby.gov.in/ [accessed April 2, 2022].

[29] Farmers' portal. Available at: https://farmer.gov.in/handbooks.aspx [accessed April 2, 2022].

[30] Sakib, S. N. (2021). A Systematic Literature Review of Blockchain-Based Crops Traceability in Agricultural Supply Chain.

[31] Xiong, H., Dalhaus, T., Wang, P., & Huang, J. (2020). Blockchain technology for agriculture: applications and rationale. frontiers in Blockchain, 3, 7. DOI:10.3389/fbloc.2020.00007.

[32] Alobid, M., Abujudeh, S., & Szűcs, I. (2022). The role of blockchain in revolutionizing the agricultural sector. Sustainability, 14(7), 4313. DOI:10.3390/su14074313.

[33] Bhamidipati,N.R.,Vakkavanthula,V.,Stafford,G.,Dahir,M.,Neupane,R., Bonnah, E., ... & Calyam, P. (2021, December). ClaimChain: Secure Blockchain Platform for Handling Insurance Claims Processing. In 2021 IEEE International Conference on Blockchain (Blockchain) (pp. 55–64). IEEE. DOI: 10.1109/Blockchain53845.2021.00019.

[34] Sheth, A., & Subramanian, H. (2019). Blockchain and contract theory: modeling smart contracts using insurance markets. Managerial Finance. DOI: 10.1108/MF-10-2018-0510.

[35] Digital transformation for reinsurance is here. Available at: https://b3i.tech/home.html [accessed May 2, 2022].

[36] Tasca, P. (2019). Insurance under the blockchain paradigm. In Business transformation through blockchain (pp. 273–285). Palgrave Macmillan, Cham. DOI: 10.1007/978-3-319-98911-2_9.

[37] Li, J., Niyato, D., Hong, C. S., Park, K. J., Wang, L., & Han, Z. (2021). Cyber Insurance Design for Validator Rotation in Sharded Blockchain Networks: A Hierarchical Game-Based Approach. IEEE Transactions on Network and Service Management, 18(3), 3092–3106. DOI: 10.1109/TNSM.2021.3078142.

[38] Xiao, Z., Li, Z., Yang, Y., Chen, P., Liu, R. W., Jing, W., ... & Goh, R. S. M. (2020). Blockchain and IoT for Insurance: A Case Study and Cyberinfrastructure Solution on Fine-Grained Transportation Insurance. IEEE Transactions on Computational Social Systems, 7(6), 1409–1422. DOI: 10.1109/TCSS.2020.3034106.

[39] Averin, A., Musaev, E., & Rukhlov, P. (2021, September). Review of Existing Blockchain-Based Insurance Solutions. In 2021 International Conference on Quality Management, Transport and Information Security, Information Technologies (IT&QM&IS) (pp. 140–143). IEEE. DOI: 10.1109/ITQMIS53292.2021.9642748.

[40] Connecting the specialty insurance market. Available at: https://insurwave.com/ [accessed March 25, 2022].

[41] Chishti, M. S., King, C. T., & Banerjee, A. (2021). Exploring half-duplex communication of NFC read/write mode for secure multi-factor authentication. IEEE Access, 9, 6344–6357. DOI: 10.1109/ACCESS.2020.3048711.

[42] Sood, K., Seth, N., & Grima, S. (2022). Portfolio Performance of Public Sector General Insurance Companies in India: A Comparative Analysis. In Simon Grima, Ercan Özen, Inna Romānova (Ed.), In *Managing Risk and Decision Making in Times of Economic Distress, Part B.* (pp.215–229). Emerald Publishing Limited,UK.

[43] Grima, S., & Marano, P. (2021). Designing a Model for Testing the Effectiveness of a Regulation: The Case of DORA for Insurance Undertakings. *Risks*, *9*(11), 206.

[44] A. Singh, R. K. Dhanaraj, M. A. Ali, B. Balusamy and V. Sharma, "Blockchain Technology in Biometric Database System," 2022 3rd International Conference on Computation, Automation and Knowledge Management (ICCAKM), Dubai, United Arab Emirates, 2022, pp. 1–6, doi: 10.1109/ICCAKM54721.2022.9990133.

[45] M. A. Ali, B. Balamurugan, R. K. Dhanaraj and V. Sharma, "IoT and Blockchain based Smart Agriculture Monitoring and Intelligence Security System," 2022 3rd International Conference on Computation, Automation and Knowledge Management (ICCAKM), Dubai, United Arab Emirates, 2022, pp. 1–7, doi: 10.1109/ICCAKM54721.2022.9990243.

[46] Singh, Anamika, et al. "Blockchain: Tool for Controlling Ransomware through Pre-Encryption and Post-Encryption Behavior." 2022 Fifth International Conference on Computational Intelligence and Communication Technologies (CCICT). IEEE, 2022.

[47] M. A. Ali, B. Balamurugan and V. Sharma, "IoT and Blockchain-Based Intelligence Security System for Human Detection using an Improved ACO and Heap Algorithm," 2022 2nd International Conference on Advance Computing and Innovative Technologies in Engineering (ICACITE), 2022, pp. 1792–1795, doi: 10.1109/ICACITE53722.2022.9823827.

16

Blockchain Technologies, a Catalyst for Insurance Sector

Rupa Khanna[1], Priya Jindal[2], and Graţiela Georgiana Noja[3]

[1]Department of Commerce, Graphic Era Deemed to be University, Dehradun, India
[2]Chitkara Business School, Chitkara University, Punjab, India
[3]Vice-Dean of the Faculty of Economics and Business Administration West University of Timisoara, Romania

Email: dr.rupakhanna@gmail.com; priya.jindal@chitkara.edu.in; gratiela.noja@e-uvt.ro

Abstract

Insurance history can be traced back to early human civilisation in the form of Granaries i.e., risk pooling early adopted by merchants for the transportation of goods through ships. In 1968, Lloyds, a formal company of insurance, came into existence. Since then, the insurance industry has evolved immensely and the insurers have witnessed a lot of disruptions. In the year 2008, blockchain technologies were thought of as the architecture for the Bitcoin cryptocurrency and are currently a burning area and the theme of various studies for various banking and financial payment-related industries. Blockchain can be a highly beneficial technology to simplify the complex processes of insurance. The insurance business runs on trust, but there are several frauds committed by people associated with the industry. Many times, insurance companies had paid the fraudulent claims due to lack of evidence and reported a loss of around 80 billion dollars, majorly 34 billion dollars reported from the property and casualty industry. Technology like blockchain is required to stop these practices and improve efficiency and trust while improvising transparency and reduction to claim costs. This chapter will discuss how blockchain technologies work, the applicability of blockchain technology in insurance, smart contracts, the type of blockchain technologies, and transforming asset management and reinsurance.

16.1 Introduction

In today's transparent world, the application of insurance is founded on a basis of faith between customers and insurers. This is an age-old idea that applies to consumer insurance as well as commercial and government entities' protection from natural disasters and specialty hazards [1]. This confidence is built on an intangible "promise to pay," as well as a one-of-a-kind mix of experience, service quality, capital, and security. This is based on the timely exchange of payment, the disclosure of accurate personal data describing the client's insurable interests, and the agreement to a contract between two legal and willing parties [2]. If a vulnerability exists in the transaction chain, especially as a result of the increased use of technology, trust is significantly harmed, resulting in a diluted brand, diminishing confidence, and the potential loss of business and shareholder value [3]. Digitisation has progressed far beyond the automation of machine-to-machine (M2M) processes. Across an integrated ecosystem, we are seeing a progressive convergence of technology, processes, data, assets, and people. The intricate web interplay between data and capital underpins smart life, mobility, farming, manufacturing, and accounting services. Business models are being driven to evolve or disrupt to survive as the digital trend continues [4]. In reaction to the connected world, insurance products, services, and infrastructure are evolving. Innovations particularly in the incorporation of disruptive technologies in the mobile industry, business analytics, big data and IoT, and reimbursement platforms will fuel progress in the insurance digital sector [5]. The main motivations for innovation in the insurance sector were cost-cutting and procedure improvement. Customer experience and service quality, particularly in claims, have recently supplanted these prized client-centric ideas [6]. Customer relationship, transparency, and safety (of personal data) are becoming increasingly important to an insurer's framework as digitisation spreads. To develop a customer-oriented culture, these fundamentals are required in place; otherwise, the firm's value can soon disappear.

New technologies like blockchain, as well as next-generation models like Ethereum, are fast-growing and supporting many facets of creativity and novelty [7]. These were originally designed to coordinate Bitcoin payment methods using distributed secure ledger systems [8]. They are now constructing industrial-strength platforms employing industry standards to protect the entire blockchain from customer purses and customer-oriented claims to deal and money exchanges, thanks to sustained development investments [9]. Customer intimacy, transparency, and security (of personal data) are becoming increasingly important to an insurer's infrastructure as digitisation spreads. Without these fundamentals in place to develop a customer-centric culture, firms' value can soon disappear as quickly as customers are attracted. They are now constructing industrial-strength platforms employing industry standards to cover the full value chain from customer wallets and client-driven applications to transaction and money exchanges, thanks to sustained development investments.

The B3i is the biggest insurance syndicate moving forward quickly in blockchain insurance investment [10]. The rapidity with which other big insurance companies are putting together B3i-style initiatives is a good indication to realise the potential of blockchain technology. Leading insurance and reinsurance companies are driving a snowball effect in the market. Regardless, each company will choose its strategy based on its vision and mission. Blockchain technology can be used in integrating proprietary information systems, which require hosting two types of exchanges simultaneously. Insurers, reinsurers, and brokers are all ardent proponents of digitising all stakeholder exchanges, and the outcomes so far have been convincing. In addition to market participants' tests, task groups are being formed within syndicates and governing agencies to assist viral marketing and adopt blockchain as a first choice in technology in insurance, reinsurance, and brokerage [11]. Insurance has a long history dating back to the dawn of civilisation. Granaries were the first kind of insurance (risk pooling), which merchants modified to carry their goods aboard ships over time. Lloyds was founded in a coffee shop in 1968. The insurance sector has been working on four essential principles of risk management since 1968. The most well-known is "utmost good faith," also known as "Uberrima fides" [12].

Even though the insurance industry is built on confidence, it has been broken multiple times, resulting in many insurance organisations paying bogus dues owing to an absence of evidence. Claims and underwriting fraud cost the property and casualty (P&C) industry $80 billion and $34 billion every year approximately, respectively [13]. As a result, a trust and efficiency engine like blockchain technology can put an end to false claims while also improving transparency and lowering claim costs [14].

16.2 Blockchain

Blockchain is considered the next technological revolution after the Internet. The availability of a decentralized, protected, and translucent ledger between

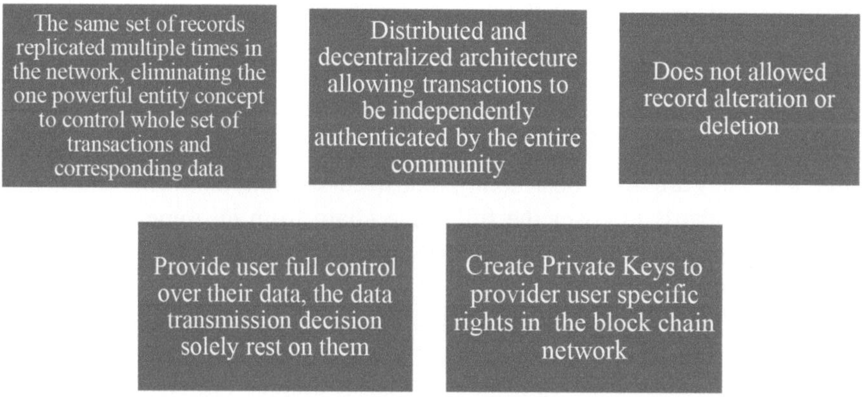

Figure 16.1 Blockchain mechanism. Source: Infosys 2018 [18].

users can be pertinent to many different fields as shown in Figure 16.1. Blockchain technology could wholly topple the insurance value chain by abstracting mediators in a unique type of procedure by:

- development of new players, products, or markets by developing new business models for customer satisfaction;
- with the help of the Internet of Things and smart contracts, there can be new methods for underwriting, contracts, and claims management;
- re-engineering of the technique of insurance agreements;
- new reinsurance methods through smart contracts;
- management of assets with automatic reimbursement and delivery of intangibles.

This technology should be used to help to reduce procurement, management, paperwork, and compliance costs. Although blockchain technology looks very promising, there can be a few challenges that cannot be ignored, such as competition with Insurtechs, a new legal framework, and infrastructure. Blockchain technologies can radically change the industry, augment faith, and disrupt the value chain. A blockchain is a perpetual and inflexible network and record of dealings. The blockchain is built on "digital ledgers," which are spread transversally with all network members and set out to be a single source of veracity. When a proceeding is completed, it is listed in the digital ledger in sequential order, and the blockchain is formed by linking the blocks as shown in Figure 16.3 [15]. It is nearly impossible to falsify the system since it relies on cryptographically secure references to other blocks inside the digital ledger. As a result, most observers believe the structure is far more

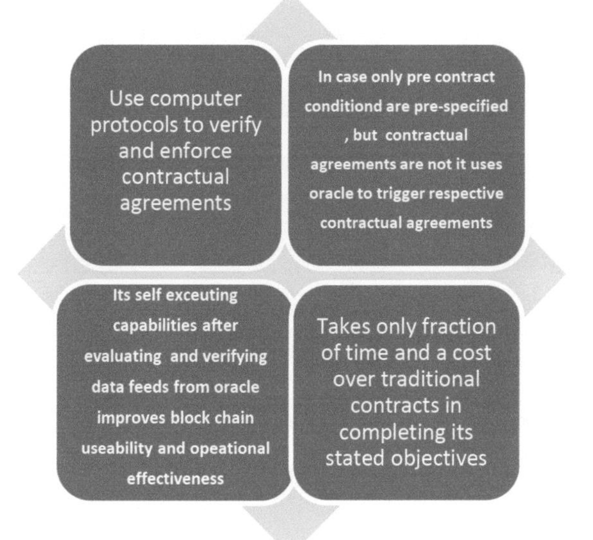

Figure 16.2 Smart contract. Source: YouHodler.com [20].

dependable and explicit as compared to traditional approaches to data sharing [16]. Blockchain establishes conditions for faster, more accurate, and more efficient procedures by establishing "truth" that is shared mutually in the present system. However, it can disrupt traditional business methods by removing the intermediaries and linking counterparties more effectively, which allows them to conduct business without a reliable central authority. Financial services firms can be given credit as pioneers in blockchain technology, even though it may be used in nearly any business. Approximately from the year 2014 onwards, 40% of financial services firms have capitalised on blockchain technologies and it is affecting other aspects of finance as well [17].

16.2.1 Smart contracts

When two anonymous parties without the help of an intermediatory do business together [19] Insurance industry experts believe Smart contracts are the future of the industry as shown in Figure 16.2.

16.2.2 How the industry is responding to blockchain

In the fourth generation industrial revolution also known as "data-driven revolution," some of the most innovative-thinking insurance companies are looking forward to blockchain technologies for their larger revolution agenda [21].

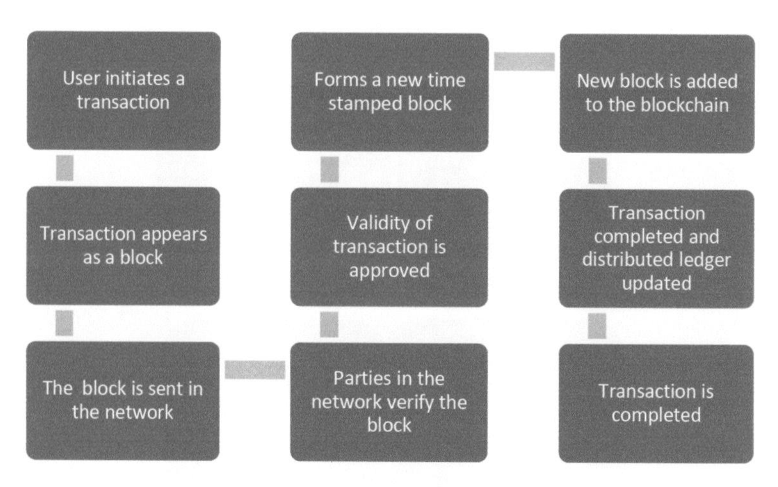

Figure 16.3 Transaction process in blockchain. Source: Investopedia.com [25].

The value of engaging in the broader financial services blockchain ecosystem is recognised by these early adopters and second-movers [22]. However, they see blockchain as a way to increase efficiency, cut transaction processing costs, ameliorate user experience, enhance data integrity, improve confidence among parties, and promote accountability, among other things as shown in Figure 16.4. A lot of people have invested money into their vision. In February 2016, AXA Strategic Ventures invested roughly 55 million dollars in a blockchain start-up [23]. In the year 2015, the USA put about 75 million dollars into a digital payment platform. In addition, Lloyd's London, as part of their strategic model upgrading plan [24], has embraced blockchain. Many other insurance companies like New York life, MSIG, and AIA are experimenting with blockchain applications in insurance as shown in Figure 16.5.

16.3 Blockchain Application

Blockchain applications may fall into two broad categories as explained below.

1. **Internal use cases:** Internal use cases do not count on network effects as much as external use cases, but they do try to enhance internal efficiency to lower service costs. However, based on experience, these measures can also result in significant revenue increases. Experience

with leading companies reveals that the revenue created by internal process generalisation may overshadow the predicted working amount savings.

2. **Industry use cases:** They count on network effects and need industry or cross-industry buy-in. The major syndicate in the blockchain in insurance is B3i, which came into effect in October 2016 by Allianz, Aegon, Munich Re, Swiss Re, and Zurich. Their effort aims to share ideas, test use cases, and pursue concepts relating to the insurance industry as a whole.

16.3.1 Categories of blockchain in the insurance sector

It can be broadly divided into three categories:

- **Public blockchain:**
  Bitcoin, which was released in 2009, was the first public blockchain in history. Any computer, regardless of location, may view the blockchain and participate in the process of approving new blocks [26]. Since the birth of Bitcoin, new blockchain concepts have evolved. These new types of distributed ledgers have some of the benefits of blockchain technology, but they limit network access and user rights [27]. There are three types of blockchains at the moment.

  Public blockchains let all members view, edit, and store the database. A public blockchain, for example, is Bitcoin.

- **Syndicate blockchain:**
  Although they are accessible to the public, not all information is available to all participants. Blocks are evaluated based on established rules, and user rights differ [28]. As a result, consortium blockchains are "partially decentralised." This category includes the R3 syndicate.

- **Private blockchain:**
  The permissions to view or edit the database are managed by a central authority. The technology can be integrated into prevailing information systems and includes a coded audit trail. The blockchain network does not encourage further use of its processing resources to miners to conduct the authentication of algorithms in private blockchains [29]. For example, Crédit Mutuel Arkéa has adopted the same system.

16.4 Advantages of Blockchain Technology

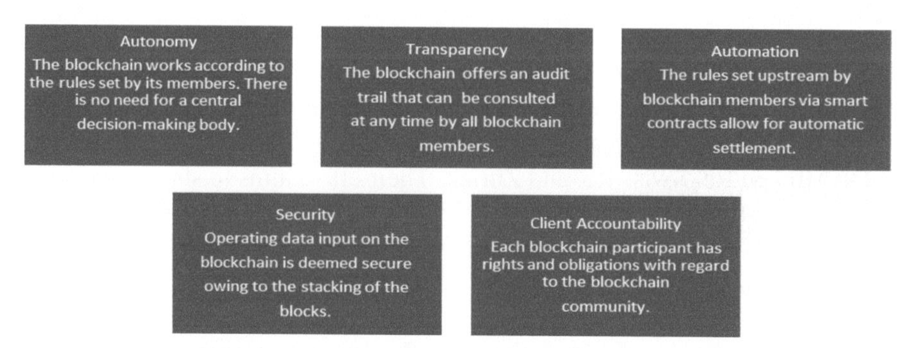

Figure 16.4 Blockchain advantages. Source: pwc.com blockchain survey [30].

16.5 Challenges of Blockchain Technology

While there are apparent benefits to having an innovative architecture, insurers must also consider the following industry challenges [31].

1. **Innovations are not without flaws**. Many of these new technologies are still in their nascent stage and are continuously evolving, and it is difficult to predict when and how they will be adopted. Insurers must consider the characteristics and advantages of blockchain, as well as the possible client adoption curve [32].

2. **Transparency is a highly sought-after offering**. Changes in commercial focus and regulation are being driven by the digital world and the demand for data protection. This has had a tremendous impact on enterprise risk management, data protection, and consumer legislation for firms [33]. It is vital to participate in proactive compliance and regulatory frameworks to fit the new model.

3. **Infrastructure is required by markets**. To deal with disruptive innovation, insurers and their ecosystem suppliers must design and create their organisations. This necessitates anticipating future client needs and growing into new areas like governance and cyber risk [34]. Today's insurance markets and the prudential system will need to maintain proper regulations and encourage the innovation curve in ways that are both feasible and scalable.

Opportunities for Insurers	Considerations
Unique Digital Identity Management	• Provides a public or private ledger with adequate personal privacy for each transaction • Digital identity verification and authentication over the web or mobile
Decentralised Infrastructure	• Reduces reliance on centralised market infrastructure • Creates a need for new regulations and controls
Eco System Scalability	• Adapts to local conditions and its globally scalable • Accelerating distribution of regionalised or personalised • Peer-to-peer insurance exists
Fraud and Security	• BT works at the technology-agnostic level • Interaction between suppliers, entities, system, and services are transparent • Reduce fraud resulting in faster settlements times • Eliminates paper works
Automation	• BT is capable of permitting time-based transactions and services • Digitally native systems that support scripted, programmable transactions
Innovation	• Capture timely and accurate big data resources • Allows introduction of new risk instruments and capital • Self-insurance and new tailored insurance products • Distributed risk mutualisation support efficient claims management and fraud reduction
Data Pooling Opportunities	• Access a single and real-time resource of data will change the ways • Enable insurers to price and govern claims recoveries

Figure 16.5 Intended consequences of blockchain in the insurance market. Source: prove.com [35].

16.6 Conclusion

The potential impact on current business models poses a slew of problems. Blockchain technology is drawing unprecedented interest from senior management. By combining security, decentralisation, and transparency, blockchains will aid in the management of increasing global complexity. They will return power to the customer and assist in the introduction of new participants to the market. Blockchains have technological restrictions that must be considered. However, regardless of whether blockchain technology or an alternative is used, the use cases for which blockchains are paving the way will be implemented. The number of potential use cases in the insurance business far exceeds those covered in this research, with diverse consequences on the value chain [36].

Certain uses appear to be easy to adopt and bring large benefits, while others appear to be riskier, especially when compared to the projected profits. In the insurance business, the blockchain's breadth of potential is enormous, but it will take some time to adapt and adjust. Regardless of industry, the key task for all stakeholders will be to determine the use case that will be most beneficial to them and to investigate other options if their first decision fails.

References

[1] Kellenberg, D., & Mobarak, A. M., The economics of natural disasters. *Annu. Rev. Resour. Econ.*, 3, 1, 297–312, 2011.

[2] Bednarz, Z., & Manwaring, K., Hidden depths: The effects of extrinsic data collection on consumer insurance contracts. *Computer Law & Security Review*, *45*, 105667, 2012.

[3] Nissim, D., Analysis and valuation of insurance companies. *CE| ASA (Center for Excellence in Accounting and Security Analysis) Industry Study*, 2, 2010.

[4] Kotarba, M., Digital transformation of business models. *Foundations of management*, 10, 1, 123–142, 2018.

[5] Latif, S., Qadir, J., Farooq, S., & Imran, M. A., How 5g wireless (and concomitant technologies) will revolutionize healthcare? *Future Internet*, 9, 4, 93, 2017.

[6] Sarkkinen, E., Designing a Digital Tool Concept for Business-to-Business Dialogue Using a Customer-Centric Approach—Case Fazer Food Services, 2018.

[7] Hsieh, Y. Y., Vergne, J. P., Anderson, P., Lakhani, K., & Reitzig, M., Bitcoin and the rise of decentralized autonomous organizations, *Journal of Organization Design*, 7, 1, 1–16, 2018.

[8] Chapman, J., Garratt, R., Hendry, S., McCormack, A., & McMahon, W., Project Jasper: Are distributed wholesale payment systems feasible yet, *Financial System*, *59*, 2017.

[9] http://www.econotimes.com/five-major-insurers-and-reinsurers-team-up-for blockchain initiative- b3i-355513

[10] Zachariadis, M., Hileman, G., & Scott, S. V., Governance and control in distributed ledgers: Understanding the challenges facing blockchain technology in financial services. *Information and Organization*, 29, 2, 105–117, 2019.

[11] Crawford, M., The insurance implications of blockchain. *Risk Management*, 64, 2 , 2017.

[12] Bennett, H., The Three Ages of Utmost Good Faith. Bloomsbury Publishing 2020.

[13] Sayegh, K., Blockchain Application in Insurance and Reinsurance. *Skema. Businessschool*, 1, 38, 2018.

[14] Akande, A. 2018. Disruptive power of blockchain on the insurance industry. *Master's Thesis, Universiti of Tartu, Institute of Computer Science, Tartu.*

[15] Chen, W., Xu, Z., Shi, S., Zhao, Y., & Zhao, J., A survey of blockchain applications in different domains. In *Proceedings of the 2018 International Conference on Blockchain Technology and Application,* 17–21, 2018.

[16] https://btcmanager.com/news/business/fintech-startup-nimbrix-partners-with-microsoft-and-kpmg-to- launch-asset-management-blockchain-consortium

[17] Ali, O., Ally, M., & Dwivedi, Y., The state of play of blockchain technology in the financial services sector: A systematic literature review. *International Journal of Information Management*, *54*, 102199, 2020.

[18] https://www.infosys.com/services/blockchain/overview.html, 2018.

[19] Borselli, A.,Smart contracts in insurance: a law and futurology perspective. In *InsurTech: A Legal and Regulatory View,* Springer, Cham. 101-125, 2020.

[20] https://www.youhodler.com/blog/smart-contract-examples-7-interesting-use-cases

[21] Signé, L., Strategies for Effective Health Care for Africa in the Fourth Industrial Revolution: Bridging the Gap between the Promise and Delivery, 2021.

[22] Dadios, E. P., Culaba, A. B., Albert, J. R. G., Paqueo, V. B., Orbeta, A. C., Serafica, R. B., & Bairan, J. C. A. C., *Preparing the Philippines for the fourth industrial revolution: A scoping study* PIDS Discussion Paper Series, 2018.

[23] https://www.axa.com/en/newsroom/news/axa-strategic-ventures-blockchain

[24] http://www.coindesk.com/lloyds-sees-blockchains-potential-insurance-markets

[25] https://www.investopedia.com/terms/b/blockchain.asp

[26] http://www.businessinsider.in/Bitcoin-Startup-Coinbase-Is-Raising-75-Million-In-New- Funding/articleshow/45956614.cms

[27] Yang, R., Wakefield, R., Lyu, S., Jayasuriya, S., Han, F., Yi, X., & Chen, S., Public and private blockchain in construction business process and information integration. *Automation in construction*, 118, 103276, 2020.

[28] Aleksieva, V., Valchanov, H., & Huliyan, A., Smart contracts based on private and public blockchains for the purpose of insurance services. In *2020 International Conference Automatics and Informatics (ICAI),* IEEE, 1-4, 2020.

[29] Hwang, S. O., & Mehmood, A., Blockchain-based resource syndicate. *Computer*, 52, 5, 58-66, 2019.

[30] https://www.pwc.in/assets/pdfs/consulting/digital-services/blockchain/blockchain-is-here-whats-your-next-move., 2018.

[31] Gatteschi, V., Lamberti, F., Demartini, C., Pranteda, C., & Santamaría, V., Blockchain and smart contracts for insurance: Is the technology mature enough?. *Future internet*, 10, 2, 20, 2018.

[32] Mohanta, B. K., Jena, D., Panda, S. S., & Sobhanayak, S., Blockchain technology: A survey on applications and security privacy challenges. *Internet of Things*, *8*, 100107, 2019.

[33] https://www.cbinsights.com/blog/financial-services-corporate-blockchain-investments/

[34] Pandey, P., & Litoriya, R., Promoting trustless computation through blockchain technology. *National Academy Science Letters*, 44, 3, 225-231, 2021.

[35] https://www.prove.com/blog/impact-of-blockchain-on-insurance-industry, 2021.

[36] Zheng, Z., Xie, S., Dai, H. N., Chen, X., & Wang, H., Blockchain challenges and opportunities: A survey. *International Journal of Web and Grid Services*, *14*, 4, 352-375, 2018.

SECTION IV

17

Protecting High-value Assets: Insurance Implications of Cybercrime for Individuals

Jasmine Kaur[1], Priya Jindal[1], and Peter J. Baldacchino[2]

[1]Chitkara Business School, Chitkara University, Punjab, India
[2]University of Malta, Accountancy Department, Faculty of Economics Management and Accountancy, Malta

Email: jasminekalra821@gmail.com; priya.jindal@chitkara.edu.in; peter.j.baldacchino@um.edu.mt

Abstract

India is one of the most famous digital markets, with huge potential for exponential growth. In terms of telephone transactions, UPI transactions exceeded INR 1 trillion. This creates the prerequisites for the huge potential of cyberattacks. Opportunities are being replaced by the threat of cyber risks and fraud, and these threats are also rising sharply. Many insurance companies have launched consumer cyber liability insurance to help customers overcome such tragedies of cyber fraud to protect individuals and their families from cyber fraud or digital risks that may cause financial or reputation damage. This product is a form of insurance that protects people from losses from online theft to unauthorised transactions. Although insurance can effectively reduce the financial impact of cyber incidents, policy responses are not always reliable. Research on the assessment of the various institutions protecting individuals through insurance against cyber frauds is missing in India, and additionally, a comparison of all the products available in the market is still shady. This chapter examines the Indian cyber insurance market and its awareness among digital marketers, as well as currently available policies and their relevance to current market conditions and scale.

17.1 Introduction

All sectors of the global economy are becoming more concerned about internet security as digital threats have advanced and online criminals have grown increasingly sophisticated. Network security incidents can hinder the ability of backup plans to manage operations, jeopardise the security of customer and employee data, and undermine faith in the region. The International Association of Insurance Supervisors (IAIS) has observed that there appears to be regional variation in the awareness of digital threats, network security within the protection domain, as well as administrative approaches for combating threats. These variables provoked the IAIS to think about the area of network safety in the protection area, remembering the association of protection bosses for evaluating and advancing the relief of digital gambling.

While large numbers of the most generally promoted network safety occurrences, including buyer information, have impacted retailers, organisations in the monetary administration's area, including backup plans, have been misled too. All backup plans pay little attention towards the gathering, storing, and offering with different outsiders such as specialist co-ops, mediators, and reinsurers significant measures of private and secret policyholder data, and this in particular given the sensitivity of wellbeing related data. Assurance of the classification, uprightness, and accessibility of safety net providers' information is of essential significance. Through cybercrime, data acquired from safety net providers might be utilised for monetary benefit through coercion, wholesale fraud, misappropriation of protected innovation, or other crimes. Accidental or deliberate openness of private information might result in expected and extreme mischief for the impacted policyholders, as well as reputational harm to backup plan area members. Likewise, vindictive digital assaults against a backup plan's basic frameworks might block its capacity to lead the business.

The IAIS evaluated the view of its members regarding the protection industry's digital gamble, their involvement in battling digital threats, and ongoing or potential administrative methods to deal with network safety in 2015 [1]. This section aims to bring issues to light for backup plans and managers of the difficulties introduced by digital gambling, including current and potential administrative methodologies for tending to these dangers. It gives a foundation, depicts current practices, distinguishes models, and investigates related administrative issues and difficulties. It centres around digital gambling to the protection area and the moderation of such dangers. It excludes extensive I.T. security gambles, digital protection (safety net providers' selling or guaranteeing that kind of protection item), or dangers emerging from online protection episodes, including management, which while being significant issues are not within the terms of reference of this chapter.

17.2 The Cyber Risk Landscape

The term "cyber risk" has no established definition. Cyber risk is the phrase used to describe the hazards associated with using electronic information, particularly specialised instruments like the internet and media communications organisations and their interchange [2]. It includes the actual harm that can be brought about by network safety occurrences, extortion submitted by abuse of information, any responsibility emerging from information capacity, and the accessibility, trustworthiness, and secrecy of electronic data, be it connected with people, organisations, or states.

A variety of circumstances explain the adverse effects brought on by cyber risk. These include "cyberattacks," which disorder, damage, or realise unauthorised access to computers, computer systems, or electronic communication networks as characterised by the U.S. Government Financial Institutions Examination Council (FFIEC) [3]. Cyberattacks may involve attacks, through the web, zeroing in on an association's use of the web to disturb, debilitate, demolish, or toxically control a processing adobe or system, or destroy the uprightness of the data or take controlled information. Relatedly, a "digital episode" is portrayed by the FFIEC as "exercises taken utilising P.C. networks that result in a certifiable or perhaps opposing effect on an information structure or the information staying in that."

At present, the expression "cybercrime case" is by and large used to encase both digital assaults and digital cases. The World Economic Forum has distinguished specialised gambles as information misrepresentation, network protection, and framework on the planet's monetary gathering, with the best 10 dangers confronting the worldwide economy [4]. In a report on identifying the dangers of industrialisation in 2015, cyber risk was distinguished as the fourth largest gamble in the U.S. Insurance Companies.

One safety net provider, Allianz, has experienced the resulting high patterns inside the digital gambling scene. The growing interconnectivity and "commercialisation" of digital wrongdoing drive more recurrences and higher seriousness of occurrences and data breaches that data insurance regulation can take into account worldwide. Extra alerts and significant fines for information breaks are, in many cases, expected. Business interference (B.I.) affects burglary, and digital coercion chances are expanding.

17.2.1 Types of cyber security incidents

Verizon estimated in a 2015 report that most data breaches resulted from at least one of the following causes: disruptions at retail locations, crimeware, insider abuse of digitally protected information, and attacks on web applications [5]. The remaining incidents involving business insurance mostly resulted from unintentional errors, genuine theft or tragedy, and payment card skimming. While information breaks are typically associated with these network safety incidents, other tragic types may also occur in the form of robbery of licensed innovation.

Digital coercion, typically cultivated through a kind of crimeware known as "ransomware," is an inexorably normal network safety occurrence "wherein programmers invade P.C.s having a place within a business or an individual, encode the information subsequently, and afterwards request an instalment to interpret it." Such ransomware is extremely convincing, and victims of such attacks cannot restore details without paying a ransom unless they have created a backup copy of the data stored on media not exposed to the ransomware attack.

17.2.2 Recurrence, severity, and cost of cyber security incidents

The recurrence of network protection episodes is expanding. PricewaterhouseCoopers' annual corporate executive's information security examination in 2014, which involved 9700 people, found that there were around 43 million recognised network wellbeing events in 2013 [6]. The amount of organisation wellbeing events paid with respect to the 2015 PwC market share extended by a further 38% over the years [7]. Likewise, the amount of unreported organisation wellbeing events is conceivably far higher. Attacks through malware are logically inescapable: it has been claimed that, across all associations over the world, five malware assaults happen the entire day, albeit not all are effective.

Network safety occurrences occur within different organisations. Independent ventures, in any case, might be especially defenceless. For example, a 2015 audit by NetDiligence noted that associations with under $50 million in livelihoods reported most episodes (29%), these being closely followed by associations with under $2 billion in pay (25%) [8]. The severity of information breaks, the most detailed type of network safety occurrence, differs by location. Among the countries and affiliations covered by a report from the Ponemon Institute (Ponemon), the ordinary number of records lost or taken per data break is approximately between 19,000 and 30,000, with the United States, India, and Arabic nations registering the most outstanding midpoints [9]. Online protection episodes are correspondingly expensive.

Lloyd's and the University of Cambridge Center for Risk Studies delivered a report breaking down the monetary misfortunes related to a theoretical digital assault on the U.S. power matrix. The report anticipated complete misfortunes costing between $243 billion and more than $1 trillion, and guaranteed misfortunes varying between $21.4 billion and $71.1 billion, such variation being based upon the seriousness of the blackout [10]. Ponemon has assumed that the overall ordinary cost of a data break in 2014 was $3.79 million, with a typical overall cost of $154 for each lost or taken record. These costs differ topographically [11]. As per Ponemon, these expenses can be seen as tending to be categorised under one of four classifications: (i) discovery and heightening; (ii) warning; (iii) ex-post reaction; and (iv) lost business. In 2013, 2014, and 2015, lost business – which incorporates startling client turnover, notorious misfortunes, and diminished generosity – was the biggest expensive part connected with information breaks.

17.3 Cyber Threats to the Insurance Sector

As a rule, the protection area faces digital gambles from both inward and outside sources. Backup plans gather, cycle, and store significant volumes of information, including recognisable data. Contingency plans are related to other financial associations through various media such as capital raising and issuance commitment. Underwriters carry out various changes in the business plans in the form of consolidations and acquisitions that can impact advantageous online protection.

Examples of cyber security flaws, including clients highlighted by IAIS members, are as follows:

- **A skewed picture of the I.T. landscape**
  While all safety net providers ought to have a stock of I.T. equipment and authorised programming, even those keeping up with such records may not perceive the information stream among those I.T. frameworks, applications, and parts. On the off-chance that information streams exist among frameworks with elevated degrees of assurance and frameworks with lower security levels, digital lawbreakers might have the option to get close enough to get a framework in any case.

- **Insufficient process for managing user privileges**
  There are normally two kinds of issues related to client character by the executives: (i) the deficiency of controls inside the designation

interaction of client freedoms, i.e., allowing clients to have higher situation rights than justified; and (ii) the inability to perceive when a record at this point does not need a specific framework. The two sorts of deficiencies could prompt insider misuse and openness to digital dangers. Programming robotised items that can perform the Board's personality become accessible.

- **Inappropriate superuser account access**
  Without sufficient safeguards, direct worker access to "superuser" accounts poses a risk to backup preparations. First, if a programmer with access to the superuser account accessed any of the employees' data, the programmer could effectively manage the entire framework through the superuser account (concealing lawbreaker acts by adjusting or erasing log documents or by debilitating location systems). Second, the normal utilisation of superuser records could prompt accidental mistakes influencing the whole framework.

17.3.1 Cyber security incidents in the insurance industry

Recently, the protection area has encountered an assortment of network safety episodes, including acclaimed information breaks at a few U.S. well-being safety net providers. The following are a few instances of these network safety occurrences:

"Anthem Blue Cross Blue Shield and Premera Blue Cross" had data breaches in 2015, exposing Mastercard information and genuinely recognised data, including health information. The breaches may have exposed the personal information of up to 91 million policyholders or almost one-fourth of the population of the United States. The safety net providers needed to respond quickly to alleviate reputational harm and to limit suit costs, even though the losses eventually caused by the backup plans because of this break are as yet unclear.

In a different new U.S. model, a North Dakota information system was hacked, revealing a trove of personal information about employees' pay records due to 43,000 Incident Reports and 13,000 Payroll Reports that had been logged online by employees and management. Clinical data and guarantee papers were allegedly not exposed, but the information in jeopardy was stated to include people's identities, government-backed retirement numbers, birth dates, injury representations, episode depictions, boss names, and business locations.

In 2012, inward extortion was discovered in a French insurance business, demonstrating that network security incidents may also include insider ones. This scam, which stemmed from an internal data breach in a simulated environment containing sensitive customer information, sparked a wave of widespread deception and fraud.

In the Netherlands, a safety net provider was liable for a particular type of phishing digital assault, termed as "President hack". Professing to be the CEO of a significant and notable business client of the safety net provider, the crook/s attempted to convince workers of the backup plan to move cash into a specific record. The crooks had investigated the specific functional subtleties of the safety net provider.

A new report from PwC distinguished a few extra online protection occurrences experienced by backup plans in various locations, including assaults against individual lines, travelling, wellbeing, and marine safety net providers.

17.4 Insurance Cyber Resilience

The changing difficulties introduced by digital gambling ought to be met with a wide reaction by guarantors. Undeniably, appropriate administration considerations are a need, this comprising a viable administration structure ready to comprehend, forestall, distinguish, answer, and address network safety occurrences. Moreover, a well-working gamble for the executives' program with digital versatility best practices should be set up and confirmed through the administrative survey. As will be shown later in this article, this degree of reaction is consistent with the insurance Core Principles.

To be powerful, network safety should be applied to all levels of an entity and take into account any possible outsider courses of action. The bulk of an effective digital gambling executives program incorporates continuous interaction and control updates, the board systems, including, for example, reaction and disaster recovery, appropriate organisation arrangements and methodology, thorough administration and control of client honours, secure design direction, appropriate malware assurance techniques, consistent control of removable media use, checking of versatile and home working strategies, and ways to progress. It is typically perceived that the prescribed procedures for digital strength include corporate governance. A genuine digital flexibility structure, in addition to the Board's and top-level management's dedication and responsibility, does contribute to the moderation of digital gambling. For instance, Senior Management ought to have the authority to access the Board, which is ultimately liable for creating and carrying out the digital versatility structure.

- **Testimony**
  I.D. implies distinguishing those business capacities and cycles that ought to be safeguarded. Data resources (including sensitive individual data) and related framework access should be important for the I.D. interaction. Standard surveys and updates are key variables, as a digital gamble is continually developing and "stowed-away dangers" can arise at any time. Associated elements are essential for the entire picture, keeping in mind that the meaning of the dangers they represent is not proportionate to the materiality of the specific help. For instance, the notable digital assault against retailer target included a section using a ventilation specialist organisation.

- **Safeguarding**
  Flexibility can be included in the plan. Thorough security involves safeguarding interconnections and different methods for admittance to insider and pariah dangers to the establishment. While planning security, the "human element" ought to be taken into account. Subsequently, being prepared is likewise fundamental in the security net against digital gambling. Controls ought to be under specialised norms, as solid I.T. controls add to security.

- **Identification**
  Persistent and extensive network safety observation is fundamental for recognising potential digital occurrences. Performing security examination additionally assists with alleviating digital occurrences [12].

- **Feedback and restoration**
  It is not generally imaginable to identify or forestall digital episodes before they occur, even with the best cycles being set up. Thus, planning for occurrence reactions is essential. Depending on the impact of the events and the urgency of the need for assistance, administrations should be effected as soon as it is reasonable. The potential for planning, organising, and business reconciliation, as well as information credibility (also in view of any agreements for information sharing) are major enabling factors for rapid business restarting. Actions taken to prevent infection can reduce additional risks. An exposure plan needs to be implemented so as to improve emergency correspondence. In conclusion, a scientific preparation is essential, particularly for deeper examinations. In particular, if one is to plan for business growth, these factors should be taken into account.

- **Examination**
  The basics of the testing stage include examination programs, weakness evaluations, situation-based testing, infiltration tests, and red group tests. Network security testing should include an understanding of the conditions when determining, creating, and coordinating frameworks. The prevalence of fear is a distinctive evidence of the digital dangers. The setting up of a dangerous knowledge process assists with relieving the digital gamble. In such a manner, backup plans ought to consider taking an interest in laid-out data-sharing drives.
- **Mastering progress**
  Backup plans ought to consistently evaluate the viability of online protection to the Board. Experiences gained from digital occasions and digital episodes add to further developments [13]. Any advancement in innovation ought to be sought after and checked.

17.4.1 Assessment of institutions providing cyber security insurance

The profiles of central market participants such as Allianz, American International Group, Inc., Aon PLC, AXA, Berkshire Hathway Inc., etc., are examined in this section. These companies have employed a range of techniques to boost their market share and solidify their position in the digital security industry.

- **Allianz**
  Allianz Global Corporate and Specialty (AGCS) has long been involved in digital protection, safeguarding associations against digital wrongdoing and computerised dangers. That is why it can offer its clients with a scope of digital insurance items varying from subject matter expert and independent digital protection to devoted digital gamble inclusion in conventional property and setback contracts. The sorts of dangers covered by Allianz incorporate first-party misfortunes (for example, business interference, reclamation, and emergency correspondences) and outsider misfortunes (for example, information breaks, network interference, and notice costs). In any case, digital protection offers much more than merely paying for possibly critical monetary misfortunes. Allianz offers its clients significant counteraction and episode reaction administrations through its worldwide network organisations.

These assist organisations with working on their digital strength and moderate adverse consequences after an episode. They incorporate all-day and everyday admittance to I.T. criminological specialists or legitimate or emergency interchanges support. The Allianz Cyber Center of Competence is inserted into AGCS and centres around group-wide coordination and arrangement of digital openings and endorsing in corporate and business protection.

- **American International Group, Inc.**
  AIG's honour-winning digital arrangements help the insured to acquire better understanding. They address digital gambles with far-reaching administrations, backing, and inclusions to assist with safeguarding their primary concern. AIG's Cyber Insurance can be composed through an independent CyberEdge strategy or embraced onto select Financial Lines, Property, and Casualty arrangements.
- **Aon PLC**
  In assessing cyber risk insurance, an understanding of the dangers and openings involved in a protection contract is essential. While various standard items are accessible in the commercial centre relating to planning cyber risk and professional liability, take-up by professional assistance firms such as law offices, bookkeeping firms, counselling firms, and planning and development firms remains difficult.

17.4.2 Digital risk protection can be intended to cover specific extra openings

Digital risk insurance contracts give direct admittance to information break advisors and boards of specialists to help firms, which endure information break occasions. Their protection addresses costs for scientific examination and advertising, warning expenses, credit observing, buyer training, and assistance costs resulting from an information break. Furthermore, a few strategies cover the expense of holding outside advice to assess the company's expected commitments for a break.

- **AXA XL**
  Essentially, all associations are in danger, and this continues to increase with new techniques, entertainers, and targets. AXA XL has a profound authoritative obligation to assist its clients with safeguarding against one of the world's most difficult dangers and one that indicates no decrease. AXA XL's digital insurance arrangements incorporate a secluded

Figure 17.1 Digital risk insurance gaps. Source: Author's compilation.

digital and information assurance contract intended to be customised to the necessities of the clients' specific industries and this for both outsider obligation and first-party misfortunes (Figure 17.1). Alongside these broad and adaptable inclusions, AXA XL can convey a variety of administrations to assist clients with forestalling such assaults and answer rapidly and successfully if their business is struck.

AXA XL adds digital and information security inclusion to correspondences, media, and innovation organisations either on an independent premise or as a more extensive expert reimbursement strategy component.

- **Outsider obligation**
  This consists of the following:
  - Information break security and protection obligation.
  - Media web correspondences .
  - Dealing with:
    - first-party misfortunes;
    - business interference and additional costs;
    - misfortune/annihilation of electronic resources;

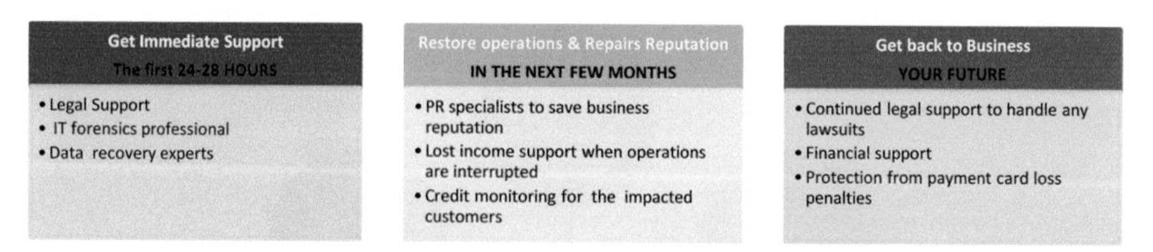

Figure 17.2 Protection and safeguarding business against cyber insurance. Source: [14].

- first-party occurrence reactions, including I.T. criminology, notice costs, call focus, and P.R. costs;
- protection of administrative safeguard expenses and inclusion of administrative fines and punishments (where insurable by regulation) emerging from a security or security unjust demonstration;
- information reclamation;
- digital blackmail.

- **Berkshire Hathway Inc.**
  Malware, cyberattacks, and information misfortune are real dangers to each business. Cyber policy gives clients insurance where it is required most – on the web. This includes insurance when:
  - handling visas;
  - tolerating advanced payments;
  - gathering client information;
  - putting away classified data;
  - putting away medical and monetary information.

- **Lloyd's of London Ltd.**
  Lloyd's of London Ltd. (Lloyds) accepts that digital protection can assist organisations by safeguarding their accounting reports not by simply giving monetary compensation after things have turned out badly but also by offering expert consultancy to further develop security and on-the-ground support during time of emergency. Lloyd's is at the forefront of this new protection arrangement. As a protection market, Lloyd's can give admittance to the joined scale, mastery, and a limit of more than 77 master digital gamble backup plans in a single spot. It is this capacity to make proper and custom-made protection arrangements from a variety of markets that distinguishes Lloyd's.

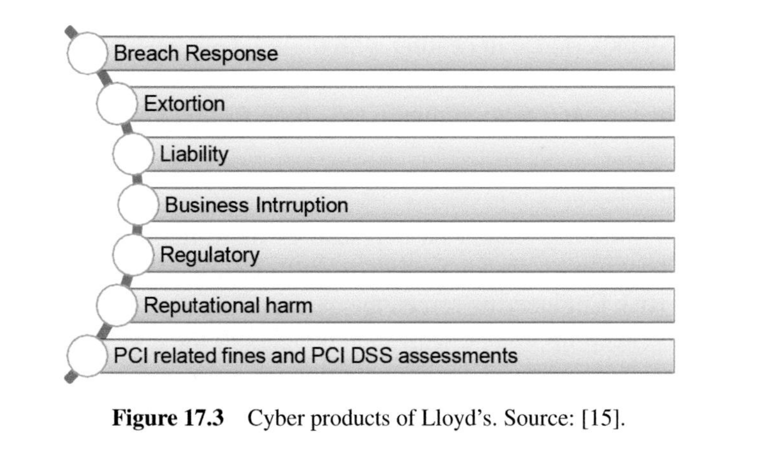

Figure 17.3 Cyber products of Lloyd's. Source: [15].

Figure 17.4 Lockton Companies Inc. products and services. Source: Author's compilation.

- **Lockton Companies, Inc.**
  In addition to the ability for in-house data security, Lockton Companies, Inc. have an extra cover for innovation blunders and exclusions to media obligations. The companies' security and privacy protection strategies are explicitly intended to match the quickly growing extent of occasions and steadily evolving innovations. This gives a more significant level of conviction for client business, disposing of large holes between existing gambling controls and customary insurance contracts.
- **Munich Re**
  Munich Re provides the solutions and services that individuals and organisations need to avoid risks, respond quickly when breaches occur, and get back on track after cybercrime has cost them time and money.
- **The Chubb Corporation**
  Extended dependence on P.C. frameworks and admittance to data can fundamentally build an organisation's weakness to network protection dangers. Blackouts, errors, or assaults on these new cycles can result in huge cash-based costs that can crush an association's primary concern. Regarding information security breaks or protection misfortunes, it is anything but a question of whether it will work out, yet rather when it will work out. To keep it from working out, clients will require complete assurance from the Chubb Corporation, a safety net provider with

practical experience in dealing with digital dangers, which offers a full set-up of coordinated protection answers to assist with limiting holes in inclusion, and who comprehends how to tailor inclusion for your business.

- **Zurich**
  Digital gambling should be overseen as more than a gambling move issue. That is why digital gamble experts in Zurich Resilience Solutions Cyber Risk Engineering screen the extent of the developing digital dangers and keep up to date with such dangers to the client organisations, offering experiences and administrations to help protect against them.
  The Cyber Risk Engineering group can assist with administration matters such as:
  - digital risk gap analysis and strategic roadmap;
  - episode response plan evaluation and tabletop exercise;
  - ransomware threat assessment;
  - representative cyber security awareness training (counting phishing social designing mindfulness);
  - framework penetration testing;
  - pre-break assessments.

17.4.3 Procedure, features, and timeline of cyber insurance

There is a sequence of steps taken for getting cyber insurance and certain eligibility criteria for getting cyber insurance done. This part appends the qualification criteria, timeline, and functionality of cyber insurance.

- **What are the steps for setting out cyber security insurance?**
  Setting out cyber security insurance:
  Setting out network safety and inclusion is a simple interaction. The following are the steps:
  - Present an application for protection.
  - Educate the client regarding the nature of hazards to be covered.
  - Investigate the dangers, the inclusion is examined by the insurance agency, and agreements are made with the guarantor.
  - Once the above steps are settled, the guarantor requests a proposition from the strategy searcher.

- ○ The proposition, together with the necessary archives, is submitted to the insurance agency.
- ○ The approach is organised and given over to the strategy holder.
- ○ If an occurrence of any possibility should arise, the protected informs the insurance agency about the equivalent.
- ○ Archives and proof are submitted to the safety net provider. An agent approves the case and sends the report to the safety net provider.
- ○ Where there is a valid case in line with the required approach, remuneration is given to the petitioner. Otherwise, the case is dismissed, and the inquirer is informed.
- ○ If the inquirer is not satisfied, further intervention can be arranged.

- **Qualification criteria:**
  Anybody over the age of 18 years can purchase a digital protection insurance contract.

  The network protection insurance claim process:
  Network protection has a fast and simple case settlement process. The following are most important details about lodging claims:

  - ○ The guarantee should be submitted in 90 days or less.
  - ○ The guarantee should be submitted in a composed design.
  - ○ Required proof of misfortune should be submitted to the organisation.
  - ○ An agent from the insurance agency checks the case.
  - ○ Where the case is indicated as valid by the approach, the remuneration is paid to the recipient account.
  - ○ Assuming that the case is dismissed, information is passed on to the inquirer.
  - ○ The petitioner might request intervention if the goal is not good.

- **Reports required for a claim process:**
  - ○ scientific reports;
  - ○ appropriately filled-in guarantee structure;
  - ○ screen captures of discoveries;
  - ○ subtleties and proof of misfortune like dates, acts, conditions, individuals included, and alleviating rehearses.

- **How long does it take to pay out a claim?**
  After documenting the case for network safety protection, the insurance agency, as a rule, requires roughly 30 days to make the payment under the case.

17.5 Supervisory Resilience to Cyber Risk

While the insurance core principles (ICPs) do not specifically address digital gambling and digital versatility, they provide a framework for executives to address the protection area related to digital gambling and digital strength by requiring the management of significant risks and related internal controls [16].

The ICPs that might generally be significant for the management of digital gambling in the protection area include the following.

- **ICP 7 related to corporate governance**
  In November 2015, ICP 7 was modified. Guarantors should be able to show the appropriateness of frameworks and controls and the corporate administration structure under this ICP. According to the guidelines for this ICP, "It is the Board's responsibility to ensure that the safety net provider has appropriate frameworks and capacities for risk management and internal controls, and to provide oversight to ensure that these frameworks and the capacities that supervise them are working successfully and as expected." Identifying and managing digital risk should be a fundamental part of a guarantee's risk management strategy [14].
- **ICP 8 related to risk management and internal controls**
  In November 2015, ICP 8 was updated. This ICP necessitates that a guarantee has an effective risk to the Board and internal controls as part of its overall corporate administration structure, including effective risk to the executives, consistency, actuarial concerns, and internal review. ICP 8 Guidance records a base arrangement of classes about the gamble that the board framework ought to cover. Concerning digital gambling, the Guidance is significant as it alludes to "functional gamble the executives," "director of business," and other gamble relief methods [17]. Additionally, the rules state that the framework for the digital gambling, board should take into account all material risks, both present and future, that are predictable and significant. The Guidance likewise depicts regular parts of a powerful inside control framework. The Guidance describes such an inward control framework as having "fitting controls for all business cycles and strategies," including "basic

I.T. functionalities," "admittance to data sets," and "I.T. frameworks by representatives" under the "arrangements and cycles" that are recorded as one of these parts. This envelops digital gambling.

At last, ICP 8 Guidance expresses that the inside review capacity ought to give free affirmation to the Board and Senior Management regarding issues including limit and flexibility of the I.T. design to give promptly bookkeeping, monetary, and risk-revealing data [18]. While going into or amending a re-appropriating course of action, the Board and Senior Management ought to consider how the backup plan's gamble profile and business congruity will be impacted by the re-appropriating [19].

- **ICP 9 related to supervisory review and reporting**
  It addresses the broad cycles and techniques managers should have for administrative surveys and disclosure. These cycles incorporate investigating the administrative system for auditing and answering in order to guarantee that it gives due consideration to the developing nature, scale, and intricacy of dangers that might be presented by safety net providers and of dangers to which backup plans might be uncovered. The administrative system should demand a backup plan under this ICP to report any substantial changes or events that might quickly affect the safety net of the provider's condition or the clients. Off-site checking and on-site investigation executives should gather enough data to analyse and break down the risks to which a backup plan and its clients are exposed, as well as audit the feasibility of the guarantor's risk management. ICP 9 is currently being audited.

- **ICP 19 related to the conduct of business**
  Security assurance arrangements under which backup plans and mediators are authorised to acquire, keep, use, or impart clients' data to outsiders are among the requirements for the direction of the protection business. ICP 19 requires safety net providers and go-betweens to have strategies and methods for the assurance of private data on clients. The Guidance portrays various measures to guarantee protection assurance and forestall security breaks, zeroing in on counteraction of the abuse or improper correspondence of individual data to outsiders.

- **ICP 21 related to countering fraud in insurance**
  Under ICP 21, bosses expect that safety net providers and mediators go to compelling lengths to hinder, forestall, distinguish, report, and cure extortion in protection. Misrepresentation in protection might happen

through digital episodes [20]. ICP 21 is currently under audit, partially to consider whether it should be reconsidered to address digital gambling all the more straightforwardly.

Furthermore, especially concerning data trade and administrative participation, the accompanying ICPs apply to:

ICP 3 "Information Exchange and Confidentiality Requirements"

ICP 25 "Supervisory Cooperation and Coordination"

ICP 26 "Cross-border Cooperation and Coordination on Crisis Management"

ICP 3, ICP 25, and ICP 26 – information sharing and supervisory cooperation

ICP 3 is significant because, given the idea of a digital gamble, it is conceivable that more than one location would be associated with the recognisable proof, the executives, and the moderation of such dangers. The ability to impart data to several parties, as defined in ICP 25, is a critical tool for supervisors. An efficient data-sharing tool will boost the ability to respond quickly to detect, mitigate, and eliminate threats [21]. As such, ICP 26 is currently being reviewed.

17.6 Conclusion

Organisations will probably embrace portable, cloud, remote access, and IoT advancements, not out of choice but out of the need to support the business during the pandemic and flourish from then onwards. Such extraordinary digitisation will likewise result in expansion. For bank chiefs, the attention will be on accomplishing business objectives even as they recalibrate procedures to address the steadily advancing digital disruptions. Organisations should therefore put resources into the digital guard to make a versatile framework for the future. This will address the ongoing network protection risks and prepare for the digital difficulties of things to come [22].

The drive must come from establishment leaders and board individuals who set objectives and designate spending plans. Speeding up digital abilities to match the speed of progressive change will require leader consideration, prioritisation, financial planning, assets, and administration. Such an initiative can drive the required change. Bank chiefs will choose the level of dexterity, speed of progress in framework, and cooperative endeavours expected towards building the network protection of things to come.

References

[1] Bharadwaj, S. K., & Bba, H. H, "Cyber liability insurance in India: Growing importance". *Imperial Journal of Interdisciplinary Research (IJIR), 2*, 1, 2454–1362, 2016.

[2] Böhme, R., & Kataria, G., Models and measures for correlation in cyber-insurance. In *WEIS* , 2, 1, 3, 2006.

[3] Camillo, M., Cyber risk and the changing role of insurance. *Journal of Cyber Policy, 2,*1, 53–63, 2017.

[4] Gajapathy, V., & Patil, R. M., Cyber Insurance–A Rising Market in India. *Global Journal of Enterprise.Information System, 10*, 3, 13–18, 2018.

[5] Franke, U., The cyber insurance market in Sweden. *Computers & Security, 68*, 130–144, 2017.

[6] Traum, A. B., Sharing risk in the sharing economy: Insurance regulation in the age of Uber. *Cardozo Pub. L. Pol'y & Ethics J., 14*, 511, 2015.

[7] Conteh, N. Y., & Schmick, P. J., Cybersecurity: risks, vulnerabilities and countermeasures to prevent social engineering attacks. *International Journal of Advanced Computer Research, 6,* 23, 31, 2016.

[8] Kuru, D., & Bayraktar, S., The effect of cyber-risk insurance on social welfare. *Journal of Financial Crime, 24*, 2, 329–346, 2017.

[9] Makanda, K., & Kim, H., Cyber security insurance status in Malawi. *International Journal of Applied Engineering Research, 12,* 17, 6983–87, 2017.

[10] Mukhopadhyay, A., Chatterjee, S., Saha, D., Mahanti, A., & Sadhukhan, S. K., Cyber-risk decision models: To insure I.T. or not? *Decision Support Systems, 56*, 11–26, 2013.

[11] Pal, R., Cyber-insurance for cyber-security a solution to the information asymmetry problem. In *SIAM Annual Meeting*, 1–13, 2012.

[12] Pal, R., & Hui, P., Cyber insurance for cybersecurity a topological take on modulating insurance premiums. *ACM SIGMETRICS Performance Evaluation Review, 40,* 3, 86–88, 2012.

[13] Kshetri, N., The economics of cyber-insurance. *I.T. Professional, 20,* 6, 9–14, 2018.

[14] https://www.cyberpolicy.com/cyber-insurance,2016.

[15] https://www.lloyds.com/about-lloyds/our-market/what-we-insure/cyber/cyber-products, 2018.

[16] Shackelford, S. J., Should your firm invest in cyber risk insurance? *Business Horizons, 55,* 4, 349–356, 2012.

[17] Yassir, A., & Nayak, S., Cybercrime: a threat to network security. *International Journal of Computer Science and Network Security (IJCSNS)*, *12*, 2, 84, 2012.

[18] Zhao, X., Xue, L., & Whinston, A. B., Managing interdependent information security risks: A study of cyber-insurance, managed security service and risk pooling. *ICIS 2009 proceedings*, 49, 2009.

[19] American Banker Association, Cyber Insurance Buying Guide. http://www.aba.com/Tools/Function/ Documents/2016Cyber-Insurance-Buying- Guide, 2016.

[20] Bartol, N; & Coden, M; Our critical infrastructure is more vulnerable than ever. It doesn't have to be that way, https://www.bcg.com/publications/2017/engineered-products-critical-infrastructure-vulnerable-doesnt-have-to-be-that-way, 2017.

[21] Lu, X., Niyato, D., Privault, N., Jiang, H., & Wang, P., Managing physical layer security in wireless cellular networks: A cyber insurance approach. *IEEE Journal on Selected Areas in Communications*, *36,* 7, 1648–1661, 2018.

[22] Rusi, T., & Lehto, M., Cyber threats mega trends in cyberspace. In *ICMLG 2017 5th International Conference on Management Leadership and Governance. Academic Conferences and Publishing Limited* 323, 2017.

18

Cybersecurity in Insurance

Jagjit Singh Dhatterwal[1], Kuldeep Singh Kaswan[2], Sanjay Kumar[3], Kiran Sood[4,5], and Simon Grima[6]

[1]Department of Artificial Intelligence & Data Science, Koneru Lakshmaiah Education Foundation, India
[2]School of Computing Science & Engineering, Galgotias University, India
[3]Computer Science Engineering, Galgotias College of Engineering and Technology, Greater Noida, India
[4]Postdoc Researcher, University of Usak, Faculty of Applied Sciences, Dept of Finance and Banking
[5]Chitkara Business School, Chitkara University, Punjab, India
[6]Department of Insurance and Risk Management, Faculty of Economics, Management and Accountancy, University of Malta, Malta

Email: Jagjits247@gmail.com; kaswankuldeep@gmail.com; skhakhil@gmail.com; kiransood1982@gmail.com; simon.grima@um.edu.mt

Abstract

It is a common assumption in policy discussions that the widespread adoption of cyber insurance would boost network security. However, this depends on potential applicants' processes to get cyber insurance. Candidates often submit proposals with self-evaluations as part of the selection process. A review of cyber insurance claim forms is presented here. According to the evidence presented in our contributions, the insurance industry will promote best practices in cybersecurity. The risk-based and proposal-based approaches to incident prevention are presented as solutions to the problem of information asymmetry. In this last section, we discuss the factors that may have led to the gap between insurance practices and informationally strong cybersecurity procedures. In this chapter, the importance of information management economics is highlighted for comprehending insurance coverage.

18.1 Introduction

The cost of cyber insurance for healthcare already surpasses $2 billion, according to a study conducted in 2015. According to the same study, demand for cyber insurance is anticipated to triple by 2020. This comes as a surprise since a Marsh-commissioned Cyber Risk Evaluation Study from 2015 found that boards of directors are gradually coming to a deeper understanding of the dangers their companies face and the gaps in their existing insurance policies (Siegel, 2002). For instance, the "cost of a security breach is $3.8 million," according to a study conducted in 2015 on 350 businesses in 11 countries. While security breaches get all the attention, there are really quite a few other risks, including cyber kidnapping and unintended outcomes. Many of these are now covered by innovative forms of cyber insurance (Kaswan, 2022; Deore, 2022).

According to a 2015 review of the trends in the cyber liability insurance market, the biggest obstacle to selling insurance is "not" comprehending customers' sensitivity to cyber risks. If you mess up this process, it might cost you a lot of money. Target's insurance paid them $90 million after the data incident in 2013. A common method used in the insurance industry is constructing mathematical models of loss experience across well-defined risk categories (Kesan, 2005). There are two key reasons why these are completely irrelevant: first, insurers lack knowledge of the traits and attributes that determine different risk profiles, and, second, insurers lack the loss statistics needed to create the computer models. Due to the ever-changing nature of cyber risk, past losses may never be repeated. To determine cyber risk, insurers now rely only on test results. However, information collected during these review processes that indicate whether or not certain types of insurance-mandated protective measures are in place might have additional implications (Shetty, 2010).

It is widely held that discussions on policy in the "United States, United Kingdom, and European Union" are influenced by insurers' security concerns. The underlying assumption of these discussions is that the insurance industry can significantly and positively impact cybersecurity management. That risk has been effectively managed by insurers for centuries is an argument supporting the approach. A deeper look into the insurance industry reveals significant mistakes made by insurers. For instance, the lawyers' mandatory insurance market saw major insurers withdraw from the industry's base during the crisis of 2010, and the Irish insurer Quinn fell into administration. This suggests there is a need for further study into the hypothesis that cyber insurance would improve the security posture of businesses (Bandyopadhyay, 2009).

Table 18.1 Statistics of the cybercrime.

S. No.	Threat by equipment	2003	2004	2005	2006
1	Virus, worms, and Trojan infection	80	85	60	63
2	Laptop theft	55	58	55	59
3	Problem inside in computer system	25	25	28	30

This chapter aims to look at how effectively existing cyber insurance planning phases adhere to standards established by organisations like the Center for Internet Security (CIS) and the International Organization for Standardization (ISO/IEC) 27002. As part of the cyber insurance online application process, applicants are required to complete self-evaluation questionnaires, which is the focus of our study. The true value of our research is that it allows us to draw attention to factors in the review process that have been neglected in the past. This may be useful for policymakers since it provides evidence of cyber insurance's ability to complement established risk management practices. Moreover, it may help cyber insurers refine the evaluation process that is based on security criteria (Kaswan, 2021).

18.1.1 Cyber insurance industry

As one author puts it, "hacker's insurance providers of the late 1990s were the first independently Internet-based insurance providers." The insurer and the IT company worked together to provide first-party damage coverage for the insurer. It became clear that standard insurance plans had significant loopholes as non-IT firms increasingly relied on their networks. For example, automated data loss is often not covered by policies that insure the physical assets of a corporation. So, insurance companies started offering cyber insurance as a different product. These strategies are broken down into a wide variety of sub-policies, each of which addresses a particular category of risk. In the event of data loss, for instance, "the cost of restoring or retrieving lost data" would be covered by first-party insurance. The coverage and hazards included were selected based on an analysis of existing insurance policies. The coverage descriptions of Table 18.1 will be used as a working definition of cyber insurance at the subgranular level (Grzebiela, 2002).

The cyber insurance industry is driven by companies that protect large organisations. While 26% of US companies with $5 billion or more in sales have cyber insurance, just 3% of companies with less than $500k in sales do. According to a poll conducted in 2015 in the United Kingdom, just 2% of large organisations have dedicated cyber insurance, while the percentage for smaller businesses is "very minor." Cyber insurance may become more important to business owners if this trend continues. There is an "increased incidence of cybercrime" against small firms. The three biggest threats they face are a disruption to corporate operations, breaches of personal data, and fraudulent activity. In addition, these threats are already covered by cyber healthcare coverage (Gordon, 2003; Sood, Seth, & Grima, 2022; Grima & Marano, 2021).

A corporation may get cyber insurance without having enough security measures to back it up. This is an example of information asymmetry, which occurs when one party in a contract uses strategic actions that benefit itself at the expense of the other side. Insurers deal with this problem by conducting a thorough ex-ante investigation, including collecting data from an application. An individual's yearly premium is based on the risk category to which the underwriter assigns them. The candidate will complete an inquiry and a proposal form that will provide most of the necessary information. Some of the data collected from the survey requested by the states are included. For instance, an insurer may inquire, "Do you retain, analyse, and transmit any Necessary Information on Your Computer System (Kaswan, 2022)?" to learn more about the applicant's data collection practices. These were selected to help the reader make sense of the questionnaire provided; a fuller perspective may be found by using all of the available options. It is common practice to supplement this questionnaire with in-person audits and interviews with C-level information technology (IT) executives. This supplementary review is on the design, implementation, and civilisation of networking security. Our chapter analyses the questions about the applicant's security measures in self-evaluation forms. We will not take into account details like the applicant's wealth, the technique of data gathering, or any previous loss records in our analysis. The personality forms provide a scalable assessment approach that may help fulfil small and medium businesses (Bolot, 2008).

18.2 Infrastructure and Facilities

Internet shopping represents a dynamic area of the ever-evolving technological infrastructure. In today's world, everyone has access to some type of information and communication technology (ICT), and the pace of digitalisation is increasing. Information technology has been integrated into previously incompatible domains, such as transportation and construction, due to the rising need for computer and data communication. Now that we have widespread access to high-speed wireless networks, information and communication technologies (ICTs) are essential to virtually every aspect of modern life, from providing basic necessities like power and transportation to more complex tasks like defence and supply chain management. Although innovative technologies are often implemented to meet the demands of Western customers, developing countries may also benefit from cutting-edge WiMAX5 and consumer electronics currently available for less than USD 200. More people in underdeveloped regions should easily access the internet and associated services and products (Gordon, 2002).

Information and communication technologies have far-reaching societal effects outside of information infrastructure. The spread of ICTs provides a foundation for developing new network-based services that are easily accessible to more people. Internet-based communication necessitates the development of services quicker than landline connections, and e-mails have almost all supplanted traditional letters. Having access to information and communication technologies (ICTs) and cutting-edge networking devices has far-reaching positive consequences for society at large, especially in underdeveloped regions (Hausken, 2006).

18.3 Merits and Risk

The contemporary notion of the knowledge economy is a direct outcome of the widespread use of ICTs in many spheres of everyday life. The information revolution's progress opens up many possibilities. As information dissemination moves away from centralised control, democracy may benefit from unfettered access to information (as has happened, for example, in Eastern Europe and North Africa). ICT integration has advanced so that it is now possible to use things like Google Wallet and buy on the move, use mobile internet services, and communicate via voice over IP (Kunreuther, 2007).

"However, the rise of the digital age has coincided with the emergence of new and perilous threats". The provision of vital public services and electricity is becoming more reliant on ICTs. To function effectively, things like cars, traffic control, elevators, HVAC systems, and phones depend on ICTs. There is a new and critical risk that attacks on electronic information and internet connections might have on society. In the past, cyber assaults on digital data and services have happened. Computers are used regularly in criminal activities such as online fraud and cyberattacks. According to the studies, cybercrime may result in very high monetary losses (Rothschild, 1976).

18.4 Cybersecurity and Cybercrime

There is no separating cybercrime and cybersecurity as distinct issues in today's linked society. The resolution on information technology adopted

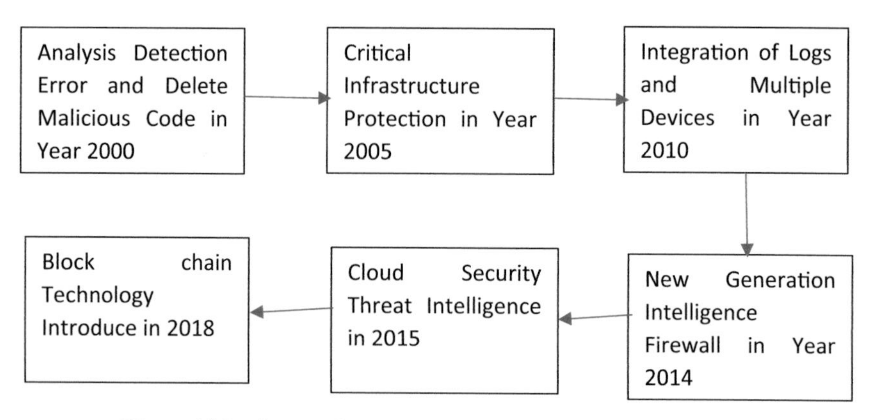

Figure 18.1 Stages of cyber errors. Source: Authors' compilation.

by the UN General Assembly in 2010 (see Figure 18.1) identifies cybercrime as a major issue. The development of new IT and web-based services depends critically on the safety of their infrastructure against cyberattacks. Every nation's economic and social well-being depends on the strength of its cyber defences and the integrity of its important information infrastructure (Kesan, 2005).

New internet companies and user protection regulations must take into account the need to make the World Wide Web a safer place. Protecting our nation's critical information infrastructure and services and information and technology is a top priority, and cybercrime prevention is essential to that strategy. To do so, it is necessary to create laws that make it illegal to use information and communication technologies (ICTs) for criminal or other purposes or in ways that endanger vital national infrastructure. This collective responsibility calls for coordinated action at the national level in readiness, response, recovery, and prevention (Anderson, 2001).

The illustration now includes text explaining the disrupter timeline in more detail:

At the global level, it is necessary to communicate and collaborate with the right partners. Therefore, a comprehensive strategy is required for creating and executing "a national framework and plan for cybersecurity." The danger of cybercrime may be mitigated by cybersecurity measures, including the construction of technical protection systems and the expansion of awareness about how to avoid being a victim. In the fight against cybercriminals, developing and enforcing cybersecurity measures is an essential component (Dhatterwal, 2022).

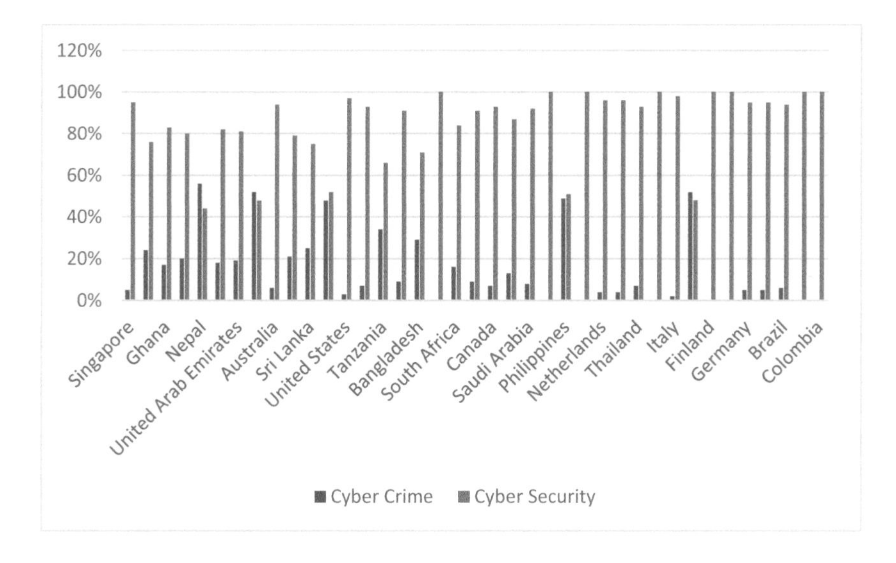

Figure 18.2 Country-wise cybercrime.

Cybersecurity's legal, technical, and institutional challenges are global in scope, but they are manageable with a coordinated approach that accounts for the many stakeholders and ongoing efforts shown in Figure 18.2 [30].

National crime data are the source for these numbers. Neither the global evolution of cybercrime nor the entire ramifications of cybercrime at the governmental level are intended to be covered by these articles. Therefore, they are merely included to offer some context for statistics on individuals in each country.

According to the US Internet Complaint Centre, claims of cybercrime increased by 22.3% between 2008 and 2009.

Statistics from Germany show a 23.6% increase in internet-related crime between 2008 and 2009.

The data may or may not be all-inclusive, and they may or may not provide reliable proof of the effects of crime. The use of crime data to assess the global cybercrime threat has several difficulties. To bring attention to an issue, statistics might be employed. As was previously said, crime data seldom break down crimes. Given the persistence and magnitude of the issue, it is crucial to recognise that a shortage of accurate data on the severity of the issue and arrests, punitive measures, and convictions is one of the important concerns associated with cybercrime.

In general, there is a lack of reliable statistics on the impact of cybercrime that would be useful for government.

18.5 Illegal Access

As one of the oldest computer-related crimes, "hacking" refers to the illegal entry into a computer system. Since the development of computer networks, this crime has exploded in prevalence (particularly on the internet). The German government, Yahoo, Google, eBay, and the United States National Aeronautics and Space Administration (NASA) are all frequent targets of cyberattacks (Böhme, 2006).

Attempting to access a protected area of a website or discovering a way around a password-protection system are both instances of hacking. However, preparatory acts related to the term "hacking" the password have included things like using faulty operating system implementations to illegally obtain a password to enter a piece of software, setting up "spoofing" websites for superpowered beings to reveal their passwords, and implementing hardware or software-based systems that detect intrusions. For example, "key loggers" are a kind of keylogging software that secretly monitors a user's keyboard activity and stores the data, including any passwords, entered into a computer or other electronic device (Ogut, 2005; Mah, 2022).

Over 250 million occurrences were reported worldwide in August 2007 alone, leading many to believe that the number of attempts to illegally infiltrate computer systems is on the rise. Poor and inadequate computer system encryption, the emergence of software programs that automate the attacks, and the growing importance of personal computers as a target of cybersecurity incidents are three major causes contributing to the rise in hacking attempts. Various keylogging techniques track all key presses and, consequently, all passwords entered into a computer or other devices.

Over 250 million attempted illegal network intrusions were registered worldwide in August 2007; this figure is expected to rise. Poor and inadequate computer system protection, the emergence of software packages that automate the attacks, and the growing importance of personal computers as a target of cybersecurity incidents have all contributed to a rise in the regularity of hacking attempts.

18.5.1 Lack of proper security

The internet connects a vast number of computers, but many of these networks lack the security measures necessary to prevent illegal access. Studies conducted at the University of Maryland found that an unprotected networked computer is vulnerable to attack within 60 seconds. While taking preventative steps might lessen the likelihood of an attack, even on a secure computer, such procedures are never 100% effective (Preety, 2021).

18.5.2 Enhancing software tools to protect attacks

Not all software tools have been used to automate attacks lately since most of them use predetermined attack techniques. It is possible for a single attacker to compromise thousands of systems "in a single day" by using only one computer with a collection of malicious software and prepared attacks. Attackers may do greater damage if they have access to more computers, such as a botnet.

Users that maintain a practice of updating their operating systems and applications reduce their vulnerability to these widespread assaults, as the companies that provide defensive software study attack patterns and plan to counteract them. High-profile attacks often use professionally prepared assaults. Methods (such as "key loggers") for recording one's keystrokes (and, by extension, one's passwords) on a computer or other device.

Over 250 million events were recorded worldwide in August 2007 alone, and many experts worry that this trend will continue. Poor and inadequate computer network cryptography, the emergence of software programs that automate the attacks, and the growing importance of personal computers as a target of cybersecurity incidents are three major causes contributing to the rise in hacking attempts. Despite the prevalence of internet-connected devices, many computer networks lack the protections to keep hackers away. Studies conducted at the University of Maryland found that an unprotected networked computer is vulnerable to attack within 60 seconds. While taking preventative steps might lessen the likelihood of an attack, even on a well-protected computer, these attacks are still possible (Dhatterwal, 2022).

18.5.3 Illegal data acquisition (data espionage)

Personal information is often stored on computers. If the offender's computer has internet access, they may access this data from almost anywhere in the world. Cybercriminals are more successful in obtaining sensitive data via online channels. Data espionage is tempting because of the high value of top-secret information and the ease with which it may be accessed across borders. Hackers from Germany were able to penetrate US intelligence and military computer networks in the 1980s, stealing and selling sensitive data to agents from other nations (Varian, 2000).

18.5.4 Interruption in data transmission

Criminals may eavesdrop on users' conversations (like e-mails) and other data transfers (like when users download files to webservers or utilise web-based

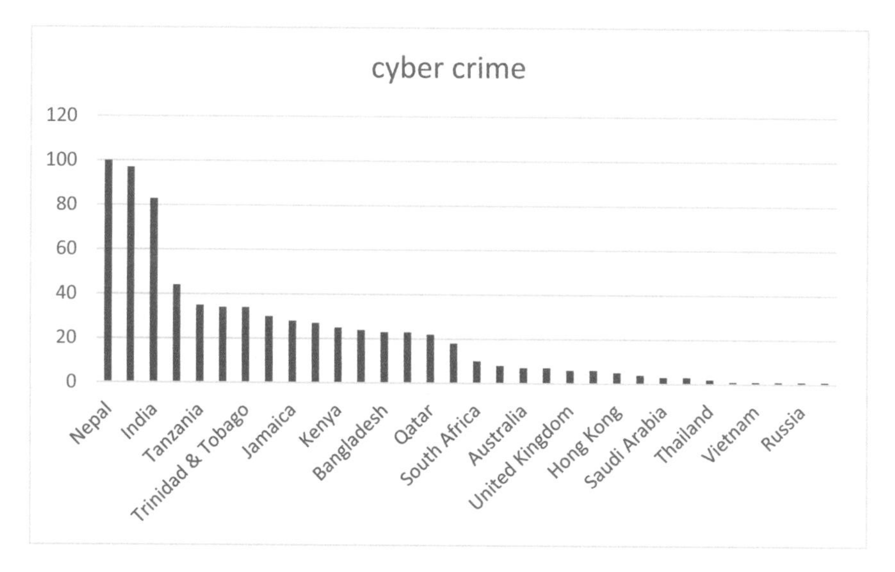

Figure 18.3 Threatening of cybercrime in developed countries.

external storage devices) to get private information. In this scenario, attackers may compromise any internet connection (including e-mail, chat, and VoIP) and any kind of telecommunications infrastructure (including fixed lines and wireless networks), as shown in Figure 18.3.

There are several sorts of cybercrime that occur in the networking world, some of which are included below:

- financial deception;
- data and other network sabotage;
- theft of confidential information;
- outside system intrusion;
- denial of service;
- insiders gaining unauthorised access;
- internet service permissions for employees;
- viruses.

18.6 Anti-cybercrime Strategies

Cybersecurity is essential to ensure the uninterrupted development of IT and internet businesses. Protecting internet users and making the internet a safer place is crucial to developing new services and regulations.

The danger of cybercrime may be mitigated by the implementation of cybersecurity measures, such as the development of technical protection systems or the dissemination of user education aimed at preventing individuals from becoming victims.

It is crucial to include an anti-cybercrime strategy in your overall cybersecurity plan. The International Telecommunication Union (ITU) Global Cybersecurity Agenda provides a structure for discussion and international collaboration to collaborate the regional and global reactivity to the evolving challenges to computer security and to boost strength and trust in the information economy (Winn, 2004).

18.7 Conclusion and Future Work

Based on the evidence presented, self-assessed proposal forms are primarily concerned with a small subset of measures in addition to malware protection, backup maintenance, and encryption usage. Insights from our study may guide the deliberate improvement of the insurance application procedure. Future presentation formats should include secure deployment, inventory preservation, administrator privilege management, and information security. Applicants should be mindful of the time commitment to writing a technical report [29].

In light of the risk consciousness of insurance companies, we think our results might be helpful to businesses in choosing strategic investments. However, misaligned motivations may generate poor security decisions since insurers only require security solutions that directly reduce financial and legal liability. After purchasing health insurance policies, businesses must consider the aforementioned considerations when allocating funds [28].

According to the authors, this is the first comprehensive analysis of cyber insurance plan proposals. Because of this, there are now several fresh avenues open to study regarding the structure of proposals. Our approach is grounded on two widely used network security frameworks and their respective issue areas. It is important to note that information security is just one of several concerns covered by cyber insurance. One example is reviewing the controls in place, such as peer review, to reduce multi-media liability, particularly considering the wide range of dangers, such as international cyber torts. We may use an inductive method to gather more controls in the future. Usability testing is another option for investigating the trade-offs between data collection depth and ease of use in web apps.

The proposal templates are just part of the issue. To better comprehend the function of telephone surveys and on-site inspections, we want to undertake further interviews with key actors in the insurance market. Perhaps the information gained from these interviews can explain the missing controls from their presentation forms. Researchers might learn more about why the insurance industry is interested in studying specific legislation if they conduct further interviews and surveys. The proportional weight of elements such claim details from insured businesses, monetary penalties from governments, changes to security regulations, and the evolution of the threats intelligence services.

References

[1] Siegel, C.A., Sagalow, T.R., Serritella, P (2002). Cyber-risk management: technical and insurance controls for enterprise-level security. *Information Systems Security*. 11(*4*). 33–49.

[2] Kaswan, K.S., Dhatterwal, J.S., Sharma, H. and Sood, K. (2022). Big Data in Insurance Innovation. Big Data: *A Game Changer for Insurance Industry*. pp.117–136.

[3] Kesan, J., Majuca, R., Yurcik, W. (2005). Cyber insurance as a market-based solution to the problem of cybersecurity: a case study. *In: Proceedings of Workshop of Economic Information Security*.

[4] Shetty, N., Schwartz, G., Felegyhazi, M., Walrand, J. (2010). Competitive cyber-insurance and internet security. *In: Proceedings of Workshop of Economic Information Security*. pp. 229–247.

[5] Bandyopadhyay, T., Mookerjee, V.S., Rao, R.C. (2009). Why IT managers go for cyber-insurance products. *Communications of the ACM*. 52(*11*). pp. 68–73.

[6] Kaswan, K.S. and Dhatterwal, J.S. (2021). The Use of Machine Learning for Sustainable and Resilient Buildings. *Digital Cities Roadmap: IoT-Based Architecture and Sustainable Buildings*. pp.1–62.

[7] Deore, S. P. (2022). Human Behavior Identification Based on Graphology Using Artificial Neural Network, Acadlore Trans. Mach. Learn., 1(2), 101–108. https://doi.org/10.56578/ataiml010204

[8] Grzebiela, T. (2002). Insurability of electronic commerce risks. In: System Sciences. *HICSS. Proceedings of the 35th Annual Hawaii International Conference, IEEE*.

[9] Gordon, L.A., Loeb, M.P., Sohail, T. (2003). A framework for using insurance for cyber-risk management. *Communications of the ACM*. 46(*3*). pp. 81–85.

[10] Kaswan, K.S., Dhatterwal, J.S., Kumar, S. and Lal, S. (2022). Cybersecurity Law-based Insurance Market. *In Big Data: A Game*

Changer for Insurance Industry. Emerald Publishing Limited (pp. 303–321).

[11] Bolot, J., Lelarge, M. (2008). A new perspective on internet security using insurance INFOCOM. *The 27th Conference on Computer Communications, IEEE*. pp. 1948–1956.

[12] Gordon, L.A., Loeb, M.P. (2002). The economics of information security investment. *ACM Trans. Inf. Syst. Security*. 5(*4*). pp. 438–457.

[13] Hausken, K. (2006). Returns to information security investment: The effect of alternative information security breach functions on optimal investment and sensitivity to vulnerability. *Information Systems Frontiers*. 8(*5*). pp. 338–349.

[14] Kunreuther, H.C., Michel-Kerjan, E.O. (2007). Evaluating the effectiveness of terrorism risk financing solutions', *NBER Working Papers 13359, National Bureau of Economic Research Inc.*

[15] Rothschild, M., Stiglitz, J.E. (1976). Equilibrium in competitive insurance markets: An essay on the economics of imperfect information. *The Quarterly Journal of Economics*. 90(*4*), pp. 630–49.

[16] Kesan J.. (2005). Cyber-Insurance as a Market-Based Solution to the Problem of Cybersecurity. *Workshop on the Economics of Information Security (WEIS).*

[17] Anderson R. (2001). Why Information Security Is Hard—An Economic Perspective. *Univ. of Cambridge Computer Laboratory working paper.*

[18] Dhatterwal, J.S., Kaswan, K.S. and Balusamy, B. (2022). Emerging Technologies in the Insurance Market. *Big Data Analytics in the Insurance Market*. pp.275–286.

[19] Böhme R and Kataria G. (2006) 'Models and Measures for Correlation in Cyber-Insurance. *Workshop on the Economics of Information Security (WEIS).*

[20] Ogut H., Menon N., and Raghunathan S.. (2005). Cyber Insurance and IT security investment: Impact of Inter-dependent Risk', Workshop on the Economics of Information Security (WEIS).

[21] Mah, P. M. (2022). Analysis of Artificial Intelligence and Natural Language Processing Significance as Expert Systems Support for E-Health Using Pre-Train Deep Learning Models, Acadlore Trans. Mach. Learn., 1(2), 68–80. https://doi.org/10.56578/ataiml010201

[22] Preety, Dhatterwal Jagjit Singh, Kaswan Kuldeep Singh. (2021). Securing Big Data Using Big Data Mining in the book entitled "*Data-Driven Decision Making using Analytics*". *Published in Taylor Francis, CRC Press*. ISBN No. 9781003199403.

[23] Dhatterwal, J.S., Kaswan, K.S., Baliyan, A. and Jain, V. (2022). Integration of Cloud and IoT for Smart e-Healthcare. In Connected e-Health. *Springer, Cham.* (pp. 1–31).

[24] H.R. Varian. (2000). Liability for Net Vandalism Should Rest with Those That Can Best Manage the Risk', *The New York Times.*

[25] J. Winn (2004). Should Vulnerability Be Actionable? Improving Critical Infrastructure Computer Security with Trade Practices Law. *George Mason Univ. Critical Infrastructure Protection Project Papers* Vol. II.

[26] Sood, K., Seth, N., & Grima, S. (2022). Portfolio Performance of Public Sector General Insurance Companies in India: A Comparative Analysis. In Simon Grima, Ercan Özen, Inna Romānova (Ed.), In Managing Risk and Decision Making in Times of Economic Distress, Part B. (pp.215–229). Emerald Publishing Limited, UK.

[27] Grima, S., & Marano, P. (2021). Designing a Model for Testing the Effectiveness of a Regulation: The Case of DORA for Insurance Undertakings. *Risks*, *9*(11), 206.

[28] Bassi, P., & Kaur, J. (2022). Comparative Predictive Performance of BPNN and SVM for Indian Insurance Companies. In Big Data Analytics in the Insurance Market (pp. 21–30). Emerald Publishing Limited.

[29] Trivedi, S., & Malik, R. (2022). Role and Significance of Data Protection in Risk Management Practices in the Insurance Market. In Big Data Analytics in the Insurance Market (pp. 263–273). Emerald Publishing Limited.

[30] Shilpa Yadav, Tanu Shree Yashika Arora (2013). Cyber Crime and Security. International Journal of Scientific & Engineering Research, 4(8), 855–861.

19

A Review of the Role of Insurance in Risk Management

Monica Gupta[1], Rajni Bansal[1], Aradhana Sharma[2], and Luke Grima[3]

[1]Chitkara Business School, Chitkara University, Punjab ,India
[2]Punjabi University, Patiala, India
[3]Department of Insurance and Risk Management, Faculty of Economics, Management & Accountancy, University of Malta, Malta

Email: monica.gupta@chitkara.edu.in; Rajni.bansal@chitkara.edu.in; aradhanasharma.as@gmail.com; grima.luke@gmail.com

Abstract

Insurance risk can be managed by spreading the risk of the events happening among insurance underwriters. It quantifies the event's happening and its financial impact on customers. The management of all of your insurance requirements, professional and personal, by a single knowledgeable vendor is known as insurance management. A significant component of risk management procedures is insurance offering various advantages, including preventing monetary loss in various circumstances, such as theft, natural disasters, or accidents influencing a business. Financial damages resulting from such dangers can be avoided with the help of insurance.

The primary function of insurance is to aid people in protecting their finances against life's unforeseen events. The paper aims to illustrate various topics related to insurance risk management, analysing the fundamentals of risk management, its significance, and the different types of insurance in risk management.

Investors are drawn to companies that manage their risks because insurance can stop or eliminate loss caused by circumstances outside the organisation's control. The risk-sharing tools insurers use to give incentives include deductibles, premium credit incentives, and contractual restrictions.

19.1 Introduction

A well-organised risk management system is essential for ensuring the optimal success of businesses in long-term operations. The main aim of risk management is to find the possible risks before they create havoc in the system and strategise to handle them in advance to negate the impact on your business objectives. It guarantees several critical aspects of business-like smooth flow of operations, enhanced productivity, and revenue generation. Business owners generally have many tasks to perform for their operations' smooth functioning. They may be looking to reach new customers, looking for financial backups, efficient employees, better workplaces, and many more. These responsibilities are more intense and riskier if you are new to the business. It can enhance your risk of safeguarding the business and complicate the process.

Insurance risk management helps create stability for a firm within a market in alignment with the dynamicity and evolution of changing business environments. For example, health insurance firms can help improve the sphere for technology enhancement and innovative practices, especially for emerging technologies, to enhance firms' competencies with the required support for implementing risky innovations [1]. Small and medium businesses fail to mitigate the demands of the market while competing with the more giant corporations due to the financial and infrastructural support acquired through the capacity of resource accessibility and allocation. At times of volatile events, insurance helps these small- and medium-sized businesses survive and access resources within a highly competitive market. The critical advantages and functionality of insurance for risk management are analysed in this chapter.

19.2 Insurance Index 2001–2020

This study thoroughly evaluated the literature on weather index insurance from 2001 to 2020. Ex-ante research on insurance based on an index that acts as a management tool of financial risk for agricultural concerns has significantly increased in recent years. To reduce basis risk in contract design, new and improved indices have been developed, and new approaches to quantifying the yield index relationship have been investigated. Our analysis revealed that temperature- and rainfall-based indices were the most common. However, indices based on vegetation, sunshine, soil moisture, and humidity hours were underrepresented despite having promise [2, 27, 28]. While methods tackling extremes like quantile regression and copulas have lately gained popularity in industrialised nations, linear regression and ordinary least squares-based correlation methods still dominate yield-index modelling. To direct future research studies, we highlighted several fresh research trends, including using the data sensed by a remote, hydrological modelling, and crop modelling to address data shortages, risks based on geographical factors, and climate change.

Agricultural climate indicator insurance may compensate farmers for financial damages sustained directly from extreme weather (IBI) [3]. We show through a comprehensive investigation that research on the efficacy and acceptance of index-based agricultural climate insurance needs to be improved in countries vulnerable to food and climatic vulnerabilities. We drew this conclusion after reviewing a number of indicators relating to the state of climate and food security in various countries and finding that many of them were vulnerable to climate and food insecurity threats but lacked any kind of research into insurance policies based on agrarian climate indices. While several options are available to these nations, they continue to be vulnerable despite being able to mitigate some of the hazards. While cereal crops and drought have received the lion's share of attention in the existing literature, there are many other crops and climate hazards that farmers can better manage if they implement agricultural climate index-based insurance. Our findings assist in filling in the holes of the existing insurance studies based on an agricultural environment index by suggesting which nations need further investigation.

The factors underlying firm lobbying actions and, in particular, where corporations concentrate their lobbying expenditures have received less attention, even though various studies have explored the association between lobbying expenditures and improved performance outcomes [4]. We offer an all-encompassing lobbying strategy to reduce the likelihood of governmental interference and official disapproval of business activity. Accordingly, we contend that the strategic risk taken by lobbying companies is directly related to the size of the target market they serve. Furthermore, the scope of lobbying activity is used as a risk management approach. In that case, this relationship should be influenced by other factors that affect decisions on the level of risk protection a corporation may require. Since the political unpredictability of enterprises and CEO ownership may increase the exposure of corporate risk-taking to governmental actions, we suggest that this strengthens the insurance link.

Making up the majority of businesses in Malaysia, SMEs today maintain a critical role in the nation's economic progress. As a result, many Malaysians, particularly the younger generations, are engaged in this activity because they think that entrepreneurship will help them raise their standard of life and income. The government's encouragement has also contributed to the desire for entrepreneurship among young Malaysians [5]. However, many prosperous businesspeople in Malaysia battled to live due to the risks and difficulties they had to deal with. Therefore, this essay aims to research earlier works that have discussed the significance of risk management for new enterprises. This study discovered that the mathematical risk index and protection of insurance/takaful are crucial tools for helping entrepreneurs manage and assess risk.

Societies are increasingly aware of the revenue costs of lengthy care as the hazards linked with ageing begin to emerge (LTC). While social insurance in many nations gives some limited coverage, there are few attractive private product offers, and there still needs to be more aware of long-term care insurance [6]. This study aims to identify the primary factors influencing people's interest in purchasing care insurance products. It is based on a survey on health, ageing, and reliance done in Switzerland. We represent the terms of potential purchasers based on important social, economic, demographic, and political issues through models incorporating features from classical statistics, including machine learning techniques. The criteria for LTC awareness and comprehension are critical. Customer and insurer interactions based on self-perceived health, conduct, and trust are crucial. The decision-making process gives socioeconomic considerations a supporting role. Our findings are important for private insurers and policymakers outside the academic community.

Using a recovery housing framework that is built in a better and long-term manner to examine strategies and impacts on rebuilding housing along with recovery, this research paper aims to understand "***community housing resilience***" along with the role of insurance. To compare insurance policies, including the results after Hurricane Katrina in New Orleans and the "***Canterbury earthquake sequence***" in "***Christchurch, New Zealand***," these incidents in affluent urban areas with robust insurance markets, a comparative case study approach has been used [7]. The community residence and insurance resilience evaluation are based on five leading indicators: community resources, governance, risk reduction, money for home rebuilding (funding along with the speed of funding), and including time compression. It is implied within the concept of "***Build Back Better***" (periods of recovery time with a built environment).

This special issue results from a conference with simultaneous submissions held in 2020 by the Corporate Finance Program and the NBER Insurance Working Group [8]. It offers a more comprehensive viewpoint on significant frictions in the insurance markets, such as the lack of faith between brokers and the policyholders, broker conflicts regarding choices, less-than-ideal behaviour of the policyholders, and capital regulation generally risk-based. New viewpoints have been sparked by several economic and intellectual developments, such as the rise of saving the products within a minimal return warranty along with the financial crisis that is spread globally and intermediate pricing of assets. We provide a summary of the research issues that show promise for additional investigation.

Businesses face risks and uncertainties as a result of the changing business environment. To improve their competitive advantage and boost their financial performance in such a setting, businesses must innovate their business models [9]. Businesses must innovate their business models to adapt to the changing business climate; in organisations operating in a dynamic business environment, enterprise risk management and strategic agility play significant roles in the innovation of a business model. An in-depth discussion of the connections between enterprise management of risks, strategic adaptability, innovation of the business model, and financial performance, particularly in the context of a changing environment of business, shall be provided in this chapter.

Actuarial science and operational research are increasingly focusing on pricing and risk modelling related to cybersecurity. This essay examines the cybersecurity risks associated with the heavily used fog computing technology in various Internet of Things (IoT) models. To this goal, a class of structural models is constructed to research the propagation of ingrain cyber risk. We can explicitly calculate the compromising probabilities of specific nodes under innovative home applications. Applications regarding the suggested structural models are thoroughly investigated regarding cyber insurance pricing [10]. We also provide an interval approach for assessing the compromised chances of fog network components to effectively detect weak nodes for cybersecurity risk management. Therefore, the evolution of insurance for risk management amplifies its advantages for customers and firms operating within a highly volatile and competitive market.

19.3 Insurance Risk Management

19.3.1 What is risk management?

Most business owners are risk-takers eager to spend money in anticipation and hope of a return, but they need more assurance that they will. However, from the perspective of insurance, "risk" is another word for "peril" and refers to potentially harmful outcomes. The list of unfavourable occurrences that might hurt your company or organisation financially goes on and on: theft, vandalism, fire, civil litigation, computer virus, failure to provide raw materials, equipment failure, and death or illness of a valuable employee [11].

The field of risk management is vast. Loss control is taking action to lessen the likelihood of something going wrong. In addition, it entails investing in insurance to lessen the financial toll that unfavourable occurrences take on a business when they do occur, despite your best efforts. Nobody enjoys contemplating what could go wrong. Nevertheless, you should be aware of the hazards your company confronts as a responsible manager. You can only effectively manage risks if you first recognise them.

19.3.2 Risk management requires leadership

Any organisation's top management should start with loss control and risk management. The rest of the organisation is more likely to adopt these values if the company's leader promotes safety, compliance, and moral and legal conduct.

Even though risk management is expensive, the long-term costs of ignoring safety issues and forgoing insurance may be much more than any short-term savings. Risk management should be kept from chance, even if small businesses often do not employ full-time risk managers. Responsibility for safety and compliance programs and insurance-related issues should fall under the purview of specific people.

19.3.3 Resources for risk management

All organisations now have quick access to a wealth of knowledge about risk management, including loss prevention techniques, safety procedures, legal requirements, and disaster preparedness and recovery [12]. Additionally, to material related to particular business kinds, extensive generic checklists and recommendations are also accessible. Examine the tools that your insurance provider offers. Your insurance agent is one resource you can use. Allowing the agent to visit your location will allow you to explain your current risk management practices. They will be able to assess your actions and make recommendations. A risk management expert could be a wise investment for your company, depending on the nature of your industry.

19.3.4 Loss control and insurance

Effective loss control can affect both the accessibility and cost of insurance by lowering the quantity and magnitude of losses. A company that does not care about loss prevention may file more insurance claims than usual. A terrible loss history can make it difficult to acquire insurance [13]. On the other hand, companies who aggressively manage the risks and consequently limit losses will see fewer claims and frequently have their efforts rewarded with minimal insurance premiums.

19.3.5 Importance of risk management

There has probably never been a time in history when risk management has been more vital than it is right now. The increasing rapidity of globalisation is accompanied by an increasing complexity of the dangers that modern enterprises must contend with. The widespread use of digital resources has resulted in the appearance of hitherto undiscovered threats. According to the terminology used by those who research risks, climate change is a "hazard multiplier."

The COVID-19 pandemic is a recent worldwide issue that initially manifested itself within a few businesses as a supply chain issue but has since evolved into an existential threat to their employees' health and safety, their ability to conduct business, their capacity to communicate with customers, and their reputations [14]. The pandemic first appeared in a few businesses as a supply chain issue but has since manifested itself in a broader range of businesses as a supply chain issue. Businesses quickly changed to address the pandemic's concerns. However, as time goes on, they shall have to face newer concerns, such as how or if to get workers back to work and what has to be done to make their supply networks more resilient to crises.

Companies, including their boards of directors, have been reevaluating the risk management measures as the globe struggles to reach terms with COVID-19. They have been reevaluating the risk exposure along with investigating risk management techniques. They are taking a second look at who should be a part of risk management. Numerous companies that now utilise a reactive strategy for risk management and providing protection against historical issues and altering procedures once a newer risk results in harm are now thinking about the advantages of a more proactive strategy from a competitive standpoint. Supporting sustainability, resilience, and enterprise agility are of greater interest. The use of advanced governance, risk, and compliance (GRC) systems and artificial intelligence technology by businesses is also being investigated.

19.4 Process of Risk Management

In the field of risk management, several academic publications have documented all of the steps that firms should take to manage risks. In terms of standardisation, moreover, ISO, which is a standard agency along with the international organisation, has created the ISO 31000 standards [15]. In addition to this, the most well-known sources are risk management guidelines.

Any entity consumes the ability to apply the five-step risk management approach of ISO, which includes the following steps:

1. identification of the risk factors;
2. analysing the likelihood along with the impact of each one;
3. prioritising the risks that are based upon numerous objectives of any business;
4. treating (or responding to) the risk factors;
5. monitoring the results along with adjusting as per necessity.

The stages are simple; however, risk management organisations should consider the effort needed to complete the procedure. It starts with having a thorough awareness of the organisational culture. The main objective is creating a set of procedures in terms of recognising the risks that are faced by the organisation, their likelihood and effects, along with the procedure of relating to the maximum risk the company is willing to receive and the measures that must be conducted to protect and improve organisational value.

19.5 Literature Survey

The chapter analyses the risk management factors by insurance policies and schemes that can be utilised to support a firm or individual customers. The sustainability and survival of enterprises are crucial factors impacting the operational efficiency of enterprises in the face of dynamic environmental changes. On the other hand, environmental crises may also be analysed for risk assessment and management. Assessment of global data for predicting food crisis issues in countries such as Ethiopia, Malaysia, Colombia, India, and England indicates the mitigation of risks on a national level [16]. On the other hand, risk management measurements may also be applied to demonstrate linear or dynamic changes. Urban agricultural projects must be analysed with the help of a flood risk management framework for building a preventive model to prioritise the building of water-resilient urban landscapes [17]. The position of insurance companies is relevant to the prospect of risk management on a larger scale, such as urban landscape building projects, as it helps to provide financial support to the firms involved in the project. Thus, insurance and risk management are correlated, which is further discussed in the study.

Analysis of risk management for both the urban and agricultural spheres is defined in this chapter, highlighting the crucial factors that can be involved in the process of claiming insurance and obtaining proper support. According to a study conducted in the Czech Republic, the necessity and willingness among farmers to consider agricultural insurance is highlighted due to predictions of production loss, insurance premium prices, and overall distrust of insurance [18]. These factors are integrated into the analysis of the chapter as the willingness and intention among consumers from different sectors contribute to the impact of insurance and risk management.

19.6 Various Types of Insurance in Risk Management

Traditional risk management vs. enterprise risk management:
It has been noticed in the current times that enterprise risk management has been gaining a better reputation than those risk management methods that have been used traditionally. Both strategies have tried to reduce those hazards that could hurt businesses. Both the risk management methods that are traditional and enterprise risk management purchase insurance to defend businesses against a wide variety of such hazards that include cyber liability along with such losses that are generated from theft and fire [19]. Both risk management methods follow the recommendations that the major standards groups generate. However, it is evident that according to the opinion of the experts, the traditional risk management methods need to have such a mentality and framework that is necessary to comprehend the risks as an essential step for their business strategies and enhance performance.

Risk "is a dirty four-letter word for many companies, and that's unfortunate," according to the opinion of Forrester's Valente. The risks are viewed as a strategic enabler for business conduction rather than a cost for business conduction in ERM.

According to Shankman of Gartner, one of the key differences between the two approaches is "siloed" vs. holistic. For instance, in conventional risk management programs, the risks that pop up have generally been the company executive's responsibility because they are in charge of such risks that are noticed by the company [20]. For instance, it is noticed that the CIO or the CTO of the company is generally in charge of the I.T. risk and the CFO is in charge of such financial risks, the COO is in charge of operational risk, etc. Shankman has even argued that in such cases, the business divisions have sophisticated procedures in such places to handle the multiple risks that they face; the organisation might still encounter problems in such cases that it fails to see the connections that exist between the risks or their influence that is combined on operations. The reactive rather than proactive nature of traditional risk management is another issue.

Regarding risk issues, Shankman said, "The pandemic is a perfect example of one that is very simple to miss if you don't consume a comprehensive, strategic view of the risks that could damage you as a company." Many businesses will reflect on the past and acknowledge that they should have been aware of this or, at the very least, considered the financial ramifications of an event of this nature before it occurred. Here is a primer on the evaluation of risk exposure and how the risks are calculated.

Risk management in an institution or enterprise setting entails teamwork, cross-functionality, and a broad perspective. The ERM team of any enterprise might consists of as few as five individuals, collaborates with business unit executives and staff to debrief them, assists them in using the proper resources to consider the risks, and compiles the data to deliver the data to the organisation's executive or the one in charge of the leadership and board. For the mitigation of risks of this calibre by the leaders, confidence in corporate executives is a need, according to Shankman.

He asserted that these professionals increasingly have a history in consulting or a "consulting mindset" and are well-versed in business principles. Contrary to the traditional methods of risk management, in which it is seen that the chief officer is in charge of the risks typically reports to the CFO, the chiefs of enterprise that are in charge of the risk management teams, whether they are holding the title of their chief officer that is in charge of the risks or some other title, report directly to the CEOs, a recognition that risk is an integral part of a business strategy.

All the "transactional CROs" commonly located in conventional risk management programs and the "transformational CROs" that are noticed to use the ERM approach are two different types of chief risk officers, according to Forrester Research's definition of the position. The former is employed by businesses where, according to Forrester, the risks are viewed as the cost centre of managing risks as an insurance policy [21]. According to the opinion of Valente, the transformational CROs are "customer-obsessed" in the Forrester sense. They pay attention to their brands' reputations, comprehend the horisontal nature or behaviour of the risk, and characterise ERM as the "right amount of risk needed to grow."

Traditional risk management companies also tend to be risk-averse. However, as Valente pointed out, businesses that consider themselves to be low-risk-taking and risk-averse can make inaccurate risk assessments.

19.7 Risk Management Best Practices

The 11 risk management principles of ISO 31000 serve as a solid place to start for any business that wants to adhere to best practices in risk management. The following goals for a risk management program should be met, according to ISO:

- generate essential valuation of the organisation;
- behave is an essential aspect of the overall processes of the organisation;
- be an essential factor in the overall decision-making process of the company;
- address all sorts of uncertainties explicitly;
- be structured and systematic as required;
- it is based on the best such information;
- be tailored to enhance the project efficiently.

According to security consultant Dave Shackleford, "digital reform" is another valuable practice for contemporary organisational risk management. To do this, manual operations that are wasteful and ineffectual must be automated utilising AI and other cutting-edge technology.

19.8 Risk Management Limitations and Examples of Failures

Failures in risk management are attributed to egregious negligence, blatant carelessness, or a chain of dire circumstances that no one would have foreseen. However, the typical risk management blunders along with the risk management failures frequently result in avoidable errors.

19.8.1 Poor governance

Even the biggest bank in the entire globe can mess up risk management, despite consuming updated policies for pandemic work conditions along with numerous controls in place, as evidenced by the 2020 tangled tale of Citigroup accidentally repaying a $900 million loan, using its own money to Revlon's lenders when a tiny interest payment was due. Although there were instances of human mistakes and clumsy software, a judge finally decided that weak governance was to blame. U.S. regulators imposed $400 million penalties on Citigroup in exchange for its agreement to update its risk management, data governance, and compliance procedures.

19.8.2 Overemphasis on efficiency vs. resiliency

When everything works out, increased efficiency can result in higher profitability. The problem with doing things the same way every time can lead to a lack of resilience, as businesses discovered during the epidemic when their supply networks collapsed [22]. "Things change all the time when we look at the nature of the planet," said Valente of Forrester. Therefore, while efficiency is important, it is equally important to plan for all potential outcomes.

19.8.3 Distrust and dishonesty

The scandal surrounding the underreporting of COVID-related mortality rates in New York nursing homes is an example of a systemic breakdown in risk management. Data concealment, data absence, and data segregation are all acts of conduct or omission that can impede openness. "Many procedures and systems were not created with risk in mind," Josh Tessaro, a risk specialist, told Lawton. As a result, several authorities end up with siloed data that does not belong to anyone. Tessaro claims that risk managers usually utilise the data they already possess because of its convenience while ignoring the procedures because of their inconvenient nature.

19.8.4 Limitations posed by different approaches to risk assessment

A large amount of data is required for many approaches to risk analysis, such as building a risk model or conducting a simulation. Extensive data collection can be unreasonably expensive and may need to be more precise. The consequences of data use in decision-making may also fall short if simple indicators signify complex risk scenarios. A wrong conclusion could be drawn about the project if a decision about a minor detail was extrapolated to include the whole thing.

19.9 In What Ways Might Risk Management Evolve Shortly?

Many businesses have gone through the motions of revisiting their risk management protocols and exploring new techniques, tools, and processes in this area due to the COVID-19 pandemic's increased focus on risk management. After reading Lawton's coverage of the innovations altering risk management, it is clear that the field is teeming with new ideas.

As threats become more integrated across the company, more organisations have been seen to adopt a risk maturity framework to assess their risk management procedures. They are reconsidering GRC technologies to, among other things, automate internal audits, administer policies, conduct risk assessments, and combine their risk management activities. The following new GRC features are being thought about:

- analytics in terms of the geopolitical risks along with the natural disasters;
- social media monitoring to track the changes within the reputation of the brand;
- security systems in terms of assessing the potential impact of breaches and cyberattacks.

A growing number of businesses are attempting to formalise the process of managing positive risks to increase business value and utilise risk management to prevent unfavourable circumstances.

They are also reexamining risk appetite declarations. Risk appetite statements previously used to communicate with staff members, regulators, and investors are now being used more dynamically for replacing "check the box" compliance exercises with risk scenarios that are more complex. The caution? A poorly written risk appetite statement could confine a business or be perceived as endorsing unreasonable risks by regulators.

In addition, technologies for assessing and reducing risks are improving, even though it is difficult to make forecasts, especially regarding the future. What improvements were made? – detecting trends and new hazards using internal and external sensing tools.

19.10 Conclusion

In the aftermath of the COVID-19 pandemic and the Ukraine–Russia war, it has become imperative that students and young professionals develop insurance and risk literacy and strengthen their capabilities in mitigating risks arising from volatile circumstances. "Whether it is the pandemic or events such as the Ukraine–Russia war or the Sri Lanka crisis, the insurance industry will have to refocus and strengthen its capability in predicting uncertainties and planning for unforeseen events. The more globally aware an insurance professional is, the better they will contribute to their function. It is vital that an insurance professional is risk-literate, financially literate, and understands global businesses and interconnected events.

19.11 Suggestions and Recommendations

- All organisations' top management should start with loss control and risk management, which will benefit the organisation.
- By consulting scholarly publications on risk management, businesses can increase their knowledge of the actions they need to take to manage risk effectively.
- One of the many benefits of risk management is increased awareness of risk within an organisation. This is only one of the many advantages. There is a rising sense of confidence inside the company that it will be able to accomplish its objectives, and this is reflected in the strategic planning that it is performing. There is the coordination of compliance with internal demands and regulatory requirements, which results in higher quality compliance.

References

[1] Nayak, Bishwajit, Som Sekhar Bhattacharyya, and Bala Krishnamoorthy. "Explicating the role of emerging technologies and firm capabilities towards the attainment of competitive advantage in health insurance service firms." *Technological Forecasting and Social Change* 170, 2021: 120892.

[2] Abdi, M. J., Raffar, N., Zulkafli, Z., Nurulhuda, K., Rehan, B. M., Muharam, F. M., ... &Tangang, F. Index-based insurance and hydroclimatic risk management in agriculture: A systematic review of index selection and yield-index modelling methods. *International Journal of Disaster Risk Reduction*, *67*, 2022.102653.

[3] Adeyinka, A. A., Kath, J., Nguyen-Huy, T., Mushtaq, S., Souvignet, M., Range, M., & Barratt, J. (2022). Global disparities in agricultural climate index-based insurance research. *Climate Risk Management*, *35*, 100394.

[4] Abdurakhmonov, M., Ridge, J. W., Hill, A. D., &Loncarich, H. (2022). Strategic Risk and Lobbying: Investigating lobbying breadth as risk management. *Journal of Management*, *48*(5), 1103-1130.

[5] Fauzi, S. N. M., Ghazali, P. L., Razak, R. A., Zain, E. N. M., & Muhammad, N. (2022). Risk Management of Start-up Business for

SMEs: A Review Paper. *The Journal of Management Theory and Practice (JMTP)*, *3*(1), 67–72.

[6] Fuino, M., Ugarte Montero, A., & Wagner, J. (2022). On the drivers of potential customers' interest in long-term care insurance: Evidence from Switzerland. *Risk Management and Insurance Review*.

[7] Hofmann, S. Z. (2022). Build Back Better and Long-Term Housing Recovery: Assessing Community Housing Resilience and the Role of Insurance Post Disaster. *Sustainability*, *14*(9), 5623.

[8] Koijen, R. S., &Yogo, M. (2022). New Perspectives on Insurance. *The Review of Financial Studies*.

[9] Wirahadi, A., &Pasaribu, M. (2022, March). Business Model Innovation: The Role of Enterprise Risk Management and Strategic Agility. In *7th Sriwijaya Economics, Accounting, and Business Conference (SABC, 2021)* (pp. 284–290). Atlantis Press.

[10] Zhang, X., Xu, M., Su, J., & Zhao, P. (2022). Structural models for fog computing based internet of things architectures with insurance and risk management applications—*European Journal of Operational Research*.

[11] Adeabah, D., Andoh, C., Asongu, S., &Gemegah, A. (2022). Reputational risks in banks: A review of research themes, frameworks, methods, and future research directions. *Journal of Economic Surveys*.

[12] Bracci, E., Mouhcine, T., Rana, T., &Wickramasinghe, D. (2022). Risk management and management accounting control systems in public sector organizations: a systematic literature review. *Public Money & Management*, *42*(6), 395–402.

[13] Jalilvand, A., &Moorthy, S. (2022). Enterprise Risk Management Maturity: A Clinical Study of a U.S. Multinational Nonprofit Firm. *Journal of Accounting, Auditing & Finance*, 0148558X221097754.

[14] McMaster, May, et al. "Risk management: Rethinking fashion supply chain management for multinational corporations in light of the COVID-19 outbreak." *Journal of Risk and Financial Management* 13.8. 2020: 173.

[15] Kajwang, B. (2022). Influence of Knowledge Management Practices on the Performance of Insurance Firms. *African Journal of Information and Knowledge Management*, *1*(1), 1–11.

[16] Bernhofen, M. V., Cooper, S., Trigg, M., Mdee, A., Carr, A., Bhave, A., ... & Shukla, P. (2022). The Role of Global Data Sets for Riverine Flood Risk Management at National Scales. *Water Resources Research*, *58*(4), e2021WR031555.

[17] Ebissa, Gizaw, and Hayal Desta. "Review of urban agriculture as a strategy for building a water resilient city." *City and Environment Interactions*. 2022: 100081.

[18] Kislingerová, S., &Špička, J. (2022). Factors Influencing the Take-Up of Agricultural Insurance and the Entry into the Mutual Fund: A Case Study of the Czech Republic. *Journal of Risk and Financial Management, 15*(8), 366.

[19] Illangakoon, A. G., Azam, S. M., &Jaharadak, A. A. (2022). Impact of Risk Management towards Sustainability of Microfinance Industry in Sri Lanka: A Case Study. *International Journal of Social Sciences and Economic Review*, *3*, 01–07.

[20] Kajwang, B. (2022). Global Challenges and their Role in Talent Management Crisis Facing Insurance Industry. *International Journal of Business Strategies*, *7*(1), 1–10.

[21] Yin, H., Chen, Z., Xiao, Y., Mohsin, M., & Liu, Z. (2022). A study on assessing risk management performance in maritime start-ups: evidence from China. *Maritime Policy & Management*, 1–15.

[22] Van Anrooy, R., Espinoza Córdova, F., Japp, D., Valderrama, D., Gopal Karmakar, K., Lengyel, P., ... & Zhang, Z. (2022). *World review of capture fisheries and aquaculture insurance 2022* (Vol. 682). Food & Agriculture Org.

[23] Shalender, K. (2021). Building Effective Social Media Strategy: Case-Based Learning and Recommendations. In *Digital Entertainment* (pp. 233–244). Palgrave Macmillan, Singapore.

[24] Gupta, M., Bansal, R., Hothi, B. S., & Shashidharan, M. (2021). Overview And Growth of Micro, Small And Medium Enterprises In Punjab. *Academy of Entrepreneurship Journal*, *27*, 1–7.

[25] Rana, A., Bansal, R., & Gupta, M. (2022). Emerging Technologies of Big Data in the Insurance Market. In *Big Data: A Game Changer for Insurance Industry* (pp. 15–34). Emerald Publishing Limited.

[26] Rana, A., Bansal, R., & Gupta, M. (2022). Big Data: A Disruptive Innovation in the Insurance Sector. In *Big Data Analytics in the Insurance Market* (pp. 165–183). Emerald Publishing Limited.

[27] Sood, K., Seth, N., & Grima, S. (2022). Portfolio Performance of Public Sector General Insurance Companies in India: A Comparative Analysis. In Simon Grima, Ercan Özen, Inna Romānova (Ed.), In *Managing Risk and Decision Making in Times of Economic Distress, Part B*. (pp.215–229). Emerald Publishing Limited, U.K.

[28] Grima, S., & Marano, P. (2021). Designing a Model for Testing the Effectiveness of a Regulation: The Case of DORA for Insurance Undertakings. *Risks*, *9*(11), 206.

Index

About the Editors

Kiran Sood is a Postdoc Researcher, University of Usak, Faculty of Applied Sciences, Dept of Finance and Banking. She is also an Affiliate Professor in the Faculty of Economics, Management, and Accountancy at University of Malta. She is currently employed as a Professor at Chitkara Business School, Chitkara University, Punjab, India. She received her Undergraduate and PG degrees in commerce from Panjab University in 2002 and 2004, respectively. She earned her Master of Philosophy degree in 2008 and Doctor of Philosophy in Commerce with a concentration on Product Portfolio Performance of General Insurance Companies in 2017 from Panjabi University, India. Before joining Chitkara University in July 2019, Kiran had served four organisations with a total experience of 19 years. She has published articles in various journals and presented papers at various international conferences. She serves as an editor of refereed journals, in particular IJBST International Journal of BioSciences and Technology, International Journal of Research Culture Society, and Journal of Corporate Governance, Insurance, and Risk Management (JCGIRM). Her research mainly focuses on regulations, marketing and finance in insurance, insurance management, economics and management of innovation in insurance. She has edited more than ten books with various international publishers such as Emerald, CRC, Taylors & Francis, AAP, Wiley Scrivener, IET, River Publishers, and IEEE.

Simon Grima, PhD, is the Deputy Dean of the Faculty of Economics, Management and Accountancy and Associate Professor and Head of the Insurance and Risk Management Department at the University of Malta. He coordinates the MA and MSc Insurance and Risk Management degrees and the undergraduate degree program in Insurance. He is also a Professor at the University of Latvia, Faculty of Business, Management and Economics and a visiting Professor at Università Cattolica del Sacro Cuore, Milan, Italy. He has served as the President of the Malta Association of Risk Management and President of the Malta Association of Compliance Officers. In addition, he is chairman of the Scientific Education Committee of the Public Risk

Management Organization. His research focuses on governance, regulations, and internal controls. He has over 35 years of varied experience in financial services, academia, and public entities. He has acted as co-chair and is a member of the scientific program committees at several international conferences. He is also a chief editor, editor, and review editor of several journals and book series. He was awarded outstanding reviewer for the Journal of Financial Regulation and Compliance Emerald Literati Awards in 2017 and 2022. Professor Grima acts as an independent director for financial services firms; sits on risk, compliance, procurement, investment, and audit committees; and carries out duties as a compliance officer, internal auditor, and risk manager.

Dr. Ganga Sharma obtained her PhD and MTech(CSE) from USICT, Guru Gobind Singh Indraprastha University, New Delhi. She also holds a Master of Computer Applications and BSc in Electronics degree. She has 10+ years of academic experience and is currently working as Associate Professor and Program Chair (BSc/MSc) at School of Computing Science and Engineering, Galgotias University, India. In the past, she has worked with various renowned institutions like PEC University of Technology, Chandigarh (formerly Punjab Engineering College), GD Goenka University, Gurgaon, etc. She has authored/co-authored various research publications in peer-reviewed reputed international journals, book chapters, and conference proceedings. Her research interests include search based software engineering, object oriented analysis and design, software metrics and quality, aspect orientation, data mining, and warehousing, machine learning, etc.

Balamurugan Balusamy is currently working as an Associate Dean Student in Shiv Nadar University, Delhi-NCR. Prior to this assignment he was Professor, School of Computing Sciences & Engineering and Director International Relations at Galgotias University, India. His contributions focus on engineering education, block chain and data sciences. His academic degrees and twelve years of experience working as a Faculty in a global University like VIT University, Vellore has made him more receptive and prominent in his domain. He has 200 plus high impact factor papers in Springer, Elsevier. and IEEE journals. He has published more than 80 edited and authored books and collaborated with eminent professors across the world from top QS ranked universities. Professor Balamurugan Balusamy has served up to the position of Associate Professor in his stint of 12 years of experience with VIT University, Vellore. He completed his Bachelors, Masters and PhD Degrees from top premier institutions from India. His passion is teaching and adapts different design thinking principles while delivering his lectures. He serves on the advisory committee for several startups and forums and does consultancy work for industry on Industrial IOT. He has given over 195 talks at various events and symposium.